David J. Mit...

*The Sustainable
Management of
Tropical Catchments*

The Sustainable Management of Tropical Catchments

Edited by

David Harper
Department of Biology, University of Leicester, UK

and

Tony Brown
Department of Geography, University of Exeter, UK

JOHN WILEY & SONS
Chichester · New York · Weinheim · Brisbane · Singapore · Toronto

Other Wiley Editorial Offices

John Wiley & Sons, Inc., 605 Third Avenue,
New York, NY 10158-0012, USA

WILEY-VCH Verlag GmbH, Pappelallee 3,
D-69469 Weinheim, Germany

Jacaranda Wiley Ltd, 33 Park Road, Milton,
Queensland 4064, Australia

John Wiley & Sons (Asia) Pte Ltd, 2 Clementi Loop #02-01,
Jin Xing Distripark, Singapore 129809

John Wiley & Sons (Canada) Ltd, 22 Worcester Road,
Rexdale, Ontario M9W 1L1, Canada

Library of Congress Cataloging-in-Publication Data

The sustainable management of tropical catchments / edited by David
   Harper and Tony Brown.
      p.   cm.
   Results of a conference held at the University of Leicester,
   Includes bibliographical references and index.
   ISBN 0-471-96914-1 (alk. paper)
     1. Watershed management—Tropics—Congresses.   2. Soil erosion
—Tropics—Congresses.   3. Land use—Tropics—Congresses.
   4. Watershed management—Mathematical models—Congresses.
   I. Harper, David M.   II. Brown, A. G. (Anthony Gavin), 1958–   .
   TC526.5.S87   1998
   628—dc21   98-26738
                                                              CIP

British Library Cataloguing in Publication Data

A catalogue record for this book is available from the British Library

ISBN 0-471-96914-1

Typeset in 10/12pt Times from the author's disks by C.K.M. Typesetting, Salisbury
Printed and bound in Great Britain by Biddles Ltd, Guildford and King's Lynn
This book is printed on acid-free paper responsibly manufactured from sustainable forestation, in which at least two trees are planted for each one used for paper production.

Contents

Contributors

Lorenzo Borselli
CNR-IGES, Instituto per la Genesi e l'Ecologia del Sudo, Piazzale delle Cascine 15, 50144 Firenze, Italy

James Brasington
Department of Geography, University of Hull, HU6 7RX, UK

Tony Brown
Department of Geography, University of Exeter, EX4 4RJ, UK

Stefano Carnicelli
Università di Firenze, Dipartimento di Scienza del Suolo e Nutrizione della Pianta, P.le Cascine 15, 50144 Firenze, Italy

John Dalrymple
Department of Soil Science, University of Reading, RG6 6DW, UK

Ian Douglas
Department of Geography, University of Manchester, M13 9PL, UK

Ahmed El-Hames
Department of Geography, University of Cambridge, CB2 3EN, UK

Pedro Fernandez
Centro Regional Andino – INCYTH, PO Box 6, Belgrano 210, 5500 Mendoza, Argentina

Giovanni Aurindo Ferrari
Università di Firenze, Dipartimento di Scienza del Suolo e Nutrizione della Pianta, P.le Cascine 15, 50144 Firenze, Italy

Ugo Galligani
Università di Firenze, Dipartimento di Scienza del Suolo e Nutrizione della Pianta, P.le Cascine 15, 50144 Firenze, Italy

David Harper
Department of Biology, University of Leicester, LE1 7RH, UK

Julian B. Hayball
Earth Observation Sciences Ltd, Farnham, Hampshire, UK

Jap Huygen
The Winand Staring Centre for Integrated Land, Soil and Water Research,
PO Box 125, 6700 AC Wageningen, The Netherlands

Luka Isavwa
Regional Centre for Services in Surveying, Mapping and Remote Sensing
(RCSSMRS), PO Box 18118, Nairobi, Kenya

Geoff Johnson
Department of Biology, University of Leicester, LE1 7RH, UK

Ingrid Jüttner
Catchment Research Group, School of Biosciences, Cardiff University, Cardiff,
CF1 3TL, UK

John Kirkby
Department of Geography, University of Northumbria at Newcastle, Newcastle
upon Tyne, NE1 7RU, UK

Clare Madge
Department of Geography, University of Leicester, LE1 7RH, UK

Kenneth Mavuti
Department of Zoology, University of Nairobi, Box 30197, Nairobi, Kenya

Massimo Menenti
The Winand Staring Centre for Integrated Land, Soil and Water Research,
PO Box 125, 6700 AC Wageningen, The Netherlands

Roger Michellon
CIRAD, BP 319, 110 Antsirabe, Madagascar

David J. Mitchell
School of Applied Sciences, University of Wolverhampton, Wolverhampton,
WV1 1SB, UK

Sam Mutiso
Department of Geography, University of Nairobi, Box 30197, Nairobi, Kenya

Dorothy Mutisya
Department of Geography, Kenyatta University, Nairobi, Kenya

Giovanni Narciso
Institute for Soil, Climate and Water, 600 Belvedere St, Arcadia, Private Bag X79,
0001, Pretoria, South Africa

Phil O'Keefe
Department of Geography, University of Northumbria at Newcastle, Newcastle
upon Tyne, NE1 7RU, UK

Steve Ormerod
Catchment Research Group, School of Biosciences, Cardiff University, Cardiff,
CF1 3TL, UK

Ambrose Oroda
Regional Centre for Services in Surveying, Mapping and Remote Sensing (RCSSMRS), PO Box 18118, Nairobi, Kenya

Nic Pacini
La Bindola 5, Stia, Arezzo, Italy

Susan E. Page
Department of Biology, University of Leicester, LE1 7RH, UK

Sylvain Perret
CIRAD 29A, Clonmore, PO Box 70793, Bryanston 2021, South Africa

Philip Polk
Laboratory of Ecology and Systematics, Free University of Brussels, Pleinlaan 2, 1050 Brussels, Belgium

Christian Prat
ORSTOM – 911 Avenue Agropolis, BP 5045, 34032 Montpellier Cedex 1, France

Paul Quantin
5, Rue Boileau, F21000, Dijon, France

Keith Richards
Department of Geography, University of Cambridge, CB2 3EN, UK

Jack Rieley
Department of Geography, University of Nottingham, NG7 2RD, UK

Carlos Schaefer
Departamento de Solos, Universidade Federal de Vicosa-Minas Geras, 36570, Brazil

Helen Schneider
Pusat Environmental Education Centre, PO Box 03, Trawas, Mojokerto 61375, East Java, Indonesia

Kapil Dev Sharma
Central Arid Zone Research Institute, Jodhpur 342003, India

Matthew Stuttard
Remote Sensing Applications Consultants Ltd, 4 Mansfield Park, Alton, GU34 5PZ, UK

Mauro Suppo
Aquater Eni Group, CP 20-61047, San Laurenzo Campo, Pesaro, Italy

Jurgen Tack
Laboratory of Ecology and Systematics, Free University of Brussels, Pleinlaan 2, 1050 Brussels, Belgium

Jacques Tassin
CIRAD Réunion, Ligne Paradis, 7, chemin de l'IRAT, 97410 Saint Pierre,
Isle de la Réunion

Alberto Vich
Centro Regional Andino – INCYTH, PO Box 6, Belgrano 210, 5500 Mendoza,
Argentina

Claude Zebrowski
ORSTOM – 32, Avenue Henri Varagnat, 93143 Bondy, Cedex, France

Preface

"Sustainable management" is the buzz-phrase of the 1990s, raised into public consciousness by the Rio Convention of 1992. In most, if not all countries of the world, it is a distant goal even though the long-term survival of our species requires that we set a course towards it in all human fields of endeavour, as soon and as completely as possible. The management of the natural resources on the surface of our planet – the soil and the water – cannot be separated since their natural processes are interlinked, and this is the reason that the past three decades have seen an increasing emphasis on catchment-scale management. River basin authorities exist in different forms in most countries of the world, at least on paper, and most practitioners would agree that catchment management is the appropriate scale for sustainable resource management. Nevertheless, even in developed countries, holistic catchment management is not yet high on the political agenda and the consequences of many land use decisions, particularly agricultural and urban development, for watercourses and water bodies, are rarely taken into account. The need for a body of knowledge in the sustainable management of catchments is great in all countries, but the supply of reference material is still poor.

This book is the result of a conference of the same name held at the University of Leicester at the end of the EU "Science and Technology for Development" Programme. The primary aim of the conference was to bring together people, representing as wide a range of disciplines as possible, working in tropical development, to discuss their research and management issues within the context of catchment-scale frameworks. This is because there was a strong feeling among the organisers that one of the many problems in promoting sustainable development was the lack of interdisciplinary collaboration; between managers and researchers, engineers and scientists, social scientists and physical scientists, and between ecologists and hydrologists. The second impetus for the conference was the completion of the Editors' work in Kenya under one of the EU STD Contracts in the mid-1990s which, although primarily academic in nature, was catchment-scale and involved close interdisciplinary collaboration and liaison with catchment engineers and water managers.

The conference was attended by over 150 delegates from 14 countries representing all the disciplines referred to above; from seven countries in the European Union and seven countries in the tropics. From the presentations at the conference, manuscripts were peer-reviewed and 21 papers have been accepted to form the chapters of this volume, from 45 authors, and covering issues in the three major tropical continents.

The book is divided into three sections reflecting the major areas of research. Inevitably, it represents only a snapshot of the research and management undertaken during this decade and it leans more heavily towards research than to management, reflecting the tendency for researchers to write more readily than managers. Nevertheless, successful management comes in large part through the application of sound research results and experience. The chapters in this book all consider the management applications of their research and together offer a number of clear case studies and messages from the 1990s to the new millennium. Chief among these is the overwhelming importance of local peoples to the successful implementation of development plans, and the need to learn from them at one scale and educate them at another. Second is the need for multidisciplinary approaches to developments at any scale, but particularly the whole-catchment and the supra-catchment scales. Third is the need for appropriate integration of the benefits from modern technological advances with the knowledge and skills of generations of tropical farmers. None of these messages are new, but the case studies in this book provide additional strong ammunition for the necessary radical changes in the world's approach to tropical development in the new century.

The Editors are grateful to Professor Anthony Parsons and Dr Jane Wellens, of the University of Leicester, together with anonymous referees, who provided external review of the manuscripts submitted.

David Harper

Tony Brown

Introduction

CHAPTER 1

Sustainability in the Context of Tropical Catchments

TONY BROWN
Department of Geography, University of Exeter, UK

DAVID HARPER
Department of Biology, University of Leicester, UK

INTRODUCTION

HISTORICAL CHANGES IN ATTITUDES TO TROPICAL DEVELOPMENT

"Development" has a history as long as that of colonialism, but its more recent manifestation, the post-Second World War concept of international development, grew from the concept of "first" and, to a lesser extent, "second" worlds assisting the economic progress of the "third" world towards global equality (Ryrie, 1995). The industrial nations of the world realised at the end of the 1940s that their own development over the past hundred years had made them both economically and technologically capable of alleviating the poverty and malnutrition with which over half the world, chiefly in the tropics, were afflicted. They established international institutions, chiefly the United Nations and the World Bank, to achieve this. At the same time individual countries in North America and Western Europe promoted bilateral assistance programmes with those tropical countries which were either colonies (and subsequently ex-colonies) or within their political spheres of influence.

Underlying the spirit of such development assistance have always been humanitarian motives, but they were also driven by the strong political and commercial motives of the donor countries, particularly during the "cold war" era of ideological conflict between the capitalist "first" and the socialist "second" worlds. Initially, belief in the ability of technology to solve agricultural and medical problems in the "third" world, as it had in the "first" world, was almost universal, but as the 1960s drew to a close there was increasing concern about detrimental environmental effects arising from blind faith in technology (such as the "green" revolution), as well as about the problems arising from the rapid population growth caused by improved medical aid which was not supported by improved agricultural production (Borgstrom, 1967;

The Sustainable Management of Tropical Catchments. Edited by David Harper and Tony Brown.
© 1998 John Wiley & Sons Ltd.

Farvar and Milton, 1972; Goldsmith et al., 1972). This concern was focused by the international scientific agencies such as the Food and Agriculture Organisation of the UN (FAO) and the UN Educational, Scientific and Cultural Organisation (UNESCO), supported by international research efforts such as the International Biological Programme (IBP) and by international environmental and development charities such as the International Union for the Conservation of Nature (IUCN) and others (Dasman et al., 1972), and culminated in the publication of *The World Conservation Strategy* (IUCN, 1980).

High-profile international conferences, such as the UN Conference on the Human Environment in Stockholm (Ward and Dubois, 1972) which established the UN Environment Programme (UNEP), reports of commissions such as those of Brandt (Brandt, 1980, 1983) and Brundtland (1987) on development, and reports on the destruction of global resources such as those of Myers (1979, 1984) progressively raised environmental issues higher up the development agenda. In 1992, at the Rio de Janeiro Earth Summit, the international community adopted the Convention on Biological Diversity, and the concept of "sustainable development" entered the dictionaries of international jargon.

Since the late 1980s the "second" world has effectively collapsed (although a handful of nations, notably China, retain the socialist style of government), and so alongside the global increase in environmental awareness has occurred a major change in the economic and political climate for development. A further important phenomenon has been the growing awareness that the conventional "capitalist" approach to development aid has been effectively a failure (Balek, 1992; Ryrie, 1995) causing levels of debt in recipient countries which themselves smother further economic growth, and that in its place a more open market is required (World Bank, 1997). The world is thus entering the third millennium with the recognised need to combine development which provides for the needs of all of its inhabitants, with conservation of the living resource base of biodiversity (Swanson, 1997), within an economic framework based upon fair and open trading between nations, with a lessened role of the state within each nation's economy.

Developed nations have appreciated the importance of a catchment approach to the management of natural resources for several decades (see below) even though political boundaries have often overridden watershed boundaries. It is without dispute that the single most important resource necessary for any development is water, and for sustainable development in regions where there is a scarcity of water, holistic management of the entire hydrological cycle is vital.

THE CATCHMENT (RIVER BASIN) AS A UNIT OF MANAGEMENT

HISTORICAL APPROACHES TO CATCHMENT MANAGEMENT

Early attempts at development frequently took a limited (single-issue) approach to the catchment (if any approach at all) with the result that there are now many examples of disastrous resource developments (Goldsmith and Hillyard, 1986) and many more examples where single-issue development occurred to the detriment of people, environment and wider resource-use options (Adams, 1992). With water as the most basic natural resource, a catchment-based approach to land management is essential in all countries, not just those of the tropics.

With some notable exceptions such as the Tennessee Valley Authority schemes, until relatively recently most major water-engineering schemes were essentially single-goal orientated, towards hydro-electricity generation (such as the Seven Forks scheme in Kenya), irrigation, or water extraction. Any other economic advantages were essentially seen as bonuses, or marginal, with costs that could be offset against any economic or social benefits or that would mitigate any economic or social costs of the scheme as a whole. Unfortunately, in most cases the schemes' benefits were to be national, or at least bring returns via the national government, whilst the disadvantages, and so some costs, would be locally or regionally borne. In many cases the subsidiary benefits did not materialise due to environmental changes, one example being the use of reservoirs for fishing which was frequently prevented by eutrophication or health considerations. There is little doubt that the environmental impacts of most large water-engineering schemes, especially dams, have been under-estimated. This holds for geophysical effects such as increased earthquakes and land instability (Keller and Pinter, 1996), increased soil erosion through the concentration of stock, and the increase in water-borne diseases (Timberlake, 1985).

TEMPERATE MULTIPURPOSE MODELS FOR DEVELOPMENT

There are some examples of temperate multipurpose schemes which have often been taken as models for development in other climatic zones. Probably the most obvious example is the Tennessee Valley Authority (TVA) scheme which dates from 1935. In many ways this scheme reveals both the complexity of demands such schemes may seek to satisfy and the complex environmental and socio-economic effects they can have (Finer, 1944). The TVA, which covers an area four-fifths the size of England and larger than Rwanda or many other small developing states, has to be seen in its historical context. It was born out of the Depression of the 1930s and it had the expressed aims of raising the living standards of the local population in one of the poorest areas of the United States. This was to be achieved by increasing productivity and increasing the level of consumption and possessions, i.e. by stimulating local demand and the local economy, thus maintaining the rural population and providing significant state/national benefits through flood control, food security and power generation. All this was to be achieved by infrastructural development of navigation, flood relief, hydro-electricity generation, irrigation, agricultural development and soil conservation (Finer, 1944). Even in the 1930s it was realised that the project had several potential negative impacts. For example, fertilisers and raw materials were not subsidised so as not to impact the open market, neither was electricity sold at below cost. In the 1940s the scheme was criticised for failing to sustain the development of co-operatives, acquire land for demonstration schemes and development, and bring forward legislation. The TVA is remarkably similar in physical terms to schemes which were part of the modernisation of agriculture in the USSR, which were often centred on major irrigation schemes, such as those on the Volga. There are of course

major differences including the collectivisation of Soviet schemes and their orienta-
tion to urban and outside demand rather than the regional economy. We can see these
large-scale multiple-use schemes as part of "Modernism", combining the white
heat of technology with a degree of social engineering. They were advocated with
proselytising zeal as major advances for both "first" (Lilienthal, 1944) and "second"
(Sorokin, 1967) worlds. By 1945 the TVA had received 11 million visitors including
civil servants from almost every nation, including China.

From the late 1930s two different forms of development evolution have occurred:
small-scale and large-scale. In the advanced New World economies, small-scale
developments included the advancement of area demonstrations set up by both
central and regional authorities and by farmers' associations forming the bases of
agricultural advice, education and research networks. In the many colonies this did
not occur in the same way, as in the twilight of the European Empires, agricultural
development was advanced through both Imperial and post-Imperial institutions
such as the Imperial Soil Bureau (British), the West Indian Technical Assistance
Agency (1897), the Imperial Institute (1893) and ultimately the Commonwealth of
Nations (1931). In most cases, this was also an extension of earlier colonial
development begun through the establishment of botanical gardens (Worboys,
1990). An example was the Agricultural and Horticultural Society of India, established by the British, and through which the Indian Government had, until the
famines of the 1860s, operated its agricultural research and development because
"those objects [agricultural innovations] will be best attained through the medium of
public societies in aid of whose funds a subscription on the part of the government
would be properly applied" (Kumar, 1990). It is this liberal concern for the freedom
of the market from the distortions of direct government control and the freedom of
commercial interests from the risks of agricultural development that probably
explains the relative lack of large-scale development projects in the remaining British
Colonies during the 1960s and 1970s, and upon independence the colonial agricultural infrastructure was generally adopted with little change by the new governments.

The other strand of agricultural development from the late 1930s was the almost
wholesale incorporation of the multiple-purpose development scheme into fascist
ideology of some developing and developed nations. Two examples are Spain and
Italy, where schemes generally included the compulsory movement of farmers and
villages for dam and canal construction. In Germany, and to a lesser extent Italy, the
large-scale approach and Imperial aspirations were combined, although further
evolution of this style of development was "interrupted" by the Second World
War. It is then not until after the post-war reconstruction of Europe and Japan
that similar large-scale schemes re-emerged as the solution to under-development, but
now with investment from the US and USSR and global organisations such as the
World Bank (1945) and the United Nations (1945). Although integrated in resource
terms (e.g. irrigation with hydro-electric power and navigation), these schemes were
rarely integrated into the local economy and became the focus of increasing criticism
through the late 1960s into the 1970s (Schumacher, 1973). Even some of the most
recent of such schemes will be remembered not for their economic benefits but for
political accusations of corruption and the problem of external dependency of
developing nations (e.g. the Pergau Dam in Malaysia). Given this history of large-

scale capital-intensive development, it is not surprising that any emphasis on the catchment approach to development must be combined with appropriate multi-scale development and socially integrated schemes. This is only likely to be the case where the aims do not contradict local development. In this respect the TVA is a better model than most others with its emphasis on the local population and their economic status. Many would argue that its limited use of economic levers is also appropriate, but this must depend upon the political and economic context. As Cooper (1973) argued, often the problem is the inappropriate use of developed technology and an uncritical notion of the value of science because it produces over-specified commodities and makes intensive use of resources that are scarce in less developed countries. The recent stress on the sustainablility of development, despite its theoretical weaknesses (Jacob, 1994), does address this major criticism of earlier integrated projects.

MULTI-SCALE SUSTAINABLE CATCHMENT MANAGEMENT: A THEME FOR THE TWENTY-FIRST CENTURY

RESEARCH ISSUES FOR CATCHMENT-SCALE APPROACHES

The catchment and drainage basin is the fundamental unit of monitoring, analysis and management in hydrogeomorphology because it is amenable to mass budget analyses of water, sediment and nutrients. However, it is not the only areal unit in dryland hydrology, as in many areas development is dependent upon regional aquifers and inter-basin transfers (Brown, 1996). Indeed, in some situations the fundamental problem may be hydrogeological in the definition and delimitation of the flux of groundwater and its susceptibility to environmental change (Brown, 1996). Examples of this include the Bils area of Bangladesh (Ahmed and Burgess, 1996) and many dunefields such as those in north-east Nigeria (Carter, 1996).

Several important areas of recent catchment research particularly relevant to the tropics can be identified. The first is sub-catchment variability of precipitation. This is important because tropical storms are often quite localised and/or occur on distinct tracks and this will affect basins of different sizes differentially.

A second area of research is the prevention of slope erosion particularly through raindrop impact and soil detachment, although tillage erosion may become more important in the tropics as mechanisation becomes more common and if field size increases. New methods are now increasingly being used in the tropics, including the use of Cs^{137} to determine site-specific erosion rates (Kulander and Stromquist, 1989). Modelling has increasingly been used to provide a basis for regionalising small watershed and slope-scale studies. An approach that began in the early 1980s of applying models of different levels of complexity has increased due to the wider availability of relatively cheap and user-friendly models which can be used on PCs or workstations. As well as work on both statistical and deterministic modelling of tropical catchment dynamics, there has been an increase in the application of geographical information systems (GISs), an example of which is Schneider and Brown (Chapter 3). Alongside theoretical research there has been a continued drive to apply research work through practical demonstrations and the evaluation of current practice. A number of texts exist, promoting practical approaches to soil and water conservation, an example being Pereira (1989), which advocates hillside ditches, bench terraces, live hedgerow-supported terraces, cut-off drains, wooden gabions, etc. Of particular value may be the identification and economic support of crops or protective cover with a particularly low erosivity, including in drier areas the dwarf fan palm (Brown, 1990) and in wetter areas Napier grass (Africa) or Brachiaria grass (Indonesia) and multi- and inter-cropping. The use of simple erosion control structures has shown to be effective, one example being the *fanyu juu* of Kenya (Figure 1.1), which is a ditch dug on the contour with the soil thrown uphill to form a bank that is then seeded with grass (Chapter 2; Thomas and Biamah, 1991). The fact

Figure 1.1 A soil conservation trench (*fanyu juu*) stabilised with Napier grass in the Kaihungu sub-catchment of the Tana basin, Kenya (see Chapters 2 and 3).

that it is more labour intensive than a simple narrow-based channel terrace, although less prone to silting, is not necessarily a problem if returns are high enough to maintain a rural labour force. In Java group demonstration projects of 10 ha have been set up incorporating all the surrounding farmers and using locally derived appropriate technology (Pereira, 1989).

A major problem highlighted by at least one chapter in this volume (Chapter 2) is the high erosion potential of roads and tracks in densely inhabited mountainous areas of the tropics. Studies have shown that roads and tracks can be the most active runoff generating components in the landscape (Harden, 1992), and so more attention is needed in relation to their location (on ridge-tops where possible) and design incorporating features such as side berms, correctly spaced culverts, rock surfacing and the use of local wood and brush in runoff control (Adams and Andrus, 1990).

At the end of a comprehensive review of tropical watershed science (Lal and Russell, 1981), Periera (1981) set out three areas of priority for future research. The first, to organise multidisciplinary practical studies of agriculture and hydrology of small eroding basins, has certainly occurred, as this volume illustrates. The second, to obtain quantitative evidence of the rate and extent to which catchment stability can be improved, seems still some way off. The third, to combine the concept of the representative basin with watershed applications, has progressed, through the routine use of remote sensing and GIS. The aim of future research must therefore remain quantifying and costing the potential improvements on water quality (and quantity) and soil fertility, i.e. showing that sustainability is achievable and proving that it has both economic and social benefits.

From the research perspective there are three fronts on which future developments will advance: firstly, a better understanding of physical and biochemical processes upon which improvements in modelling will be based; secondly, new techniques especially in the area of monitoring and modelling; and, thirdly, in the area of new technologies for impact minimisation. However, in the short to medium term, management will have to proceed with inadequate data. In both the short and the long term no management can be successful in environmental terms without sensitive and appropriate social and economic policies. This last constraint requires inter-disciplinary work and a willingness of all the expert parties to broaden their intellectual horizons in order to work with the dynamics of the existing social systems rather than in isolation.

PROCESSES AND MODELLING IN TROPICAL CATCHMENTS

The hydrology of tropical soils has received rather less detailed attention than that of temperate soils. It has, for example, often been assumed that bypassing flow is less common; whilst this may be the case in nitisols, it is certainly not true for vertisols which are common in the drier parts of tropical catchments (Smaling and Bouma, 1992). The determination of the hydrological properties of tropical soil *in situ* is rare – an exception being the field determination of hydraulic conductivity (K) for fersiallitic soils in Zimbabwe by Twomlow (1994). The few results from tropical instrumented catchments that there are suggest that stormflow (quickflow) is a high but seasonally variable component of total discharge (typically $> 50\%$ annually). Studies have shown that soils of moderate permeability can impede vertical percolation producing ponding in the upper 10–20 cm (Gilmour, 1980). The causes of this include swelling clays and high rainfall intensity. The result of this is short lag times and high rates of soil detachment, frequently with a risk of mudflows.

One area that has received considerable study is the effect of logging on sediment yields and the research has reached a point at which management can be provided with clear guidelines (Douglas, 1996; Greer et al., 1996, Chapter 9). The increasing agricultural demands in tropical areas will undoubtedly lead to agricultural intensification with an increased use of fertilisers. This presents the spectre of an increasing risk of eutrophication. It is not clear that buffer-strips and floodplains will act in the same manner in tropical as in temperate catchments, given that the processes of nitrogen stripping are microbiological in temperate catchments and are related to the organic content of floodplain soils (Burt and Haycock, 1996).

An area bound to receive increased attention over the next decade is the sensitivity of tropical catchments to climate change, as trends in sediment yield have already been identified from temperate and semi-arid zones (Walling and Webb, 1996). This area is particularly important and will involve both the re-interrogation of long-term records and work on the relative contributions of sediment and nutrient sources and sinks.

NEW TECHNIQUES IN TROPICAL CATCHMENT MANAGEMENT

Given the size and heterogeneity of tropical catchments, the gathering of data for resource models such as land use and soil data has been revolutionised by remote sensing. Several chapters in this volume show how remote sensing can be used to generate input data such as crop type, indices of plant cover (e.g. improved NDVI type measures), proxy data such as lake level changes and land use changes such as deforestation (Chapter 3, Chapter 20). There remain, however, serious limitations to the use of remote sensing, the most significant of which are the complex relationships between total photosynthetic area, reflectance and the hydrological characteristics of plants. Further advances will require a greater understanding of the relationship between crop geometry, viewing angles and reflectance, and combining these data with remotely sensed meteorological data will make it possible to predict raindrop impact and runoff intensities. Since antecedent moisture conditions play an important role in tropical catchments, a second area of advance is in the use of microwave remote sensing in order to measure soil moisture through the vegetation canopy.

Digital numerical modelling of catchments is increasingly becoming a practical alternative to empirical and statistical modelling. This is largely because of the ease of coupling such models to GISs. The most rapid advances are being made in the semi-arid and arid zones (MEDALUS, 1993; Sharma, 1993), due to a relative simplicity in soil/vegetation processes, and transport rather than supply limitations on sediment.

An as yet unresolved problem in tropical catchments is the prevalence of terracing, which, as illustrated in Chapter 2, means that digital slopes derived from maps or SPOT data bear little resemblance to ground slopes nor to ground shear-stress patterns, the danger being overestimation of both runoff and sediment yield if risers are stable. This may be partially offset by the increase in land area.

However, given the need of immediate help for managers, the use of empirical and pragmatic models will continue. A short cut to the modelling of different crops and the effect of land use practices such as terracing may be the use of knowledge-based rules in GIS erosion models (Moore et al., 1991; Skidmore et al., 1991) and improvement to the spatial and temporal components of existing empirical models such as the USLA or CREAMS (Rudra et al., 1986). The use of natural tracers or radioisotope fallout from nuclear testing to measure erosion rates, well established in temperate and semi-arid zones (Quine et al., 1992, 1993; Quine et al., 1996, is just starting in the more humid tropics (Kulander and Stromquist, 1989).

NEW AND APPROPRIATE TECHNOLOGIES FOR TROPICAL CATCHMENT MANAGEMENT

Traditional techniques can be enhanced (e.g. *tassa*, which are planting pits to capture water around sorghum and millet or the use in East Africa of *fanyu juu* (Figure 1.1)). These measures can be highly effective if maintained. Their maintenance is a function of an adequate population working the land and sufficient returns, either in product or cash, for that work.

Whilst most emphasis has been placed on the use of traditional and well-tested soil conservation techniques, there is a role for new technology here. The use of soak-aways, biogas stoves (Chikomba, 1993), check-dams, erosion control mats, geocells, natural and synthetic mats, and soft engineering approaches such as biodegradable rip-rap, gabions made from old mattresses, etc. is undoubtedly a way forward. Agroforestry has also become a major area of research and trials in tropical catchments. In East Africa, agroforestry ranges from the planting of shrubs or hedgerows of plants such as *Cassia siamea* (a leguminous shrub from Asia) forming natural micro-terraces, to local trees such as *Croton megalocarpus* interplanted with maize. However, given that one of the most significant sources of both runoff and sediment in tropical catchments can be roads, paths and tracks, the use of permeable metalled road surfaces and associated drainage systems should be evaluated and improved design criteria formulated.

A LOSS OF INNOCENCE: TROPICAL CATCHMENTS AND PEOPLE

In the extreme, and hopefully rare, cases of widespread social disintegration and conflict, environmental management obviously takes a back-seat in the face of immediate humanitarian concerns. However, even in these cases, such as in Rwanda, it is not long before relief efforts must be directed to sustainable agricultural productivity often in the face of greatly reduced infrastructural support. As Lewis (1991) pointed out, there were degrees of freedom in the Rwandan catchments for agriculture to evolve in both a highly productive and sustainable direction. The intellectual division between the sciences and social sciences has all-too-often in the past led to a degree of naivety on both sides as to the real causes of land degradation, most specifically the belief that population pressure is inevitably the root cause of degradation. On the other hand, Whitlow (1988) has shown that physically based mapping of erosion hazard can be a poor predictor of the extent of eroded land. Assuming that the estimation of erosion severity from aerial photographs is a reliable measure of current erosion rates (and there is some doubt about this as it is not always the case, see Thornes and Gilman, 1983), then in the case of Zimbabwe the best predictor of erosion by far is the population density (Whitlow, 1988, 1990). This may be explained by at least two arguments. The first is the argument that it is land with a moderate hazard that has seen the greatest increases in population and this has produced the most visible forms of erosion although not necessarily the highest rates. The second is that this relationship is characteristic of semi-arid rather than tropical catchments. Even if this is a causal relationship, as claimed by Whitlow (1990), given that the increase in population in less developed countries is a function of their age structure, population density is not a potentially changeable variable in the soil-loss equation. Not only is the high population argument a hypocritical position for experts from the developed world, but it ignores social processes and is fundamentally wrong in the sense that high population does not necessarily or inevitably lead to environmental degradation (Tiffen et al., 1994, and see Chapters 3, 10 and 13). Indeed, the statistical association of high population densities at the district or regional scale with high erosion and hence the view that it is people's mismanagement of land that causes erosion is a classic example of "the ecological fallacy". Instead it is the pattern and practices of agriculture and land use that create an increased erosion hazard and this may or may not be linked to population growth, decline or to external economic forces. Ideally, if human factors are to be included in hazard-mapping, a variety of variables related to current land use need to be included (Millington et al., 1982). It is therefore just as naive to regard high population as the cause of erosion as it is to proclaim the opposite as a universal generalisation. In a detailed and important study of the Machakos district in Kenya, Tiffen et al. (1994) have shown how, over a 60 year period, population increased fivefold but, due to terracing and tree management, erosion decreased and agricultural production per head and per hectare increased. This and similar studies (Blaikie, 1985; Blaikie and Brookfield, 1987) demand that we revise our theoretical framework of population growth and environmental degradation. It may be time to consider expressing erosion not in absolute terms but as erosion per head or per unit of agricultural output, especially in the tropics where deep soils can have a relatively high soil-loss tolerance. The form that erosion takes is also important, as the same rate from gullies, as opposed to sheet- and rill-erosion, has different socio-political implications.

Advances can be made: one example is the development of community based land use planning in which land use evaluation and zonation only takes place after community meetings. In the case of Ngamiland West (South Africa) these meetings took place both in the village meeting places (*kgotla*) and in remote cattleposts (Van der Sluis, 1994). A similar programme in Lesotho involved individual villages drawing up plans for the annual work programme and "Headman village work-shops" (Khatiwada, 1993). Indeed the explosion of information concerning soil erosion through organisations such as the Southern African Development Community (SADC) is of significance both for the education of agricultural advisors and aid workers and increasingly the researchers in developed countries who have until recently had far too little knowledge of indigenous practices and social systems. An area of particular importance is research on indigenous soil and water conservation and harvesting (Reij, 1991). This concerns such questions as why are indigenous soil and water conservation techniques abandoned and, if so, under what pressures?

It is clear that sustainable development has to occur at a variety of spatial levels and most particularly at the local level. Many projects have illustrated that hydrological engineering, such as irrigation, changes social relations. One example where this was studied was in the Niger river valley near Timbuktu, North Mali. Small-scale pump irrigation technology was introduced to the Songhai and Bella peoples in 1984. The traditional farming system of these peoples was of low inputs and not labour intensive, with little dependence on external market relations. Social relations were changed by an emergence of land claims by men and particularly the Songhai men, worsening the position of both women and the Bella peoples as a whole. From their study, Ton and de Jong (1991) suggested that engineers have a responsibility to try to determine whether any scheme is suited to the specific characteristics of the social group(s) in question and more particularly to assess any adverse effect on any particular sub-group within the population. Of particular importance here is the role of women, who are often the rural labour force but have been excluded from decision-taking by a combination of traditional gender roles and strong colonial reinforcement (Warren and Bourque, 1991). The area of women's role in development and changing gender roles is extremely important, particularly in Africa, and will undoubtedly inform future development schemes. In Kenya this was recognised some time ago in the Women's Soil Conservation Project of the Muranga District. Projects should be conceived as open reactive systems (Figure 1.2) where local people's views and desires are built in from the start (Melkote, 1988), or indeed even earlier, as many would now doubt whether the imposition of a development scheme which has not arisen from local needs and desires can be justified ethically or is likely to succeed.

It is at the small to medium basin scale that these principles may be attainable and maximum efficiencies may be realisable. An obvious area to tackle is evaporation loss, as it has been estimated that around 70% of all water abstracted for irrigation never

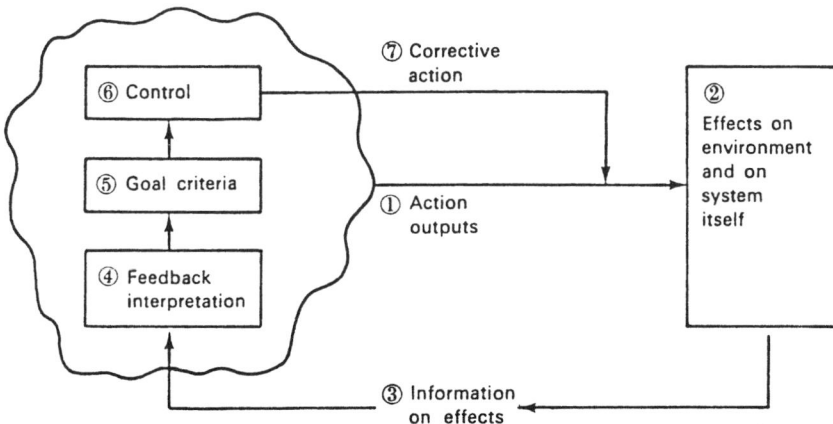

Figure 1.2   An open reactive system for project communication. From Melkote (1988).

reaches the crop (UNWC, 1978), and at the small- to medium-scale project this can be reduced by appropriate design and practices.

At the large catchment scale there are both the most serious obstacles to sustainable management but also undoubtedly the most potential hydrological benefits. At least 214 major river basins are multinational and this includes most African rivers (Falkenmark, 1986). The difficulties of managing the water resources of a cross-national catchment are illustrated by the Ganges, where there is a 40-year-old dispute between India and Bangladesh. The dispute currently focuses on the Farakka Dam (barrage) which, it is claimed, reduces dry-season flow in Bangladesh. United Nations mediation has been required and yet there is still no agreement, with India and Bangladesh proposing their own basin management plans (Falkenmark, 1986). There is, however, evidence that during the later 1980s the major international aid organisations started taking a broader, more socially and ecologically informed approach. In the Dumoga Irrigation and Water-Catchment Protection Project in Indonesia, the World Bank supported the protection of 300,000 ha of tropical rainforest in conjunction with the Indonesian Government and the World Wildlife Fund, who created a National Park. Irrigation increased rice yields fourfold, and farmers' incomes doubled (Wind and Sumardja, 1988). Whilst there were clear lessons to be learnt from the project, particularly concerning the involvement of local inhabitants which might have reduced illegal invasion of the protected forest, the joint inclusion of both the catchment and the irrigated area, and the collaboration of ecological agencies and development agencies must surely be a pointer to future sustainable development schemes in the tropics.

From the above discussion and the chapters in this book, the message for the future would seem to be that small- to medium-scale approaches to catchment management are likely to be most successful in both ecological and economic/social terms and therefore sustainable. In a recent collection of studies in Africa, Reij et al. (1997) illustrated this theme, with the most successful projects being of intermediate scale (i.e. sub-regional, 10^3–10^4 km^2), based on holistic (integrated), village-based local participation and management, and largely using locally derived materials and low or intermediate technology. This is fundamentally the *raison d'être* for the wide coverage of this volume, with studies from indigenous practices (Chapter 14) to modelling mangrove swamp hydrology (Chapter 21). The breadth of techniques and specialisms included in this volume point the way to a more sophisticated and sensitive approach to the management of tropical catchments in both environmental and human terms.

REFERENCES

Adams, P.W. and Andrus, C.W. 1990. 'Planning secondary roads to reduce erosion and sedimentation in humid tropic steeplands'. In *Research Needs and Applications to Reduce Erosion and Sedimentation in Tropical Steeplands*, pp. 318–346. International Association of Hydrological Sciences Publication 192, IAHS, Wallingford.
Adams, W.M. 1992. '*Wasting the Rain*'. Earthscan, London.
Ahmed, K.M. and Burgess, W.G. 1996. 'Bils and the Barind aquifer, Bangladesh'. In *Groundwater and Geomorphology* (ed. A.G. Brown), pp. 143–156, Wiley, Chichester.
Balek, J. 1992. '*The Environment for Sale*'. Carlton, New York.

Blaikie, P. 1985. '*The Political Economy of Soil Erosion in Developing Countries*'. Longman, London.

Blaikie, P. and Brookfield, H. 1987. '*Land Degradation and Society*'. Methuen, London.

Borgstrom, G. 1967. '*The Hungry Planet*'. Macmillan, New York.

Brandt, W. 1980. '*North–South: A Programme for Survival*'. Pan, London.

Brandt, W. 1983. '*Common Crisis*'. Pan, London.

Brown, A.G. 1990. Soil erosion and fire in areas of Mediterranean type vegetation: results from Chaparral in southern California, USA and Mattoral in Andalucia, southern Spain. In *Vegetation and Erosion* (ed. J.B. Thornes), pp. 269–287, Wiley, Chichester.

Brown, A.G. (ed.) 1996. '*Groundwater and Geomorphology*'. Wiley, Chichester.

Brundtland, G. 1987. '*Our Common Future*'. Oxford University Press, Oxford.

Burt, T.P. and Haycock, N.E. 1996. 'Linking hillslopes to floodplains'. In *Floodplain Processes* (eds M.G. Anderson, D.E. Walling and P.D. Bates), pp. 461–492. Wiley, Chichester.

Carter, R.C. 1996. 'Groundwater recharge and outflow patterns in a dunefield of North East Nigeria'. In *Groundwater and Geomorphology* (ed. A.G. Brown, pp. 157–176. Wiley, Chichester.

Chikomba, A.B. 1993. 'The biowaste gas stove'. *Splash*, **9**, 13.

Cooper, C. (ed.) 1973. '*Science, Technology and Development*'. Frank Cass, London.

Dasman, R.F., Milton, J.P. and Freeman, P.H. 1972. '*Ecological Principles for Economic Development*'. Wiley, Chichester.

Diaz-Bordenave, J. 1977. '*Communications and Development*'. UNESCO, Paris.

Douglas, I. 1996. 'The impact of land use changes, especially logging, shifting cultivation, mining and urbanization on sediment yields in humid tropical Southeast Asia: a review with special reference to Borneo'. In *Erosion and Sediment Yield: Global and Regional Perspectives*, pp. 463–472. International Association of Hydrological Sciences Publication 236, IAHS, Wallingford.

Falkenmark, M. 1986. 'Fresh water as a factor in strategic policy and action'. In *Global Resources and International Conflicts* (ed. A.H. Westing), pp. 85–113. Oxford University Press, Oxford.

Farvar, M.T. and Milton, J.P. 1972. '*The Careless Technology: Ecology and International Development*'. Doubleday & Co, New York.

Finer, H. 1944. '*The T.V.A: Lessons for International Application*'. International Labour Office, Montreal.

Gilmour, D. 1980. '*An Investigation of Storm Drainage Processes in a Tropical Rainforest Catchment*'. Australian Water Resources Council Technical Paper 56, Canberra.

Goldsmith, E. et al. 1972. '*Blueprint for Survival*'. The Ecologist, Wadebridge, Cornwall.

Goldsmith, E. and Hilyard, N. 1986. '*The Social and Environmental Effects of Large Dams, Vol. 2, Case Studies*'. Wadebridge Ecological Centre, Cornwall.

Greer, T., Sinum, W., Douglas, I. and Bidin, K. 1996. 'Long term natural forest management and land use change in a tropical catchment, Sabah, Malaysia'. In *Erosion and Sediment Yield: Global and Regional Perspectives*, pp. 453–462. International Association of Hydrological Sciences, Publication 236, IAHS, Wallingford.

Harden, C.P. 1992. 'Incorporating roads and footpaths in watershed-scale hydrologic and soil erosion models'. *Physical Geography*, **13**, 368–385.

IUCN, 1980. '*The World Conservation Strategy*'. IUCN, Gland, Switzerland.

Jacob, M. 1994. 'Towards a methodological critique of sustainable development'. *The Journal of Developing Areas*, **28**, 237–252.

Keller, E.A. and Pinter, N. 1996. '*Active Tectonics*'. Prentice-Hall, New Jersey.

Khatiwada, Y.P. 1993. 'Participatory village development planning'. *Splash*, **9**, 7–9.

Kulander, L. and Stromquist, L. 1989. 'Exploring the use of topsoil ^{137}Cs content as an indicator of sediment transfer in a small Lesotho catchment'. *Zeitschrift für Geomorphologie*, N.F., **33**, 455–462.

Kumar, D. 1990. 'The evolution of colonial science in India: natural history and the East India company'. In *Imperialism and the Natural World* (ed. J.M. MacKenzie), pp. 51–66. Manchester University Press, Manchester.

Lal, R. and Russell, E.W. 1981. '*Tropical Agricultural Hydrology*'. Wiley, Chichester.

Lewis, L.A. 1991. 'Relations between crops, topography and degradation in Western Rwanda: some strategies for sustainable agriculture. *Zeitschrift für Geomorphologie Suppl.*, **83**, 23–28.

Lilienthal, D.E. 1944. '*TVA: Democracy on the March*'. Penguin, Harmondsworth, London.

MEDALUS, 1993. '*Executive Summary*'. European Union, Environment Programme.

Melkote, S.R. 1988. Agricultural extension and the small farmer: revealing the communication gap in an extension project in Kenya. *Journal of Developing Areas*, **22**, 239–252.

Millington, A.C., Robinson, D.A. and Browne, T.J. 1982. 'Establishing soil loss and erosion hazard maps in a developing country: a west African example'. In *Recent Developments in the Explanation and Prediction of Sediment Yield*, pp. 283–292. International Association of Hydrological Sciences, Publication 137, IAHS, Wallingford.

Moore, D.M., Lees, B.G. and Davey, S.M. 1991. 'A new method for predicting vegetation distributions using decision tree analysis in a geographical information system'. *Journal of Environmental Management*, **15**, 59–71.

Myers, N. 1979. '*The Sinking Ark*'. Pergamon, New York.

Myers, N. 1984. '*The Primary Source*'. Norton, New York.

Pereira, H.C. 1981. 'Future trends in watershed management and land development research'. In *Tropical Agricultural Hydrology* (eds R. Lal and E.W. Russell), pp. 465–468. Wiley, Chichester.

Pereira, H.C. 1989. '*Policy and Practice in the Management of Tropical Watersheds*'. Westview Press, Boulder and San Francisco.

Quine, T.A., Walling, D.E., Zhang, X. and Wang, Y. 1992. Investigation of soil erosion on terraced fields near Yantang, Sichuan Province, China, using caesium-137. In *Erosion Debris Flows and Environment in Mountain Regions*, pp. 155–168. International Association of Hydrological Sciences, Publication 209, IAHS, Wallingford.

Quine, T.A., Walling, D.E. and Mandiringana, O.T. 1993. An investigation of the influence of edaphic, topographic and land use controls on soil erosion on agricultural land in Borrowdale and Chinamora areas, Zimbabwe, based on caesium-137 measurements. In *Sediment Problems: Strategies for Monitoring, Prediction and Control*, 185–196. International Association of Hydrological Sciences, Publication 217, IAHS, Wallingford.

Quine, T.A., Walling, D.E. and Govers, G. 1996. Simulation of radiocaesium redistribution on cultivated hillslopes using a mass-balance model: an aid to process interpretation and erosion rate estimation. In *Advances in Hillslope Processed* (eds M.G. Anderson and S.M. Brooks), pp. 561–588. Wiley, Chichester.

Reij, C. 1991. '*Indigenous Soil and Water Conservation in Africa*'. IIED Gatekeeper Series 27, International Institute for Environment and Development, London.

Reij, C., Scoones, I. and Toulmin, C. 1997. '*Sustaining the Soil: Indigenous Soil and Water Conservation in Africa*'. Earthscan, London.

Rudra, R.P., Dickinson, W.T., Clark, D.J. and Wall, G.J. 1986. 'GAMES – A screening model of soil erosion and fluvial sedimentation on agricultural watersheds'. *Canadian Water Resources Journal*, **11**, 58–71.

Ryrie, W. 1995. '*First World, Third World*', Macmillan, London.

Schumacher, E.F. 1973. '*Small is Beautiful*'. Blond & Briggs, London.

Sharma, K.D. 1993. '*Distributed Numerical Modelling of Runoff and Soil Erosion using Thematic Mapper Data and GIS*'. Report 75.4, The Winand Staring Centre for Integrated Land, Soil and Water Research, Wageningen, The Netherlands.

Skidmore, A.K., Ryan, P.J., Dawes, W., Short, D. and O'Loughlin, E. 1991. 'Use of expert systems to map forest soils from GIS'. *International Journal of Geographical Information Systems*, **5**, 431–445.

Smaling, E.M.A. and Bouma, J. 1992. 'Bypass flow and leaching of nitrogen in a Kenyan vertisol at the onset of the growing season'. *Soil Use and Management*, **8**, 44–48.

Sorokin, G. 1967. '*Planning in the USSR*'. Progress Publications, Moscow.

Swanson, T. 1997. '*Global Action for Biodiversity*'. Earthscan, London.

Thomas, D.B. and Biamah, E.K. 1991. 'Origin, application, and design of the Fanyu Juu terrace'. In *Development of Conservation Farming on Hillslopes* (eds W.C. Moldenhauer, N.W.

Hudson, T.C. Sheng and L. San-Wei), pp. 185–194. Soil and Water Conservation Society, Iowa.

Thornes, J.B. and Gilman, A. 1983. 'Potential and actual erosion around archaeological sites in South-east Spain'. *Catena Supplement*, **4**, 91–113.

Tiffen, M., Mortimore, M. and Gichuki, F. 1994. '*More People Less Erosion*'. Wiley/Overseas Development Agency, Chichester/London.

Timberlake, L. 1985. '*Africa in Crisis*'. Earthscan, London.

Ton, K. and de Jong, K. 1991. 'Irrigation technology and social change: an analysis of the social variables of technology'. *Journal of Developing Areas*, **25**, 197–206.

Twomlow, S.J. 1994. 'Field moisture characteristics of two fersiallitic soils in Zimbabwe'. *Soil Use and Management*, **10**, 168–173.

Van der Sluis, T. 1994. 'Community-based land use planning – Ngamiland West'. *Splash*, **10**, 8–10.

UNWC, 1978. '*Water Development and Management*'. Pergamon Press, Oxford.

Walling, D.E. and Webb, B.W. 1996. 'Erosion and sediment yield: a global overview'. In *Erosion and Sediment Yield: Global and Regional Perspectives*, pp. 3–20. International Association of Hydrological Sciences Publication 236, IAHS, Wallingford.

Ward, B. and Dubois, R. 1972. *Only One Earth*. Pelican, Harmondsworth.

Warren, K. and Bourque, S.C. 1991. 'Women, technology, and international development ideologies: analysing feminist voices'. In *Gender at the Crossroads of Knowledge: Feminist Anthropology in the Postmodern Era* (ed. M. di Leonardo), pp. 278–311. University of California Press, Berkeley.

Whitlow, R. 1988. '*Land Degradation in Zimbabwe: A Geographical Study*'. Department of Natural Resources, Harare.

Whitlow, R. 1990. 'Potential versus actual erosion in Zimbabwe'. *Splash*, **6**, 8–12.

Wind, J. and Sumardja, E.A. 1988. 'World Bank irrigation and water-catchment protection project, Dumoga, Indonesia'. In *The Greening of Aid* (eds C. Conroy and M. Littvinoff), pp. 57–63. Earthscan, London.

Worboys, M. 1990. 'The Imperial Institute: the state and the development of the natural resources of the Colonial Empire, 1887–1923'. In *Imperialism and the Natural World* (ed. J.M. MacKenzie), pp. 164–186. Manchester University Press, Manchester.

World Bank, 1997. '*The State in a Changing World*'. IBRD/World Bank, Washington, DC.

SECTION A

Soils, Erosion and Land Use

Introduction

TONY BROWN

This section includes contributions relating to the spatial dimension of soil erosion and particularly the role of soil type and vegetation. Whilst this subject area has been the focus of much work in the semi-arid zones, particularly research into the desertification process, it has received far less attention in the tropics and particularly the humid tropics. Tropical soils are spatially variable, and in particular climatic and geomorphic zones may have fundamental agronomic limitations for sustainable agriculture. As Chapter 7 shows, the soils of the Roraima district of Amazonia are fundamentally unsuited to intensive agriculture for structural reasons, whereas Chapter 8 illustrates how the indurated soils of Mexican plateau, despite severe limitations, can be used for maize under a suitable management (low tillage frequency) regime. Due to the common deep weathering of tropical soils, soil depth may never be a limiting factor in productivity, but this presents a danger in that accelerated erosion may be sustainable in purely temporal terms, i.e. there will be no exhaustion of the supply of weathered regolith even under accelerated erosion rates. This deep weathering and high clay contents also mean that tropical soils often have more extreme structural problems than temperate soils, particularly structural degradation and compaction, reducing infiltration capacity and increasing surface erosion. The rapid turnover times for organic matter, whilst promoting nutrient retention, also reduces aggregate stability and so can increase erosion. The spatial variability of tropical soils also has implications for nutrient losses. Phosphorus loss from agricultural soils is not only a waste of a resource, from whatever source, but also a considerable problem for downstream ecosystems and fisheries potential (Chapter 5). Soil variability in phosphorus content and retention and phosphorus losses are related to lithology, with volcanic areas having high background levels. Where phosphorus is closely bound to soil particles, nutrient retention programmes can be based on physical measures. It is clear that far more attention needs to be given to soil-suitability mapping linked to agricultural advice and other management tools.

The provenance of sediment also reflects soil as well as land use variation in the catchment. Catchment size is an important variable and the scale problem discussed in Chapter 2 hides a number of generation and storage processes which are very poorly understood in tropical catchments. Few studies have been conducted of floodplain storage, bank erosion or channel storage, which in temperate and sub-tropical catchments can be of overriding importance in the sediment budget. A persistent problem has been that the sediment budget database for tropical catchments is poor in comparison with temperate catchments and inadequate monitoring as well as inherent climatic variability has caused considerable variation in sediment yields from the same basin, which clearly complicates the formulation of management policies. A partial but less expensive and practical alternative to the drive for ever more comprehensive monitoring systems (desirable though this is) is the identification of the main sediment-generating processes in the catchment and management to minimise their effects. Data presented in Chapter 2 and in Chapter 9 suggest that off-plot erosion of roads, tracks, paths, ditches, drains and infrastructural areas are disproportionately high sediment production areas. Given the steep slopes of many tropical catchments, and high population densities, there will always be small plot sizes and a high density of such features. As Chapter 2 indicates, methods need to be developed which can measure and integrate infrastructural areas and point sources with the standard inputs to GIS-based erosion potential modelling such as digital elevation data and remote sensing of vegetation cover. The most appropriate erosion-control methods for most small tropical catchments are small-scale structures, slope modifications and agronomic practices. Chapter 6 shows how a form of hedge planting can be multi-functional, not only reducing runoff but also improving soil quality and producing fodder. Likewise Chapter 10 suggests the retention of mature fruit-producing trees and underplanting can both increase yields and reduce erosion during crop establishment. It is surely no accident that these techniques are to some extent re-inventions of traditional land use practices (e.g. under- and inter-planting), many of which in the original form have been lost due to colonialism or the incorporation of tropical areas into national and global markets.

We need to re-evaluate the degrees of freedom available to those involved in development and this should be based upon both a sound and more particularly, appropriate, scientific basis, and a re-examination of the past and present relationships between land use practices and social structure. Soils have different degrees of freedom for different crops and at different yields/returns, likewise erosion management or mitigation measures have degrees of freedom related to the social system and the desires and aspirations of the local inhabitants. Re-focused agronomic research into sustainable production should have the aim of manipulating the former rather than the latter.

CHAPTER 2

From Plot to Basins: The Scale Problem in Studies of Soil Erosion and Sediment Yield

TONY BROWN
Department of Geography, University of Exeter, UK

HELEN SCHNEIDER
Pusat Environmental Education Centre, East Java, Indonesia

INTRODUCTION TO THE SCALE PROBLEM

This chapter addresses the problems of integrating the results of erosion and sediment data derived from investigations at different spatial scales from within the same catchment. This is a common and inevitable sampling problem for the management of large tropical catchments. The scale dependency of geomorphic processes was recognised in the 1960s and 1970s in the classic work of Schumm and Lichty (1965), Chorley and Kennedy (1971), and others. The lack of spatial continuity evident at different scales, from the experimental plot to large drainage basin is measured, but not explained, by the sediment delivery ratio (Roehl, 1962; Brown, 1987). An example of this is the common lack of fit between soil erosion/area relationships derived from plots and those derived from catchment monitoring. The possible causes are numerous, including systematic inaccuracies in the plot or catchment data, the lack of representivity of the plot sites and sediment storage.

It is therefore necessary, even in basin-scale projects, to investigate processes both at the field scale and at the catchment scale. The advantage of multi-scale studies is that the understanding of both local and regional processes and trends can feed into management at diffcrent scales (i.e. national, district and village). The disadvantage is that the field-scale studies cannot be undertaken everywhere and it is notoriously difficult to aggregate results up to the basin scale. This is particularly true in the humid tropics where the processes of erosion are still less well understood than in the temperate or semi-arid zones (Chapter 1). The fundamental reason for these difficulties is because, as Baron et al. (1980) have observed, "The very concept of hydrological uniformity, even in small areas, is suspect".

The Sustainable Management of Tropical Catchments. Edited by David Harper and Tony Brown.
© 1998 John Wiley & Sons Ltd.

THE UPPER TANA BASIN

The problems of multi-scale data were inevitably encountered in the EU-funded project "The Sustainable Management of the Tana River Catchment", other aspects of which are described in Harper (1993) and Chapters 3, 4, 5 and 15. The aim was to re-estimate suspended sediment and nutrient fluxes to the upper impoundment in a cascade, Masinga Dam (see Figures 3.1, 5.1 and 15.1 for catchment maps), and to identify the key sediment-producing areas and mechanisms. This necessitated a multi-scale approach with sediment collection at the basin-wide scale (major tributaries) and studies of a small catchment in the high-sediment producing zone (Table 2.1).

At the basin scale, sampling during the rainy seasons of 1991 and 1992 permitted a recalculation of the rating curves and a check on temporal trends (Figure 2.1). The rainy season had, as is usual, been relatively under-represented in the sediment data. The re-estimated yield for 1981–88 was 113–178 t km^{-2} yr^{-1}, which produces an estimate of approximately 10^6 t yr^{-1} input to Masinga Dam; less than the highest previous estimates but more than the lowest (Gibb & Partners, 1959; ILACO, 1971; Dunne, 1974, 1975; Dunne and Ongweny, 1976; Ongweny, 1978; Wooldridge, 1984; Brown et al., 1996). The rates suggest that the life expectancy of Masinga Dam will not be significantly reduced at the current erosion rates. However, the inputs of nutrients, especially phosphorus, associated with the sediment and turbidity, are a cause of concern (Harper, 1993; Pacini, et al., 1993; and Chapter 5).

There is wide variation in the sediment yield of the sub-catchments from 50–100 t km^{-2} yr^{-1} in the lower semi-arid part of the catchment to over 300 t km^{-2} yr^{-1} in the Aberdare foothills (Figure 2.2). At this scale there is no relationship between basin area and unit yield as intra-basin variation in erosivity, relief and land use is too great. It is at this scale that remote sensing and GIS provide useful tools for the designation of basin differences in erosion hazard and erosion potential (Chapter 3). One of the high-yield basins, the Maragua, was chosen for further study. Suspended sampling from first- to third-order streams showed a fall in concentration, but yield remained approximately constant, due to increasing discharge. This suggested that small catchments within the basin would be representative. Monitoring through 1991 and 1992 provided an estimate of the yield for the 1991/1992 hydrological year of 139 t km^{-2} yr^{-1} for the Maragua and 156 t km^{-2} yr^{-1} for the Kaihungu sub-catchment.

Table 2.1   The spatial scales used in the upper Tana Project

Areal unit	Data used	Area (km^2)
The upper Tana river basin	Landsat, agroecological maps, Masinga Dam data	8000
Tana main tributaries	Landsat, agroecological maps, additional suspended sediment sampling	150–2400
Maragua basin	Landsat, additional suspended sediment sampling, air-photographs	357
Kaihungu catchment	Landsat, air-photographs, topographic maps, field survey	24

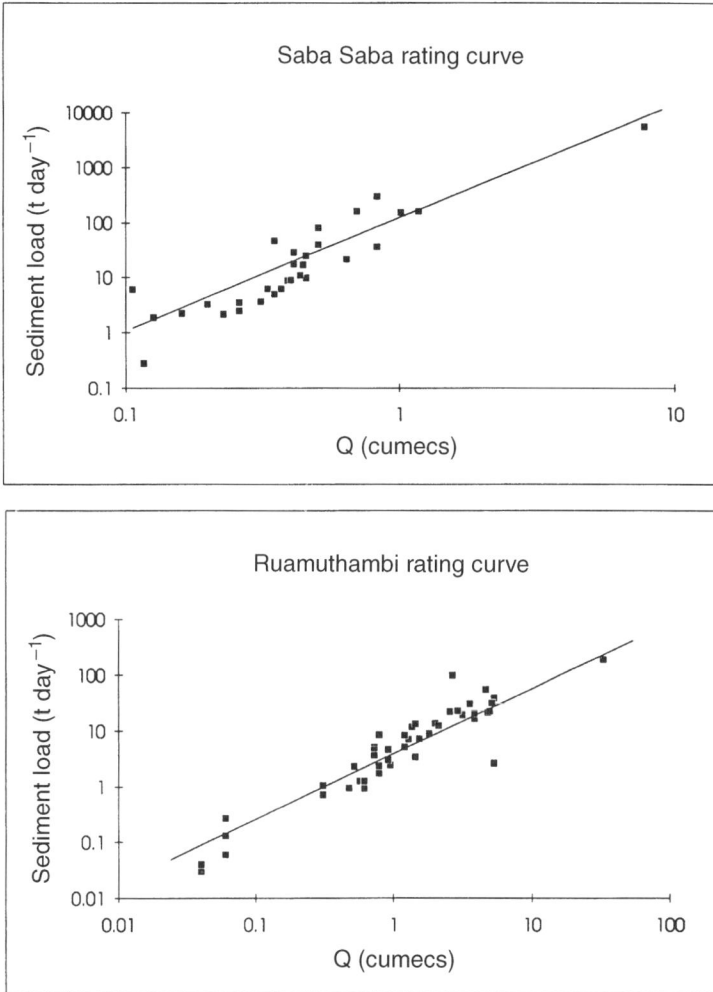

Figure 2.1   Example of the additional data collected for the Saba Saba and Ruamuthambi sub-catchments for 1991–93

As with many studies, investigations moved to the hillslopes as the presumed sediment source. The Kaihungu catchment was instrumented with simple Gerlach-type soil traps which are known to have a relatively high catch rate (Lewis, 1988). Irrespective of relationships with slope and cover, the rates suggested by the traps were unrealistically high, varying from $400 \, t \, km^{-2} \, yr^{-1}$ to $130,000 \, t \, km^{-2} \, yr^{-1}$. Even the lowest catches from traps from the most reliable farmers were over twice the erosion rate inferred from the sediment yield from Kaihungu catchment. The catchment is linear with rectilinear slopes right to the valley floor and it has little, if any, floodplain and so very little sediment storage capacity. Observations during storms suggested that detachment by splash- and saturation-induced overland flow caused the high movement of soil on the plots but most was redeposited at the breaks

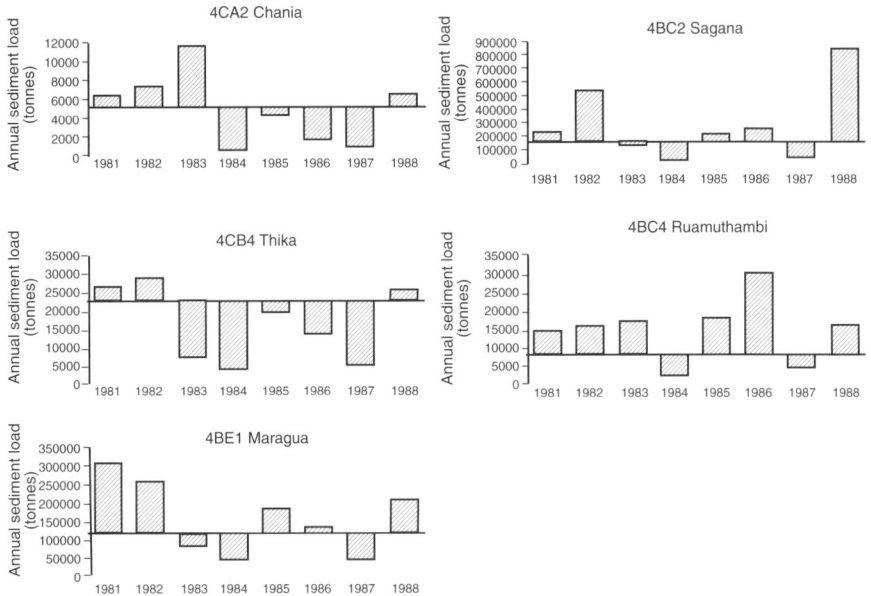

Figure 2.2 Variation in suspended sediment loads about the long-term mean for 1981–88 (note the different scales)

of slope, i.e. curvature-dependant deposition (Brown, 1992). The problem is that the traps included soil that would not have left the plot and the sediment yield included only soil that had left the plot and hillslope. Traditional soil conservation measures are designed to trap already mobilised sediment and prevent it leaving plots rather than to prevent detachment or entrainment.

Around these small plots there is a dense network of paths and tracks, many of which lead directly downslope to the river. Overland flow occurred sporadically all over the catchment during storms and suspended sediment concentrations in over-land flow were extremely high but variable ranging from 1000 to 68,000 mg l^{-1} (Table 2.2). This variation is too much to be explained by entrainment variations alone. Instead it suggests, as did observations, hyper-concentrated point injections from plots and gullies. These processes are difficult to model as they do not equate with natural flow lines. This is partly because of the slope problem. The slopes derived from topographic maps (1:50,000 scale) and used for the production of a Digital Elevation Model (DEM) are not the slopes on the ground due to the almost total terracing of the landscape. A 30 × 30 m cell/area would generally include a horizontal/gently sloping component and a smaller near-vertical component. This is also why application of the Universal Soil Loss Equation (USLE) is inappropriate and likely to produce overestimates of the erosion rate. The result of the paths and terraces is a grid-like stream/path network of conduits which can be treated as point sources to the river. Gerlach traps placed in footpath, gully and homestead locations all had high sediment catches. However, the difference between these and those on the cultivated plots is the connectivity to the river.

Table 2.2   Sediment composition (mg l^{-1}) of runoff sampled from tracks during rainstorms

Date	Maximum instantaneous track runoff (K3)(mg l^{-1})	Maximum instantaneous track runoff (K4)(mg l^{-1})
19 Oct. 91	47,000	9,607
25 Oct. 91	–	4,924
02 Nov. 91	–	2,856
03 Nov. 91	1,000	–
07 Nov. 91	6,100	–
16 Nov. 91	68,200	–
20 Nov. 91	–	42,744
26 Nov. 91	57,895	–
10 Dec. 91	12,763	12,442

Paths, tracks and communal (bare) areas were measured from aerial photographs (see Chapter 3) as covering 19.7% of the catchment. This was composed of sealed roads (0.3%), unsealed motorable roads (3.4%), footpaths (4.6%), homesteads (5.6%, of which approximately half were buildings and half bare ground around houses), and other bare ground (5.7%) associated with communal areas such as schools, market centres and churches. A typical slope is shown schematically in Figure 2.3. It is composed of small plots or *shambas* often subdivided into small rectangular sub-plots divided by paths, and small terrace risers.

Observations during storms suggested that the path and track network acted in a way analogous to the storm water system of an urban area. This was tested using a simple engineering method, the rational or time–area method. Using an impermeable area of 20% and the input precipitation measured by an automatic weather station, this method was applied to the event illustrated in Figure 2.4. The rational method predicted a peak discharge of 40 $m^3 s^{-1}$ and the observed was 38 $m^3 s^{-1}$. This, along with the short lag time to peak, suggests that the basin is acting in a way analogous to an urban catchment. It is suggested that we can conceptualise the sediment-producing mechanism as two pathways: firstly, the standard discontinuous movement of soil down hillslopes and through various stores to the streams; and, secondly, that of direct contribution of sediment through storm-activated conduits. The result at the basin scale would be an aggregate curve (Figure 2.5). This is part of the problem of relating small-scale and large-scale studies as, in the former, the effect of point sources will be much greater and stores insignificant whilst at the basin-wide scale the opposite is true. There are other scale-dependant factors in the Tana basin including slope, land use and geophysical parameters.

The lack of continuity between plot- or trap-measured rates and rates estimated from catchment sediment yield has been noted in many studies and is frequently plotted as a power function of area. An example is the relationship estimated for the Awash basin in Ethiopia from suspended sediment data and erosion plots from the Hulet Wenz (Figure 2.6, Newson, pers. comm.). There are two interpretations (not mutually exclusive) of this type of data: firstly, that the fall in rate as area increases is a function of sediment storage and, secondly, that either the sediment yield or the plots

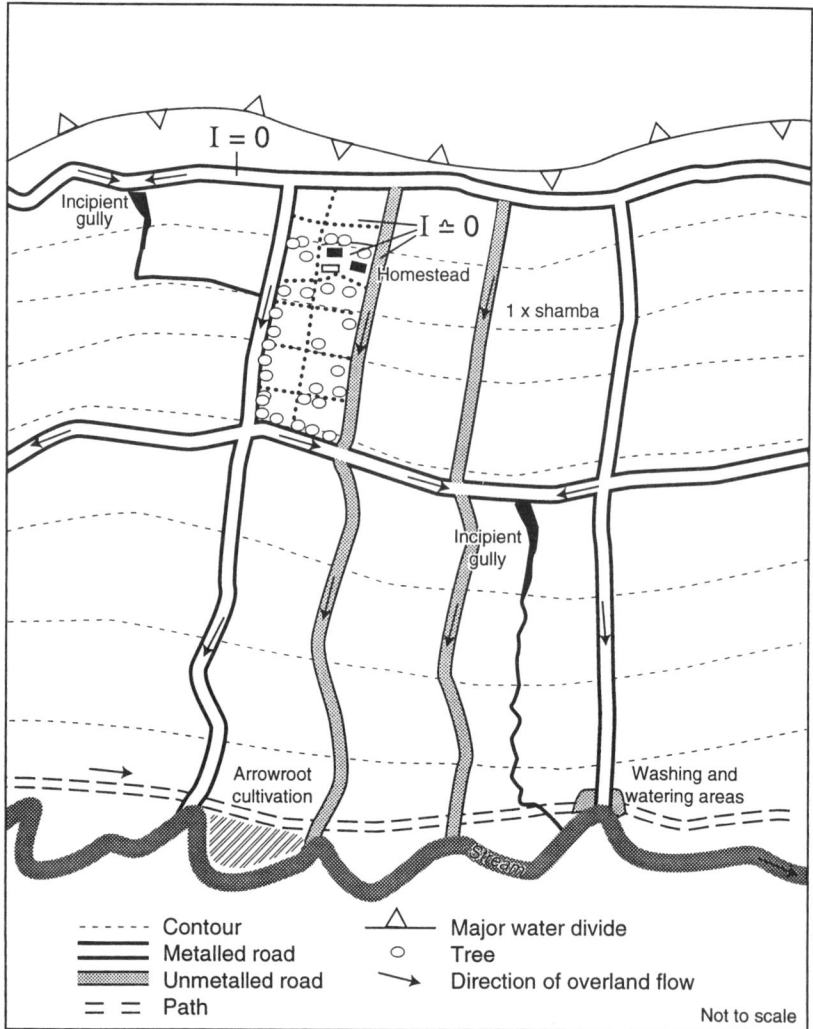

Figure 2.3   A schematic representation of a typical hillslope in the Kaihungu catchment

are significantly and systematically mis-estimating the true rate. In both the Kaihungu and Awash studies, the trap/plot rates seem to be an order of magnitude too high and, as discussed below, it is believed that the location, design and operation of traps inevitably overestimates the hillslope erosion rate. Alternative methodologies may avoid these problems, most notably the use of ^{137}Cs (Quine et al., 1992) and in some areas soil magnetic properties (Brown, 1990). Traps and plots remain useful for the observation of processes and the estimation of rates of detachment and relative rates of potential erosion (and nutrient depletion) assuming a high degree of connectivity with the stream network.

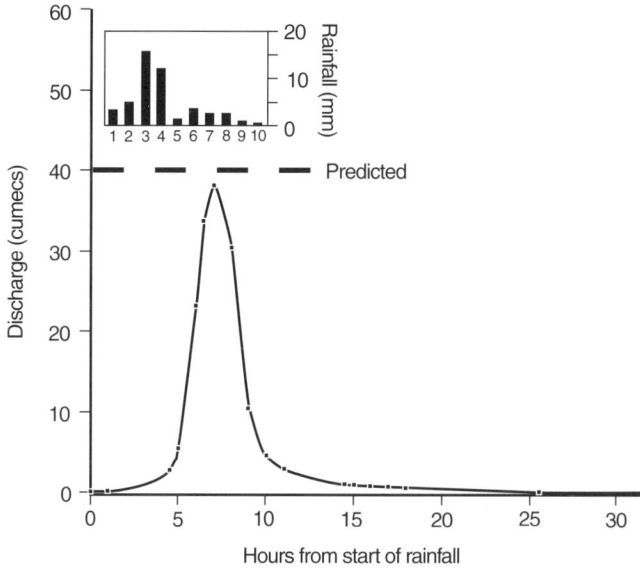

Figure 2.4 The storm hydrograph of 14–15 November 1992 in the Kaihungu catchment

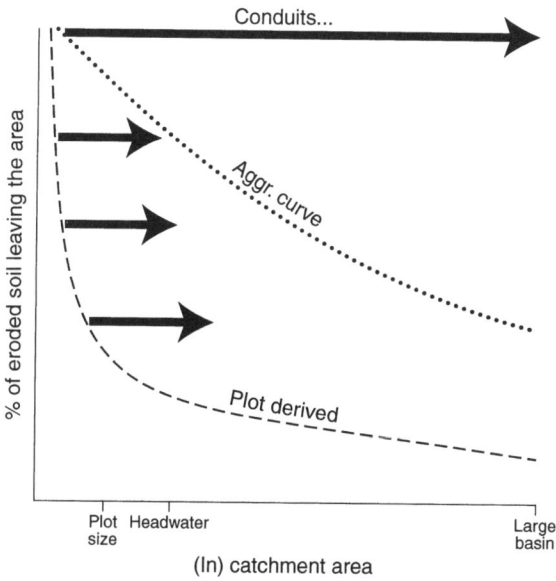

Figure 2.5 A conceptualisation of the sediment delivery ratio and mechanisms in the Upper Tana high-sediment producing basins

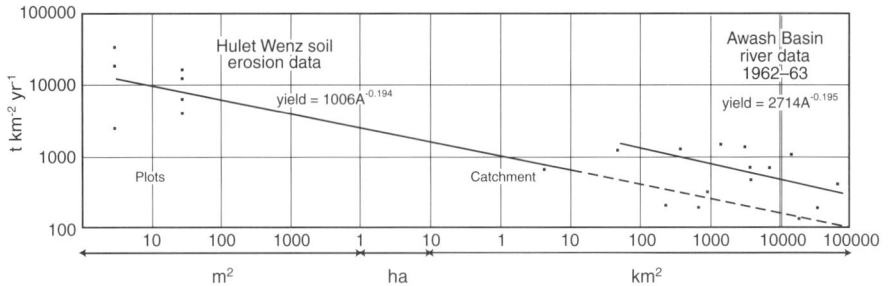

Figure 2.6   Soil erosion/area relationships based on data from the Hulet Wenz and the Awash basin Ethiopia (Newson, pers. comm.)

IMPLICATIONS FOR MANAGEMENT

The multi-scale approach taken on the Tana Project, and particularly the attempt to link sediment production and transport, has identified both the areas of major sediment generation and some of the key processes. This has four implications for management:

1. It remains desirable to optimise cover on cultivation plots in order to minimise splash detachment. When questioned in the socio-economic part of the study, farmers said that this was not achieved due to lack of fertilisers, resulting largely from a decline in incomes due to the 1980s crash in coffee prices (Chapter 4). The continued construction and maintenance of soil conservation measures (such as 'fanya juu' trenches and Napier grass) is also dependent upon a strong local economy and sufficient labour supply. There is opportunity for intervention here.
2. Of more importance is runoff management, the re-vegetation of road verges, tracks, etc., the surfacing of the larger tracks but only with adequate drainage to a watercourse, and a greater use of French drains, open half-lined drains and tile drains. This could have the additional advantage of stimulating the tile industry. In non-lined drains the current practice of using check-dams and sediment traps should be extended.
3. There is a fundamental problem in the small plot size and the fragmentation of holdings through inheritance. Not only does it have undesirable environmental effects but it is one of the factors forcing out-migration, as the plots are too small to feed a family.
4. In order to establish reliable estimates of inputs to as large a reservoir as Masinga it would be better to have fewer high-quality suspended sediment monitoring stations located just above the reservoir. In the case of the Tana, four are urgently needed: on the Tana itself, Maragua, Saba Saba and Thika.

IMPLICATIONS FOR RESEARCH

There are several implications for future catchment research:

1. It is necessary to estimate basin response characteristics before instrumentation for suspended sediment monitoring. It seems likely that tropical catchments will vary from pseudo-urban systems, dominated by stormflow and point sediment sources, to baseflow-dominated catchments which can be modelled using distributed basin models from humid studies.
2. It is necessary to examine a range of sediment sources and measurement techniques (not just plots) as this is crucial in the attribution of the responsibility for soil erosion, i.e. whether it is the farmers' or others' responsibility. The assumption that it is the cultivated land that is the most significant sediment source is not only socially unacceptable but unhelpful, as, even if it is, it may not be the component most easily reduced. Communal and other bare areas may well suffer from a "tragedy of the commons" syndrome where no one is responsible for maintenance. Some would argue this is, at least partially, a legacy of the colonial regime.
3. There is a need for intermediate-level models, which can integrate both the distributed field-scale erosion and the sediment production and conveyance of linear features in the landscape.
4. There is a need for full cost–benefit analysis. The population of Kaihungu in 1991 was 12,000 people which produces a per capita erosion rate of $13\,\mathrm{kg}\,\mathrm{person}^{-1}\,\mathrm{yr}^{-1}$. It is this figure which can be set against the productivity and the downstream costs, given that the catchment is covered with deep soils with probably a high soil-loss tolerance.

ACKNOWLEDGEMENTS

The authors acknowledges a grant from the European Community, Science and Technology for Development, Tropical and Subtropical Agriculture Programme (Contract No. TS2-A-0256-UK), held jointly with the University of Nairobi, which enabled the monitoring to be undertaken and supported one of us (HS). We would also like to thank J. Warburton and B. Hickin for advice and help, and M. Newson for the unpublished data in Figure 2.6.

REFERENCES

Baron, B.C., Pilgrim, D.H. and Cordery, I. 1980. '*Hydrological Relationships Between Small and Large Catchments*'. Australian Water Resources Council Technical Paper 54, Canberra.

Brown, A.G. 1987. 'Long-term sediment storage in the Severn and Wye catchments'. In *Palaeohydrology in Practice* (eds K.J. Gregory, J. Lewin and J.B. Thornes), pp. 307–332. Wiley, Chichester.

Brown, A.G. 1990. 'Soil erosion and fire in areas of Mediterranean type vegetation: results from chaparral in southern California, USA, and matorral in Andalucia, southern Spain'. In *Vegetation and Erosion* (ed. J.B. Thornes), pp. 269–287. Wiley, Chichester.

Brown, A.G. 1992. Slope erosion and colluviation at the floodplain edge. In *Past and Present Soil Erosion* (eds M. Bell and J. Boardman), 77–88. Oxbow Monograph 22, Oxford.

Brown, A.G., Schneider, H. and Harper, D. 1996. 'Multi-scale estimates of erosion and sediment yields in the Upper Tana Basin, Kenya'. In *Erosion and Sediment Yield: Global and Regional Perspectives*. Publications of the International Association of Scientific Hydrology (IASH), 49–54.

Chorley, R.J. and Kennedy, B.A. 1971. '*Physical Geography: A Systems Approach*'. Prentice-Hall, London.

Dunne, T. 1974. 'Suspended sediment data for the rivers of Kenya'. Unpublished report to the Ministry of Water Development, Government of Kenya.

Dunne, T. 1975. 'Sediment yield of Kenyan rivers'. Unpublished report to the Ministry of Water Development, Government of Kenya.

Dunne, T. and Ongweny, G.S.O. 1976. 'A new estimate of sedimentation rates on the Upper Tana river'. *Kenyan Geographical Journal*, **2**, 109–126.

Gibb & Partners, 1959. '*Upper Tana Catchment Water Resources Survey 1958–59*'. Report to the Kenyan Government.

Harper, D.M. (ed.) 1993. '*Tana River Project – Development of Ecologically Sustainable Catchment Land Uses*'. Final Report, EU Contract No. TS2-A-0256-UK (SMA).

ILACO, 1971. '*Upper Tana Catchment Survey*'. Report to the Government of Kenya, ILACO, The Netherlands.

Lewis, L.A. 1988. 'Measurement and assessment of soil loss in Rwanda'. In *Geomorphic Processes in Environments with Strong Seasonal Contrasts Volume I: Hillslope Processes* (eds A. Imeson and M. Sala), pp. 151–165. Catena Supplement 12.

Ongweny, G.S.O. 1978. 'Erosion and sediment transport in the upper Tana catchment with special reference to the Thiba basin'. Unpublished PhD Thesis, University of Nairobi.

Pacini, N., Harper, D.M. and Mavuti, K.M. 1993. 'A sediment-dominated tropical impoundment: Masinga Dam, Kenya'. *Verhandlungen Internationale Vereinigung Limnologie*, **25**, 1275–1279.

Quine, T.A., Walling, D.E., Zhang, X. and Wang, Y. 1992. 'Investigation of soil erosion on terraced fields near Yanting, Sichuan Province, China, using caesium-137'. In *Erosion, Debris Flows and Environment in Mountain Regions* (*Proceedings of the Chengdu Symposium, July 1992*) (eds D.E. Walling, T.R. Davies and B. Hasholt), pp. 155–168. IAHS Publication 209, Wallingford.

Roehl, J.W. 1962. 'Sediment source areas, delivery ratios and influencing morphological factors'. *Publications of the International Association of Scientific Hydrology*, **59**, 202–213.

Schneider, H. 1993. 'Soil loss and sediment yield in a tropical agricultural catchment: the upper Tana river basin, Kenya'. Unpublished PhD Thesis, University of Leicester.

Schumm, S.A. and Lichty, R.W. 1965. 'Time, space and causality in geomorphology'. *American Journal of Science*, **263**, 110–119.

Wooldridge, R. 1984. '*Sedimentation in Reservoirs – Tana River basin, Kenya III – Analysis of Hydrographic Surveys of Three Reservoirs in June/July 1983*'. Hydraulics Research Report OD46, Wallingford.

CHAPTER 3

Remote Sensing and GIS Studies of Erosion Potential for Catchment Management: A Densely Populated Agricultural Catchment in Kenya

HELEN SCHNEIDER
Pusat Environmental Education Centre, East Java, Indonesia

TONY BROWN
Department of Geography, University of Exeter, UK

INTRODUCTION

The size of many catchments and the multi-scale approach needed in projects makes remote sensing and the use of Geographical Information Systems (GIS) the only practical way to achieve project aims and objectives. The work described here was focused upon the Tana catchment in Kenya (7,950 km^2, Figure 3.1). A major aim was to assess the spatial distribution of potential erosion (and erosion factors) in the basin. The effects and consequences of soil erosion, both on-site and off-site, are a continuing cause for concern in many parts of the world and particularly in the tropics where a combination of physical and socio-economic factors often increases the impact of erosive processes (Greenland, 1977; Stocking, 1980; Pereira, 1989; El-Swaify, 1990). Unfortunately, the erosion processes operating within sedentary agricultural systems in the humid tropics are far less well known than those in the temperate or semi-arid zones (Chapter 1). It is, however, widely recognised that soil erosion in the tropics is serious and likely to become an increasing threat to the sustainable management of catchments in the face of increasing population pressure (Greenland and Lal, 1977).

There has been previous research on erosion in the upper Tana catchment, largely in order to estimate siltation rates for the reservoirs of the Tana hydro-electric scheme, which now provides over 40% of Kenya's hydro-electric potential (Gibb &

The Sustainable Management of Tropical Catchments. Edited by David Harper and Tony Brown.
© 1998 John Wiley & Sons Ltd.

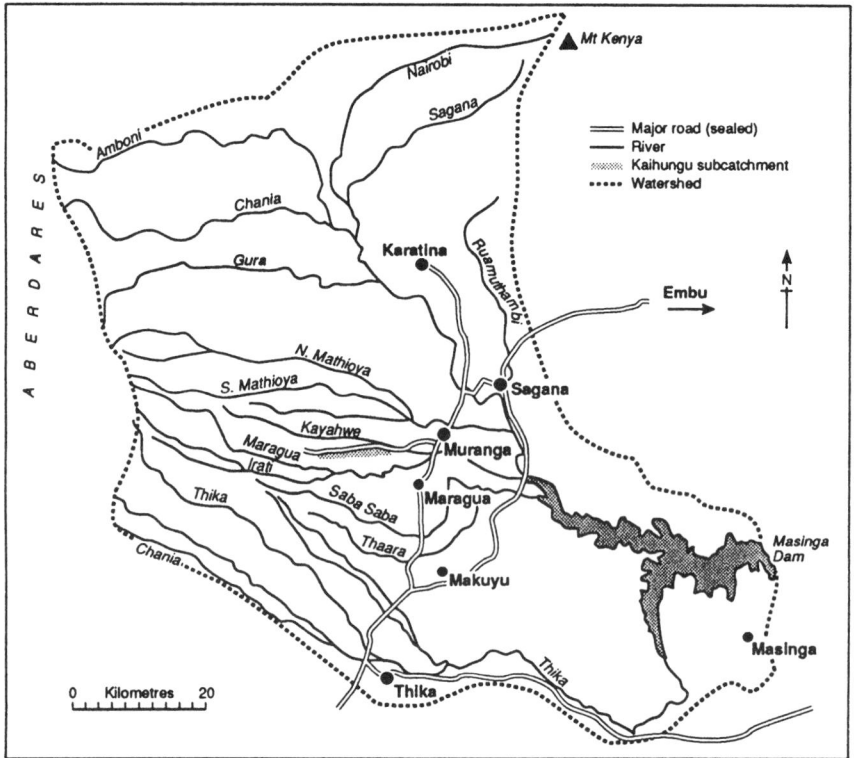

Figure 3.1   The upper Tana catchment, Kenya, showing the Kaihungu sub-catchment (stippled)

Partners, 1959; ILACO, 1971; Dunne, 1974, 1975; Dunne and Ongweny, 1976; Ongweny, 1978; Wooldridge, 1984). Parts of the basin had also been subject to a baseline survey (Odingo, 1979) prior to reservoir construction.

The geology of the basin can be divided into volcanic rocks to the north and west (forming the highlands) and the Precambrian basement complex forming the lower plateau to the south-east. The highest uplands are covered by andosols whilst the basalt foot ridges are covered by deep, well-drained, red–brown friable clay-rich nitisols. The basement complex soils are acrisols, luvisols and ferralsols (Sombröek et al., 1982). Further details of the basin can be found in Schneider (1993) and Harper (1993). The basin is divided into eight agroecological zones which are systematically distributed by height and from the south-east to the west and north-west (Figure 3.2). The Kaihungu sub-catchment extends from the tea zone of the Aberdare uplands through the coffee and mixed cropping belt (Figures 3.3 and 3.4) to the marginal coffee/maize zone.

METHODOLOGIES FOR REMOTE SENSING OF EROSION

There are three fundamental ways in which remote sensing and digital elevation data can be used in studies of soil erosion:

Major physical characteristics of land units in the Upper Tana River Basin

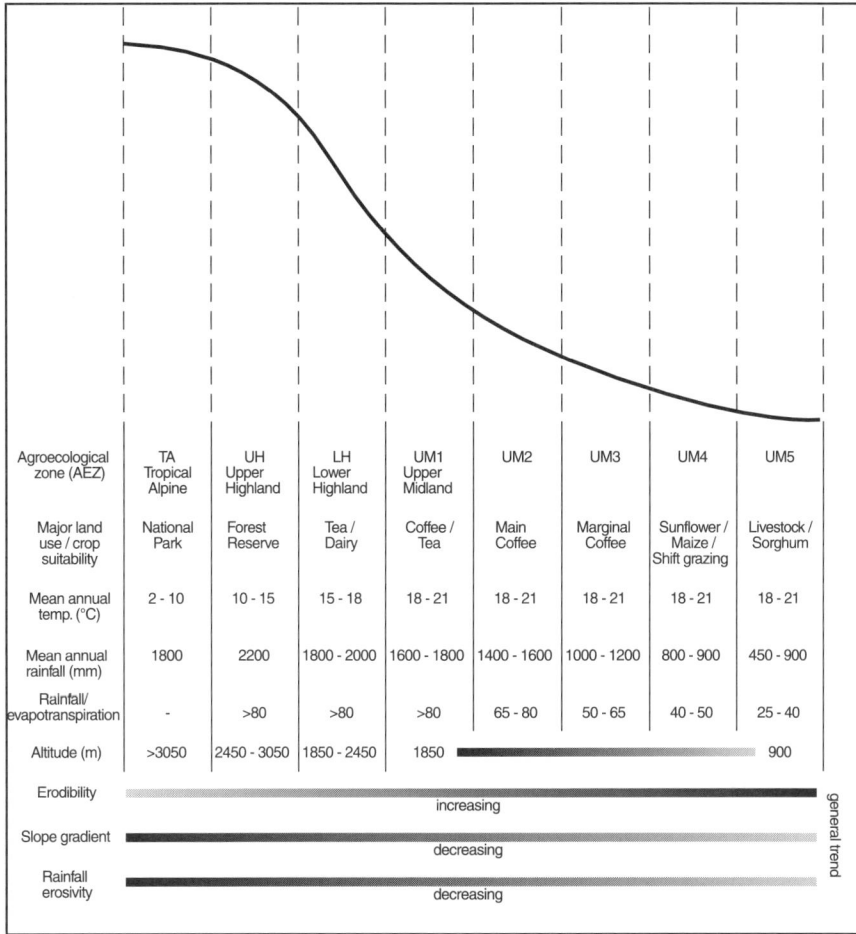

Agroecological zone (AEZ)	TA Tropical Alpine	UH Upper Highland	LH Lower Highland	UM1 Upper Midland	UM2	UM3	UM4	UM5
Major land use / crop suitability	National Park	Forest Reserve	Tea / Dairy	Coffee / Tea	Main Coffee	Marginal Coffee	Sunflower / Maize / Shift grazing	Livestock / Sorghum
Mean annual temp. (°C)	2 - 10	10 - 15	15 - 18	18 - 21	18 - 21	18 - 21	18 - 21	18 - 21
Mean annual rainfall (mm)	1800	2200	1800 - 2000	1600 - 1800	1400 - 1600	1000 - 1200	800 - 900	450 - 900
Rainfall/ evapotranspiration	-	>80	>80	>80	65 - 80	50 - 65	40 - 50	25 - 40
Altitude (m)	>3050	2450 - 3050	1850 - 2450	1850				900
Erodibility					increasing			
Slope gradient					decreasing			
Rainfall erosivity					decreasing			

general trend

Figure 3.2   The agroecological zones of the Tana catchment (from Sambröek et al., 1982)

1. The direct measurement of erosion.
2. The measurement of one of the main controls on erosion (such as vegetation cover).
3. The geographical information system (GIS) approach where remotely sensed data is combined with other digital data in order to produce a potential erosion or erosion hazard map.

   The direct method may involve either the identification of characteristic landforms such as gullies associated with accelerated erosion or estimation of soil properties indicative of a high erosion rate. In the former case the assumption is made that the forms are in equilibrium with the contemporary erosion rate, and in the latter case a model of the relationship between the soil profile and erosion rate is needed (Brown

Figure 3.3   Coffee terraces in the Kaihungu sub-catchment (photo H. Schneider)

Figure 3.4   Underplanted bananas and coffee in the Kaihungu sub-catchment (photo H. Schneider)

et al., 1990; Payne et al., 1994). Since vegetation cover is the most important factor in reducing soil erosion, the remote sensing of vegetation can be used to estimate cover values which can be used as inputs into a distributed erosion model. Vegetation cover may also be combined with other digital data in order to map the erosion hazard and it is this approach that is used here (Figure 3.5).

REMOTE SENSING AND GIS

A digital image (the Nairobi scene from the Landsat Thematic Mapper (TM) for 16 October 1988) which covered the southern half of the upper Tana basin including the Kaihungu sub-catchment and Masinga Dam was obtained. No suitable, recent, cloud-free scenes could be found for the northern section, including Mt Kenya which is normally cloud-covered. Since this scene did not cover the entire basin, it could not be directly classified to produce a land use or vegetation cover map. In addition, the volume of digital data covering even the southern part of the basin was too large for automated manipulation. To overcome these problems, subsets of the imagery, which covered the major vegetation and agroecological zones of the basin (as mapped by Trapnell and Brunt, 1987), were taken along a transect from west to east (Figure 3.1). The National Park and forested areas of the Aberdare (Nyandarua) mountains could not be included due to extensive cloud cover. Although sensors are designed to avoid the wavelengths of maximum atmospheric absorption, atmospheric scattering (Mie scattering) can affect the use of band ratios (Mather, 1987). The most commonly used method of atmospheric correction, the dark level offset method, could not be used in this study because of the absence of any large enough areas of uniformly dark pixels. The largest water body, Masinga Reservoir, had very variable reflectance due to high turbidity. The variable nature of aerosols and water vapour over such a large area also suggests that the use of a single correction factor for the whole scene would do little to improve spatial differentiation of vegetation.

In this study only the bands covering the red and near-infrared (NIR) were used as these are the most useful for vegetation detection (Holben and Justice, 1981), and the Normalised Difference Vegetation Index (NDVI; Curran, 1985) was computed to further enhance the spatial variation in vegetation type and cover both within and between these zones. Due to the very variable slopes in the basin and the difficulties of atmospheric and ground calibration, no attempt was made to establish an empirical relationship between NDVI and percentage cover; instead, as NDVI has been shown to correlate well with biomass (Curran, 1985), it was used to generate a relative indication of vegetation cover/biomass at both the basin and the sub-basin scale and ground radiometry was used to aid qualitative interpretation of the imagery.

The Kaihungu sub-catchment was identified on the Landsat imagery, geometrically corrected to a 1:50,000 topographic map and the watershed was digitised in order to cut the sub-catchment from the full scene. Attempts to use mixture modelling in order to quantify vegetation cover and bare soil radiance were confounded by an inability to identify any pixels which could act as end-members (100% bare soil and 100% vegetation), a problem related to the land use characteristics of the area (discussed later in this chapter). Estimation of land use in the basin were made from random-point analysis of aerial photographs taken in 1983/84 as part of the baseline survey and 1:22,500 scale aerial photographs taken in February and June of 1992 by the Kenya Rangeland Ecological Monitoring Unit (KREMU; Epp et al., 1983). The methodology followed that used by KREMU in the baseline survey (Muchoki, 1985).

Ground radiometry was performed on a variety of surfaces within the sub-catchment using a portable Milton Multiband Radiometer (MMR) in correspondence with the recommendations of Milton (1987) and Jackson et al. (1987). This radiometer has dual sensor heads, enabling simultaneous capture of both target and reference radiance and allowing the calculation of bi-conical reflectance. The filters used were chosen to correspond as closely as possible to bands 1–4 of Landsat TM. The sensor was mounted on a tripod giving a ground target circle of 60 cm in diameter.

The digital processing involved in this work was performed using the ERDAS System for image processing and the ARC/INFO, UNIRAS and IDRISI systems for GIS work and display.

EROSION FACTORS

A number of digital datasets for the upper Tana were obtained which had been scanned in 1987 by the Canada Land Data Division of the Canadian Department of the Environment. These included the mean annual rainfall map of East Africa (1:2,000,000; 1965) and the Exploratory Soil and Agroclimatic Zone Map of Kenya (1:1,000,000; 1980). Additional data were available from the vegetation and climate maps of south-west Kenya published by Trapnell and Brunt (1987). The boundary of the upper Tana basin was digitised from 1:250,000 topographic sheets and referenced to the same co-ordinate system as the rainfall and soil maps. Although the maps are rather large scale they are the only published data which covered the entire basin. The rainfall map was used to generate classes of rainfall erosivity based on the R factor of the USLE using the method of Moore (1979) whereby mean annual rainfall is converted to kinetic energy according to the equation:

$$KE = 3.96P + 3122 \qquad (1)$$

where P = mean annual precipitation in mm, and KE = kinetic energy in $J\,m^{-1}$, and:

$$R = 0.029KE - 26.0 \qquad (2)$$

where R = rainfall erosivity factor.

It was felt that this method was appropriate because it was derived from intensity measurements made at 35 stations in East Africa with, in most cases, 10 or more years of data (Moore, 1979).

Soil erodibility classes were assigned to the soil map using the classification of Kassam et al. (1991) which is based upon the nomograms of Wischmeier and Smith (1978). Data on the texture, organic matter content, structure and permeability were used to derive the erodibility values. Modifications were made for three soil groups (nitisols, vertisols and chernozems) which appeared from plot experiments to be more erodible than indicated by the nomograms (Kassam et al., 1991; Table 3.1). A range of erodibility (K) values were given to each class. Where the soil classification of

Table 3.1 Soil erodibility classes of soil units by soil texture (from Kassam et al., 1991)

Soil unit	S	LS	SL	L	CL	SCL	SC	C	SiC	SiCL	SiL	Si
Acrisol			4		4	3	2	3				
Cambisol			4	5	4	3	2	3	4	5		
Ferrasol			4			3	2	2				
Lithosol			5	5	4	3	2	3			6	
Nitisol								4				
Arenosol	3	3	4									7
Regosol			4	5	4	3	2	3				
Andosol				5	4	3	2	3	4	5		
Vertisol						5		5				

S = sand, Si = silt, C = clay, L = loam

Sombröek et al. (1982) contained more than one soil type per classification unit, the erodibility class to which the dominant soil of that unit belonged was assigned to the whole unit. In addition the soil map was coded with dominant slope classes derived from field survey by Sombröek et al. (1982).

Subsequently a digital version of the 1:25,000 maps was used in an attempt to produce a digital elevation map for the entire basin. After error identification and correction, a triangular irregular network (TIN) was created by the subsampling of points along the contour lines. An area larger than the basin was used as input to lessen anomalies associated with boundaries and the maximum number of points was used (50,000). The subsampling used a user-defined weed tolerance (i.e. all points lying within a defined distance of another point on the same contour are ignored) of 300 m. The resulting TIN was assessed for accuracy in two ways: firstly, the number of triangles with zero slopes were determined and compared to the contour map and, secondly, a contour map was regenerated from the TIN and overlain on the original. Both methods indicated that the TIN had failed to model the surface accurately. An alternative method was used which involved the creation of a lattice file, representing the surface using regularly spaced sample points, and then filtering these data by one or more passes, followed by extraction of those points considered most significant in relation to the general morphology of the surface. This was done on a subset but failed to improve the model significantly. The principal reason for this would seem to be that the program poorly samples areas with closely spaced contours, such as river valleys. Several parts of the catchment, such as the foothills and highlands, are characterised by such closely spaced steep-sloping valleys. This meant that the only reliable digital slope map was the map derived from the soil map slope classes which were reclassified from dominant slope class in percentages to degrees from which the sine of each set of class limits was then taken, since it is known that soil loss does not vary linearly with gradient.

It was, however, possible to derive an accurate DEM for the Kaihungu subcatchment. The contours of an area larger that the basin were digitised from the 1:50,000 topographic maps with a contour interval of 20 m, and the resulting file ungenerated to a file containing x, y and z values which were input to an interpolation program on the UNIRAS system (UNIRAS, 1986). The resulting gridded data was converted to

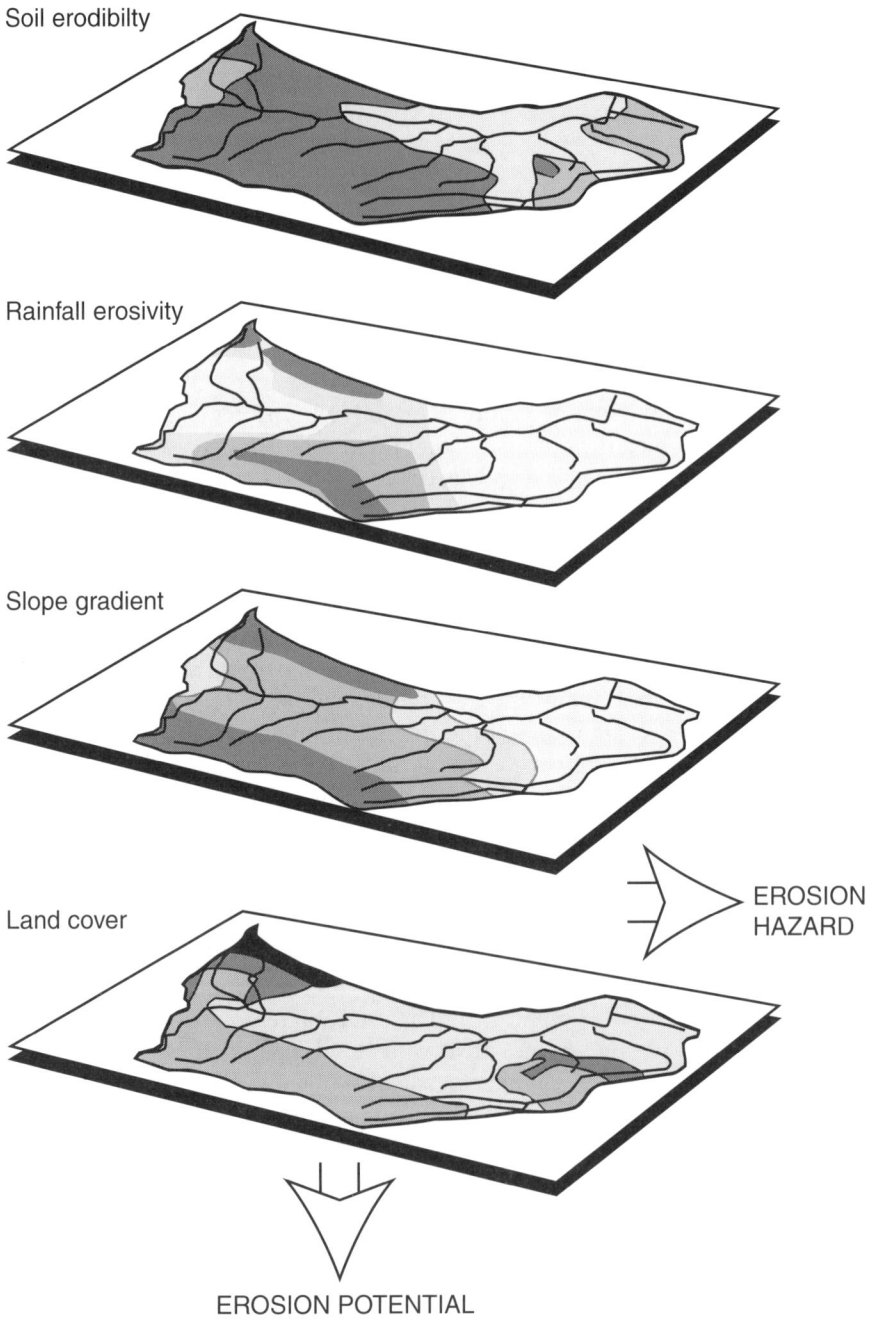

Figure 3.5 The methodology of GIS-based assessment of erosion hazard and erosion potential

IDRISI format and then subset with the basin boundary to create the final DEM to which the remote sensing imagery was co-referenced. A slope map was derived from the DEM by calculating the maximum slope around each cell from the local slopes in the x and y directions (i.e. Rook's case where diagonals are ignored). This allowed the creation of an aspect image. Watershed analysis was permitted using a vector file created from the digitisation of the stream network (blue line method) which was then converted into a target cell image. Then those cells which could transmit flow in the direction of the target cells based on the aspect of each cell in the DEM were identified. Flat areas were assumed to be capable of transmitting flow in all directions. This analysis shows only those cells theoretically capable of contributing flow to adjacent cells towards a target cell or group of cells (in this case a stream network).

EROSION HAZARD AND EROSION POTENTIAL MAPPING

The digital coverage of erosivity, erodibility and slope classes were overlain and classes of erosion susceptibility derived. This was done by multiplying the median value of each class for each factor and then applying an equal interval classification (Figure 3.5).

The result gives an indication of erosion hazard (i.e. the physical hazard irrespective of land use). In reality the erosion rate is largely determined by vegetation cover and therefore land use. The only available land use map for the whole of the basin was at 1:1,000,000 scale produced by KREMU from satellite imagery covering the period 1971 to 1980. Field checks were carried out during 1991–93 to establish that major changes had not occurred in the intervening period. Land use classes from this map were then reclassified to crop-cover classes using C-values for the dominant crop (Table 3.2) as given in Lewis (1988), taking into account known changes in land use that had occurred since the compilation of the map. Incorporation of these data in order to produce a potential erosion map involved the multiplication of crop-cover values by the median value for classes of erosivity, erodibility and slope gradient. No account could be taken of physical conservation measures (common in the area, Figure 1.1), such as terracing, due to the great spatial variability of their implementation, maintenance and effectiveness at this scale. Intercropping of subsistence food crops was assumed since observations indicated that this was by far the most common cropping system in both subsistence crop zones.

Table 3.2   USLE crop factors used in reclassification of the land use map of the upper Tana

Land use class	C-value
National Park/forest/tea	0.02
Irrigated coffee	0.02
Improved grazing	0.10
Cash vegetables	0.22
Subsistence maize/beans (intercrop)	0.30
Subsistence maize/pigeon peas (intercrop)	0.30

In order to assess relative population pressure in the basin a digital map was derived from the district boundaries and results of the 1979 and 1989 population censuses (Central Bureau of Statistics, 1991).

GROUND RADIOMETRY AND THE VARIATION IN VEGETATION COVER ACROSS THE BASIN

All vegetation plots used showed a characteristic vegetation signature with high positive NDVIs which also varied with canopy structure and geometry. Differences in the radiance caused by sun–sensor–surface geometry was particularly strong for several of the crops including banana and arrowroot. Although the normalisation used in the NDVI reduces this problem, it is still evident that canopy structure and illumination angle affects the reflectance of the crops in the area.

Bare ground is easily differentiated whatever its moisture status. Experiments suggest that in a mixed vegetation/soil pixel the absorbance feature in the red band

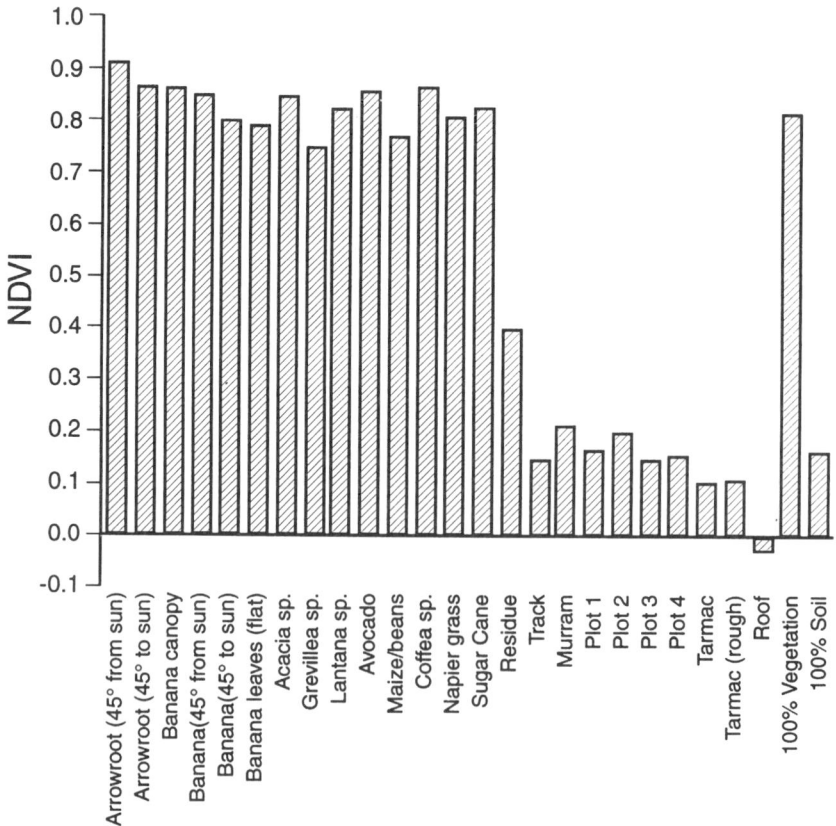

Figure 3.6   NDVI values from ground radiometry for different land uses in the Kaihungu catchment

and high NIR reflectance are both still apparent but dampened due to the presence of soils with a high ferric iron content and the percentage reduction in vegetation cover. Dead vegetation presents a potential problem of misclassification, but mulching is not a common practice in the area. Similarly, both metalled and unmetalled roads would be likely to be classified along with bare soil, although this is not a problem for erosion potential mapping since they are both sediment and runoff producing areas. The NDVI results (Figure 3.6) suggest that a clear distinction can be made between well-vegetated areas, with their low erosion potential, and bare soil, roads and tracks with their high erosion potential. This work suggested that mixed pixels are likely to show a spectral response which will vary with percentage vegetation cover and soil type and therefore even if it is possible to identify individual pixels of 100% soil and 100% vegetation cover, a simple linear model is unlikely to produce an accurate estimation of the percentage cover for all pixels in a scene.

The NDVI image of the Landsat TM transect shows clear variation from east to west with particular features such as irrigation schemes and river valleys prominent. In the east the area around Masinga Dam shows NDVI values around zero, indicating low vegetation cover, whilst NDVI values increase towards the west and the foothills of the Aberdare Range. Significant differences can be seen between the frequency distribution of NDVI in the different agroecological zones (Figure 3.7), representing an increase in cover from the eastern millet/livestock zone to the western upper coffee zone. The bimodality of the distribution in the millet/livestock zone may reflect the increasing prevalence of large bare areas in this, the driest, area of the basin.

EROSION HAZARD AND EROSION POTENTIAL MAPS

The erosion hazard map (using physical factors only, Figure 3.8) shows a low hazard in the south-east increasing with altitude to the west with the highest hazard being in the foothills of the Aberdares where the steepest slopes and highest rainfall erosivities occur together with moderate soil erodibility. There is also an increase in the foothills of Mt Kenya due to a similar erodibility and only slightly less steep slopes and lower erosivity. The erosion potential map shows that the addition of a land-cover factor dramatically alters the classification for some areas of the basin. The protective nature of the vegetation of the National Park, forest and tea zones means that potential erosion is far less than the erosion hazard (unless this vegetation was removed). The lowland zone around Masinga Dam shows little difference between the relative erosion hazard and potential erosion despite low vegetation cover. This is explained by the lower slopes and low rainfall erosivity of this zone. The erosion potential map (Figure 3.9) suggests that the highest erosion rates should occur in the areas of intensive smallholder cultivation in the upper midland (UM1, UM2) zone of the foothills of the Highlands in Muranga, Nyeri and Kirinyaga Districts. This is the zone of second highest population density within the basin (at 450–650 people km^{-2}), but not the highest which lies to the south around Thika, and with less than national average growth rates population density is unlikely to increase significantly in the short to medium term (Figure 3.10) within this zone. A more detailed discussion of the population distribution and trends can be found in Schneider (1993).

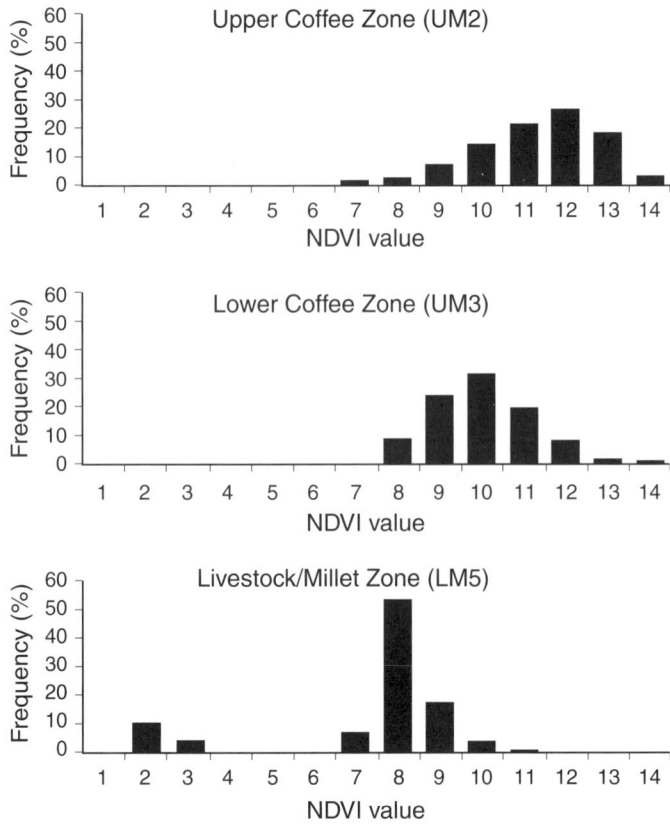

Figure 3.7   Comparison of NDVI values for the major agroecological zones of the upper Tana river basin

These maps have been compared to the revised estimates of sediment yield produced by the project (Harper, 1993; Schneider, 1993; Brown, et al., 1996; Chapter 2). Figure 3.11 shows that there is a positive correspondence between the area of high erosion potential in each sub-catchment and the estimated mean annual sediment yield. However, there are also some significant departures, particularly in the case of the Saba Saba and Ruamuthambi basins. In each case there are local factors, such as the location of the high erosion potential areas and stream gradients, which help to explain these discrepancies (Schneider, 1993).

The Kaihungu sub-catchment lies in class 4 on the potential erosion map. However, the DEM and NDVI show that considerable spatial variation exists and that it increases downstream (Figure 3.12). Land use is dominated by coffee in small plots, and subsistence cropping of a wide variety of fruit and vegetables, again in small plots and terraces (see Figures 3.3 and 3.4). Analysis of aerial photographs suggests that much of this variation is due to pixels with a mixture of bare ground/infrastructure and cultivation plots. The infrastructure comprises sealed roads (0.3%), unsealed motorable roads (3.4%), footpaths (4.6%) and homesteads (5.6%, of which approxi-

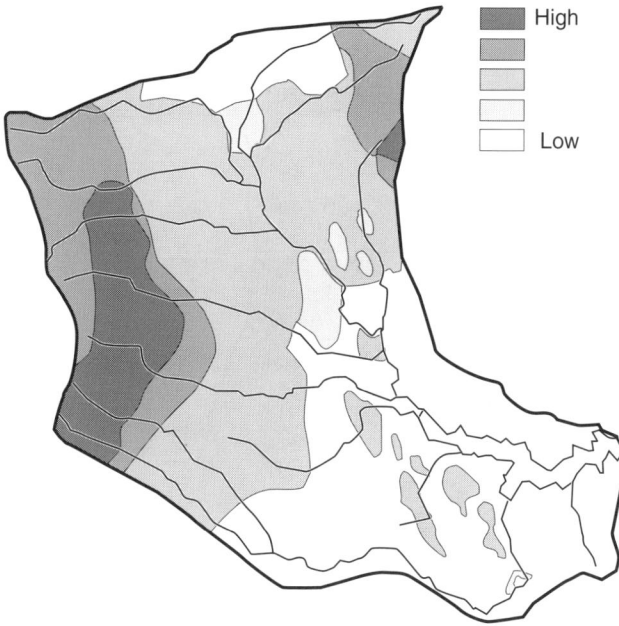

Figure 3.8   Relative erosion hazard in the upper Tana sub-basins

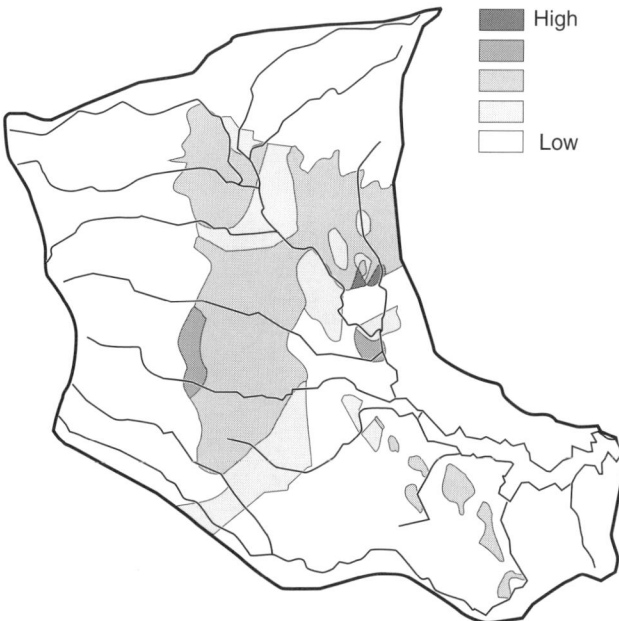

Figure 3.9   Potential erosion hazard in the upper Tana sub-basins

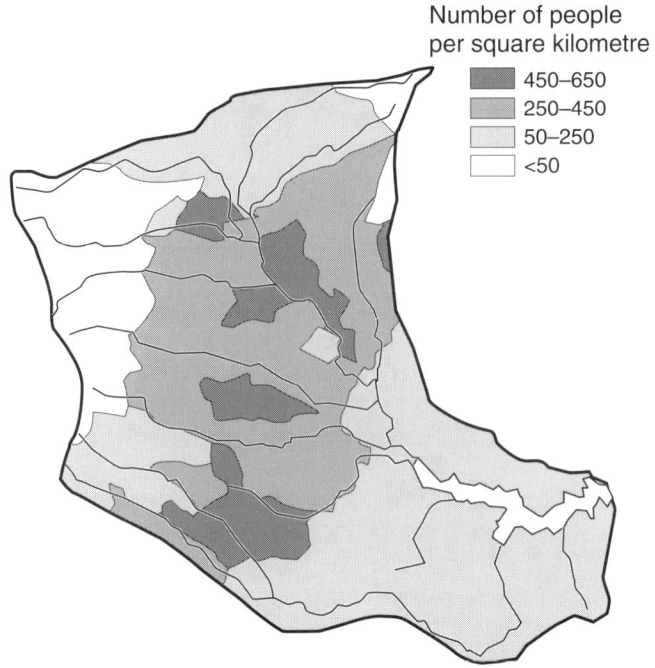

Figure 3.10   Population density distribution in 1989

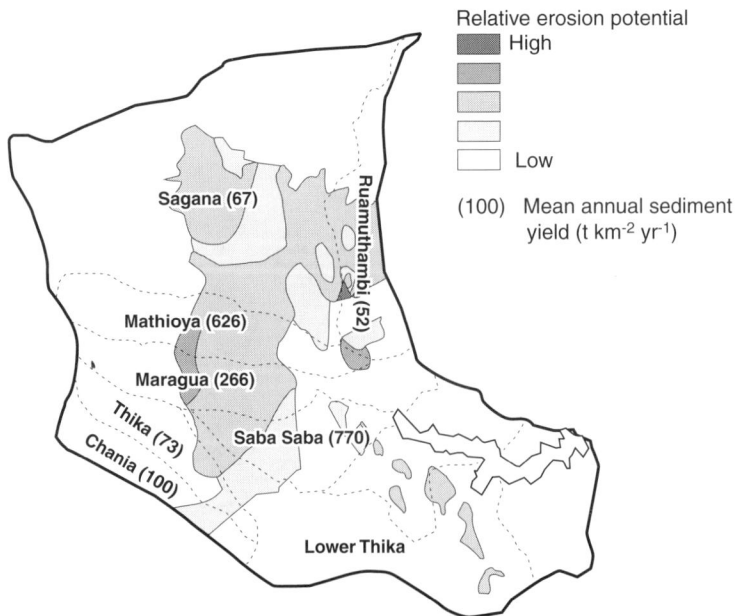

Figure 3.11   Comparison of erosion potential and suspended sediment yield of the major sub-basins of the upper Tana

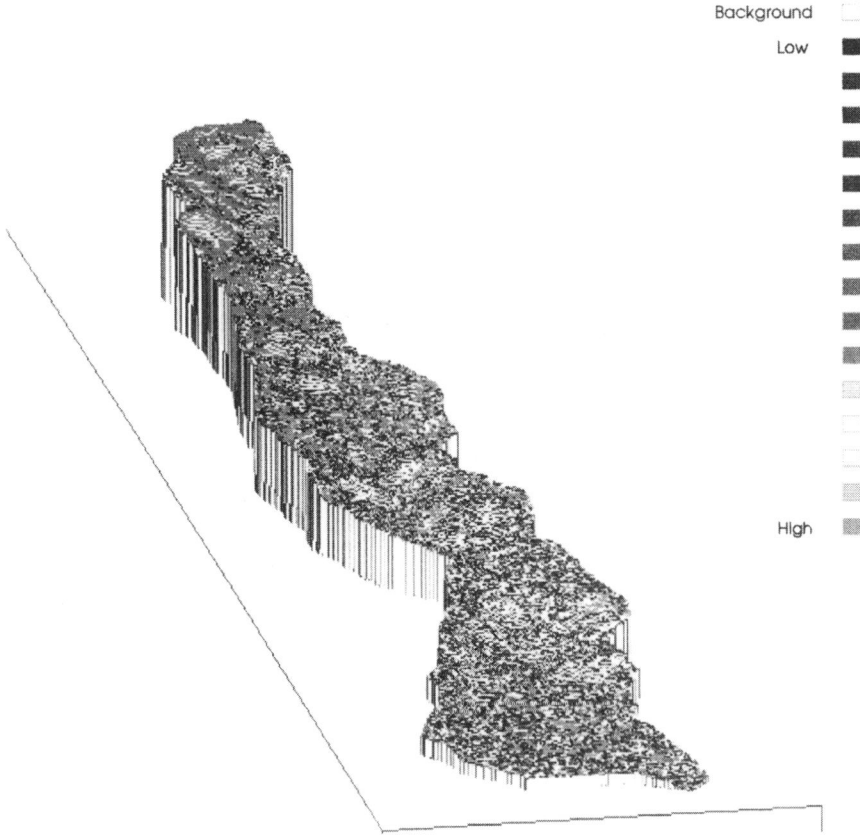

Figure 3.12   NDVI image overlain on the DEM of the Kaihungu sub-basin

mately half were buildings and half bare ground around houses). Other bare ground (5.7%) was associated with communal areas such as schools, market centres and churches. This gives a combined ground area of just under 20%. Given the nominal ground area of one pixel ($900 \, \text{m}^2$) and the typical size of plots (under 5×20 m), then it would be rare for a pixel not to include a significant proportion of infrastructural area. This high coverage of infrastructure is due to the high population, intensive cultivation in small plots and terracing; it is therefore indirectly related to the relatively steep slopes of the foothills zone.

CONCLUSIONS

The foothill zones of the Aberdare Mountains, and to a lesser extent Mt Kenya, have both the highest potential erosion (determined using the GIS methodology) and the highest sediment yields (from revised estimates). The prime causative factors in the distribution of potential erosion appear to be topography, rainfall erosivity and population density rather than vegetation cover. Analysis of the Kaihungu sub-catchment suggests that topography and population may be acting through land use, and in particular the high amount of infrastructural area, and it is not the density of cultivation *per se* which is responsible for the high sediment yields. This suggests, as do soil trap results (Harper, 1993; Brown et al., 1996 and Chapter 2) that future soil conservation efforts should be addressed to infrastructural and communal areas rather than to the cultivation plots themselves.

There are as yet unresolved problems in both the remote sensing of vegetation cover and GIS modelling of erosion associated with the small-scale land use parcels (plots and terraces) which are typical of river valleys in the tropics. Nevertheless, widely available remote sensing data (Landsat TM) and relatively inexpensive commercial remote sensing/GIS packages can be used in combination to generate maps of large tropical catchments which accord with soil erosion theory and observations of sediment yield with relatively small labour input (equivalent of 4 or 5 person-years).

ACKNOWLEDGEMENTS

We must thank the rest of the Tana Project team for all their help and assistance. In addition help was given by several members of the Geography Department at Leicester University, especially Jeff Warburton, Bill Hickin, Kate Moore and David Orme. The staff of the Kenyan Ministry of Water Development provided much useful data, as did KREMU and the Tana River Development Authority. This work would not have been possible without the kind assistance of staff of UNEP-GRID in Nairobi, Danielle Arigoni and Laura Meszaros. This project was funded by the European Community's STD2 programme, contract TS2-A-0256-UK.

REFERENCES

Brown, A.G., Schneider, H., Rice, R.J. and Milton, E.J. 1990. 'Remote sensing soil erosion: airborne thematic mapper data on soil variation in Mediterranean arable land in Southern Spain'. In *Proceedings of the NERC Symposium on Airborne Remote Sensing 1990*, pp. 7–18. British Geological Survey, Keyworth, Nottingham.

Brown, A.G., Schneider, H. and Harper, D.M. 1996. 'Multi-scale estimates of erosion and sediment yields in the Upper Tana basin, Kenya'. In *International Symposium on Erosion and Sediment Yield: Global and Regional Perspectives* (eds D.E. Walling and S. Webb), pp. 49–54. International Association of Hydrological Sciences Publication 236, IAHS, Wallingford.

Central Bureau of Statistics, 1991. '*Economic Survey*'. Ministry of Finance and Planning, Nairobi.

Curran, P. 1985. '*Principles of Remote Sensing*'. Longman, Harlow.

Dunne, T. 1974. 'Suspended sediment data for the rivers of Kenya'. Unpublished report to the Ministry of Water Development, Government of Kenya.

Dunne, T. 1975. 'Sediment yield of Kenyan rivers'. Unpublished report to the Ministry of Water Development, Government of Kenya.

Dunne, T. and Ongweny, G.S.O. 1976. 'A new estimate of sedimentation rates on the Upper Tana river'. *Kenyan Geographical Journal*, **2**, 109–126.

El-Swaify, S.A. 1990. 'Research needs and applications to reduce erosion and sedimentation in the tropics'. In *Soil Erosion and its Countermeasures* (ed. S. Jantawat), Soil and Water Conservation Society of Thailand, Bangkok.

Epp, H.A., Killmayer, A. and Peden, D.E. 1983. *'Land Use in the Kisii District'*. KREMU Technical Report 97, KREMU, Nairobi.

Gibb & Partners, 1959. *'Upper Tana Catchment Water Resources Survey 1958–59'*. Report to the Kenyan Government.

Greenland, D.J. l977. 'Soil structure and erosion hazard'. In *Soil Conservation and Management in the Humid Tropics* (eds D.J. Greenland and R. Lal), pp. 17–23. Wiley, Chichester.

Greenland, D.J. and Lal, R. 1977. *'Soil Conservation and Management in the Humid Tropics'*. Wiley, Chichester.

Harper, D. (ed.) 1993. *'Tana River Project – Development of Ecologically Sustainable Catchment Land Uses'*. Final Report, to the European Community, Contract No. TS2-A-0256-UK SMA.

Holben, B. and Justice, C. 1981. 'An examination of spectral band ratioing to reduce the topographic effect on remote sensing data'. *International Journal of Remote Sensing*, **1**, 115–133.

ILACO, 1971. *'Upper Tana Catchment Survey'*. Report to the Government of Kenya, ILACO, The Netherlands.

Jackson, R.D., Moran, M.S., Slater, P.N. and Biggar, S.F. 1987. 'Field calibration of reference reflectance panels'. *Remote Sensing of the Environment*, **22**, 145–158.

Kassam, A.H., van Velthuizen, H.T., Fischer, G.W. and Shah, M.M. 1991. 'Agroecological land resources assessment for agricultural development planning: A case study of Kenya'. Unpublished Postgraduate Project Report, Department of Agricultural Engineering, University of Nairobi.

Lewis, L.A. 1988. 'Measurement and assessment of soil loss in Rwanda'. In *Geomorphic Processes in Environments with Strong Seasonal Contrasts* (eds A. Imeson and M. Sala), Volume 1: Hillslope Processes. *Catena Supplement*, **12**, 151–165.

Mather, P.M. 1987. *Computer Processing of Remote Sensing Imagery: An Introduction*. Wiley, Chichester.

Milton, E.J. 1987. 'Review article: principles of field spectroscopy'. *International Journal of Remote Sensing*, **1**, 1807–1827.

Moore, T.R. 1979. 'Rainfall erosivity in East Africa'. *Geografiska Annaler*, **61A**, 147–156.

Muchoki, C.H.K. 1985. *'Land Use in the Murang'a District'*. KREMU Technical Report 120, KREMU, Nairobi.

Odingo, R.S. (ed.) 1979. *'An African Dam: Ecological Survey of the Kamburu/Gtaru Hydroelectric Dam Area, Kenya'*. Ecological Bulletins No. 29, Swedish Natural Research Council.

Ongweny, G.S.O. 1978. 'Erosion and sediment transport in the upper Tana catchment with special reference to the Thiba basin'. Unpublished PhD Thesis, University of Nairobi.

Payne, D., Brown, A.G. and Brock, B. 1994. 'Factors affecting remotely-sensed soil variation and near-harvest crop variation near Antequera, Southern Spain'. In *Proceedings of the NERC Symposium on Airborne Remote Sensing 1993*. NERC, Swindon.

Pereira, H.C. 1989. *'Policy and Practice in the Management of Tropical Watersheds'*. Belhaven Press, London.

Schneider, H. 1993. 'Soil loss and sediment yield in a tropical agricultural catchment: the upper Tana river basin, Kenya'. Unpublished PhD Thesis, University of Leicester.

Sombröek, W.G., Braun, H.M.H. and van der Pouw, B.J.A. 1982. *'Exploratory Soil Map and Agro-climatic Map of Kenya, 1980'*. Exploratory Soil Survey Report No. E1, Kenya Soil Survey, Nairobi.

Stocking, M.A. 1980. 'Conservation strategies for developing countries'. In *Soil Conservation: Problems and Prospects* (ed. R.P.C. Morgan), Wiley, Chichester.

Trapnell, C.G. and Brunt, M.A. 1987. *'Vegetation and Climate Maps for South-Western Kenya'*. Land Resources Development Centre, Overseas Development Administration, Surbiton.

UNIRAS, 1986. *'Geopak Interpolation Version 5'*. European Software Contractors, Denmark.

Wischmeier, W.H. and Smith, D.D. 1978. *'Predicting Rainfall Erosion Losses'*. USDA, Agricultural Research Service Handbook 537.

Wooldridge, R. 1984. *'Sedimentation in Reservoirs – Tana River Basin, Kenya 111 – Analysis of Hydrographic Surveys of Three Reservoirs in June/July 1983'*. Hydraulics Research Report OD46, Wallingford.

CHAPTER 4

Socio-economic Aspects of Subsistence Farming and Soil Erosion in Tropical Catchments: The Tana Catchment, Kenya

DOROTHY MUTISYA
Department of Geography, Kenyatta University, Nairobi, Kenya
SAM MUTISO
Department of Geography, University of Nairobi, Kenya

THE SOIL EROSION PROBLEM IN THE TROPICS

Soil erosion is one of the major land management problems threatening the economic productivity of the agricultural land and rivers in the tropics (Stocking, 1984). Many tropical countries characterised by high and rapidly growing populations are facing a population–agricultural resource imbalance. The pressure of rapidly increasing population has brought about increased activity in water catchment areas. More and more people are moving up to the sloping and marginal areas in search of land for cultivation and grazing purposes. More forests, grasslands and scrublands are being opened up for agricultural production and exposed to the hazards of accelerated soil erosion due to a variety of causes including, in places, unsuitable crop and soil management methods. Also, people are migrating from the overcrowded areas of high agricultural potential to the marginal (dry) lowlands, thus increasing the erosional risk on these more fragile environments.

The problem of a poor land–population ratio is already being experienced in Kenya where, out of 44.6×10^6 ha of land, only approximately 8.6×10^5 ha are of medium to high agricultural potential. These areas support over 80% of the Kenyan population and over half of the livestock in the country (Kenyan Government Statistics). The result is land subdivision, land fragmentation, landlessness, over-cultivation, overgrazing and deforestation, which have serious implications for soil erosion.

The Sustainable Management of Tropical Catchments. Edited by David Harper and Tony Brown.
© 1998 John Wiley & Sons Ltd.

One manifestation of soil erosion is the removal of topsoil leading to loss of plant-available water capacity, loss of applied and native plant nutrients and the non-uniform removal of topsoils from farmland (Kelly, 1983). This culminates in reduced land productivity and local food shortages. As a result, large areas in the tropics have undergone a decrease in productivity per unit area. Rehabilitation of farm land usually increases the cost of production to a level beyond the financial means of the ordinary small-scale subsistence farmer.

Soil erosion also impacts on reservoir and irrigation systems (Chapters 13 and 15). This is because a large proportion of the sediments produced from arable, degraded pasture land and open areas such as homesteads and footpaths are transported downstream into rivers and dams, potentially causing a siltation problem (Chapter 2).

Solutions to the problems associated with soil erosion are often costly at both the household and national levels. At the former, farmers are forced to spend part of their meagre incomes on the purchase of additional food and farm inputs, while at the national level the government has to spend foreign exchange on food imports, fertilisers and on equipment for dredging silted dams and irrigation channels. Mellor and Rennam (1986) estimated the damage resulting from soil erosion at US$26 US billion annually in Africa. The damage is likely to be proportionately worse in Kenya where there is still a low level of technical knowledge, and a pressing demand for more land and food with limited allocation of resources to combat the problem (Wolde and Dunne, 1986).

On many occasions, the rural poor have been blamed for land degradation and soil erosion. For instance, Hawuck (1985) stated that "in general, the main cause of land degradation and soil erosion stem from indiscriminate human interference in the natural ecological balance, from abuse and mismanagement of the soil and water resources and from trying to farm land beyond its capability", and Kelly (1983) stated that "it is not high wind or rain that is the cause of accelerated erosion ... it is the people who destroy the soil by demanding more from the land than it can provide". These authors and many others do not seek to understand and address the socio-economic and cultural problems that rural people face.

This chapter discusses the socio-economic and cultural aspects of soil erosion in a subsistence agricultural economy in the Masinga Dam catchment, Kenya. This is important because this catchment represents one of the many tropical catchments facing serious soil erosion problems in Kenya. According to Wooldridge (1984), the Masinga Dam catchment's sediment yield was estimated to be increasing at the rate of 305 m^3 km^2 yr^{-1} between 1968 and 1974; 357 m^3 km^2 yr^{-1} in the period between 1974 and 1981; and 1099 m^3 km^2 yr^{-1} between 1981 and 1987; and although revised estimates have more recently been produced (Chapter 2) the problem is nonetheless serious. This has two major implications for the Kenyan government. Firstly, the siltation process affects the aquatic life and, potentially, even hydro-electric power production. Secondly, this catchment, particularly the upper areas, forms one of the most agriculturally productive regions of the country. Therefore, if this potential is reduced through soil erosion, food shortages may occur both at the household and national level.

Moreover, socio-economic and cultural aspects of soil erosion are normally treated as separate phenomena. However, despite the fact that erosion is a physical process, it is becoming increasingly evident that the problem has a socio-economic dimension which has more than often been ignored. The central argument here is that, in their endeavour to satisfy basic needs, people are faced with many complex socio-economic problems which may force them to over-utilise and mismanage their limited, non-renewable, land resources.

THE STUDY AREA

The catchment of Masinga Dam, the uppermost in a chain of hydro-electric reservoirs on the Tana river (Chapter 15), is about 7,950 km^2 lying to the east of the Aberdare Mountains and south of Mt Kenya (Figure 4.1). It covers a large proportion of the Central Province and a small proportion of the Eastern Province. The catchment rises gradually from the east to the west ending in the slopes of the Aberdare Mountains. The highest areas in the west have a deeply dissected topography and are drained by several rivers including the Maragua, Saba Saba and Mathioya (Chapter 5). The topography of the lower eastern portion ranges from gently sloping to flat topography. The altitude varies from 3,353 to 914 m a.s.l.

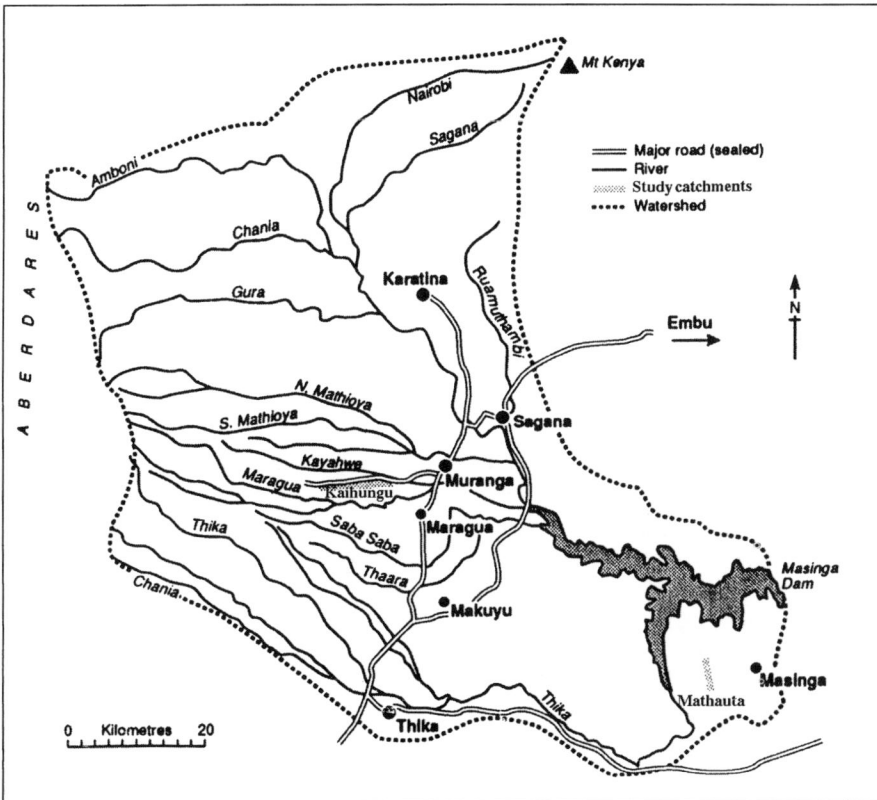

Figure 4.1   The upper Tana basin with the two study areas indicated

Catchment geology is characterised by volcanic rocks in the north and west. These include Tertiary and Quaternary basalts, phonolites and trachytic tuffs. A Precambrian basement complex terminates in the south-eastern part of the catchment. Other geological formations include igneous intrusions of granite and dolerites into the basement system and an area of Quaternary sandstones. Soils in the northern and western parts are of volcanic origin, locally referred to as kikuyu friables, and are richer and deeper than those developed over the rocks of the basement complex in the south-eastern part. The latter are poorer and shallower and rich in quartz. They overlie a deeply weathered granitoid gneiss which has decomposed to kaolin, quartz and vermiculite.

Rainfall is influenced by orographic effects. It varies from 600 mm yr^{-1} in the low-lying semi-arid areas in the east to over 2000 mm yr^{-1} in the high-altitude humid areas in the west. Four fairly distinct climatic seasons are recognised:

- the long rainy season (March–May)
- the cool dry spell (June–July)
- the hot dry spell (August–September)
- the short rain season (December–early March).

Moisture availability also varies according to climatic zone, as determined by Sombröek et al. (1982):

- humid zone (zone 1) receiving 1100–2700 mm rainfall per annum
- sub-humid (zone 2) receiving 1000–1600 mm
- semi-humid (zone 3) receiving 800–1400 mm
- semi-humid to semi-arid (zone 4) receiving 600–1100 mm
- semi-arid zone (zone 5) receiving 450–900 mm

The population distribution in the Masinga Dam catchment is, to a large extent, influenced by the physical environment. The high-potential areas of the catchment have dense population (sometimes exceeding 600 km^{-2}) resulting from an annual population growth rate of 3.2%. The increasing population is exerting pressure on the limited available agricultural land. This is indicated, firstly, by small farm holdings (usually less than 2 ha) resulting from land subdivision among family members which suggests that the farmers are financially unable to purchase land outside their home district and/or are unwilling to migrate to new areas due to cultural ties. Secondly, pressure on land is indicated by use of marginal land: in order to solve the problem of landlessness and small farm sizes, subsistence farmers with minimal resources move into fragile marginal areas such as steep slopes, roadsides, riversides and marshy lands, which already have a high erosion risk, in search of land for cultivation.

On the other hand, the lower altitude, drier areas with a low agricultural potential are less densely populated with some areas having a population density of below 100 persons km^{-2}. The annual population growth (less than 2%) is small but likely to have an important impact on soil erosion because of the fragile nature of this zone. Population movement into these marginal areas from regions with high population density is contributing greatly towards their annual growth rate and consequently to the soil erosion problem.

THE SOCIO-ECONOMIC PERSPECTIVE

SAMPLING FRAMEWORK

Investigations were conducted in the Masinga Dam catchment area between September 1991 and December 1992. Due to the large size of the catchment, only two sub-catchments were investigated: Kaihungu, lying between 1400 and 1750 m a.s.l. and Mathauta, lying between 1000 and 1400 m a.s.l. Kaihungu and Mathauta sub-catchments were considered representative of the high- and low-altitude areas of the catchment respectively in terms of the climatic regime, topography, soil characteristics and the soil erosion hazard.

Socio-economic data were then collected from selected farmers within this sampling framework. In this study, farm households formed the basic observational units. A household was defined as "a group of people who have one or more sources of income and make independent decisions on the use of that income. The members of the household are related through kinship and have a common place of residence and cook in one pot" (Bernard and Anzagi, 1979). These were selected along transects, developed along the footpaths traversing across and into the interior of the sub-catchments.

In the densely settled Kaihungu sub-catchment, 10 transects, each 1 km long, were developed. Sample households were then selected along the transects on an interval of 100 m. On the other hand, in the sparsely settled Mathauta sub-catchment, five transects, each 5 km long, were developed and households for inclusion in this study were selected every 0.5 km; the first two households on each side of the transect at this interval. This yielded 20 households along each transect – a total of 200 in Kaihungu sub-catchment and 100 households in the Mathuata sub-catchment. The heads of these households provided the socio-economic information used in this study to explain the soil erosion problem. Heads of household were self-selected but in many cases it was simply who was present at the time of the study (see below).

DATA COLLECTION

A questionnaire containing both structured and unstructured questions was used to elicit information from the farmers. The questionnaire covered three main areas, which were:

1. Farmers' socio-economic background including gender, age of the head of the household, family size, farm size, land tenure, income and employment.
2. Agricultural practices including crop types, cropping patterns, purpose for farming, crop performance, types of livestock, watering and grazing places.
3. Soil erosion and practised soil conservation measures.

At the end of each questionnaire the observer made her own assessment of the extent of the soil erosion problem in each visited household.

The chi square test was used to analyse the socio-economic data as the data were nominal (Ebdon, 1985).

FARM SIZE AND SOIL CONSERVATION

The results of this study show that the majority of the farmers did not practice adequate soil conservation measures (Table 4.1). One reason that was advanced for inadequate conservation was the small nature of the farms. There was a significant difference in farm size between farms suffering and those not suffering from accelerated soil erosion – large farms suffered less from the erosion problem than the small ones. Sixty-four per cent of the 266 farmers whose farms suffered from accelerated soil erosion and 41% of the 34 farmers whose farms did not suffer from accelerated soil erosion had farms (less than 1 ha). On the other hand, 27% of the farmers whose farms did not suffer from accelerated soil erosion and 17% of the farmers whose farms suffered from accelerated soil erosion had larger farms (exceeding 2 ha).

The main explanation is that some conservation techniques, particularly terracing and bund building, are associated with loss of valuable land. The land taken for terraces, especially the land occupied by the ditch from which the soil was excavated during the time of constructing the embankments, is considered as lost land by the farmers (Table 4.2). This explains the presence of widely spaced terraces particularly in the food crops in the study area. Similar problems have been noted in other parts of the tropics by Millington et al. (1989) and Thomas and Biamah (1991).

FARM ECONOMIC RESOURCES

There is a relationship between lack of off-farm employment and the presence or absence of accelerated soil erosion. About 75% of farmers whose farms suffered from

Table 4.1   Signs of erosion observed in the study farms grouped by size

	Farm size (ha)				
Observed erosion	0–1	1–2	2–3	3+	Totals
None	14 (41%)	11 (32%)	5 (15%)	4 (12%)	34 (11%)
Present	170 (64%)	51 (19%)	6 (2%)	39 (15%)	266 (89%)
Total	184 (61%)	62 (21%)	11 (4%)	43 (14%)	300 (100%)

Table 4.2   Typical land losses associated with commonly adopted soil conservation techniques

Conservation technique	Land loss (%)
Bench terracing	0.07–5.0
Grass strips	0.2
Contour bunding	0.1
Stick/stone bunding	0.02
Broad based bunding	0
Contour cultivation	0
Conservation tuleze	0

soil erosion are self-employed in their farms and 25% have an off-farm employment. On the other hand, 12% of the farmers whose farms did not suffer from soil erosion relied solely on farm employment whilst 88% had off-farm employment (Table 4.3). Off-farm employment is an additional source of income which makes the farmers financially superior compared to the farmers relying entirely on the farm employment. A significant difference also exists between the level of incomes of the farmers whose farms suffered or did not suffer from soil erosion. Table 4.4 shows that 52% of the 59 farmers whose farms suffered from soil erosion had low monthly incomes (less than 1800 Kenya Shillings (60 Ksh = 1 US $).

AGRICULTURAL EXTENSION ADVICE AND EDUCATION

Agricultural extension provides techniques for enhancing chances for increasing production as it offers practical education to the farmers. It plays an essential role in promoting soil conservation. There was a significant difference in farms that were or are not visited by extension officials in terms of soil erosion; 56% of the farmers whose farms did not suffer from accelerated soil erosion had received one or more visits while 85% of the farmers with eroded farms were never visited at all (Table 4.5). This indicates that farms receiving extension advice are better conserved compared to

Table 4.3   Nature of employment of farmers compared with observed evidence of erosion

Observed erosion	Farm employment	Off-farm employment	Total
None	4 (12%)	30 (88%)	34 (11%)
Present	200 (75%)	66 (25%)	266 (89%)
Total	204 (68%)	96 (32%)	300 (100%)

Table 4.4   Farm income classes compared with observed evidence of erosion

Observed erosion	Income in Ksh			Totals
	< 1800	1801–5000	> 5000	
None	3 (10%)	16 (53%)	11 (37%)	30 (34%)
Present	31 (52%)	25 (42%)	3 (5%)	59 (66%)
Totals	34 (38%)	41 (46%)	14 (16%)	89 (100%)

Table 4.5   Farm extension visits compared with signs of erosion

Observed erosion	Never visited by extension worker	Visited by extension worker at least once	Total
None	15 (44%)	19 (56%)	34 (11%)
Present	227 (85%)	39 (15%)	266 (89%)
Total	242 (81%)	58 (19%)	300 (100%)

those not receiving any advice. This is because the extension officers offer practical on-farm technical advice on soil conservation. For example, they physically measure and design the mechanical structures, thus reducing the incidence of having wrongly laid terraces. However, the ratio of extension workers to the farmer is very low. There is only on average one officer to every 5,000 farmers in Kenya and therefore the majority of farmers never benefit from such advice. For example, about 81% of the farmers in the Masinga Dam catchment asserted that they had never seen any agricultural extension officer on their farms in their lifetime.

Although the majority of the farmers have never received any agricultural extension advice, most of them appeared aware and knowledgeable about the soil erosion problem and its implications. This indicates the presence of other sources of soil erosion and conservation information besides the extension education workers. These include: local administrative chiefs, church meetings, friends/neighbours, mass media and agricultural shows. There is a considerable fund of relevant and appropriate indigenous knowledge (e.g. Chapter 14) concerning the soil erosion hazard. The local chiefs, who are members of the rural farming community, appeared to be an important source of soil conservation information. This is explained by the fact they call meetings (usually on Saturdays) which are mandatory for every one to attend. Therefore, chiefs have used those meetings to increase the farmers' awareness of the soil erosion problem and the need for conservation.

Agricultural extension is, however, likely to be more successful with an educated community. Indeed the relatively low level of education in particular among the rural poor is to an extent to blame for the low levels of rural and national development. There is a significant difference in the level of education between the farmers whose farms are or are not affected by the soil erosion problem. In the Masinga Dam catchment, about 85% (Table 4.6) of the farmers with secondary and/or post-secondary education had well-conserved farms, but only 15% with primary education. Conversely, 32% of the farmers in possession of secondary and/or post-secondary education had farms that showed signs of erosion, compared with 67% who only had primary education. This implies that farmers with higher education are able to adopt and maintain good conservation structures, thus minimising soil erosion in their farms. Education *per se* will not make good and efficient farmers. It is, however, a positive step towards solving some of the agricultural problems. This is because the higher educated one is, the higher the chances of obtaining a high-salaried employment. The majority of the rural population who have only attained primary level of education cannot join the labour market effectively, thus adding to the prevailing unemployment and underemployment problems. Most of them turn to

Table 4.6   Farmers' level of education compared with observed erosion

Observed erosion	Primary	Secondary	Post-secondary	Total
None	5 (15%)	14 (41%)	15 (44%)	34 (11%)
Present	179 (67%)	70 (26%)	17 (6%)	266 (89%)
Total	184 (61%)	84 (28%)	32 (11%)	300 (100%)

their small farms as the only alternative for employment. This contributes to the low standards of living and the increasing poverty level among the rural poor.

DISCUSSION

Socio-economic constraints facing the majority of the rural poor farmers are a major force behind the soil erosion problem. These farmers, most of them being small-scale subsistence farmers, lack resources and are therefore not in a position to assume responsibility for costly earthworks such as terracing. Due to the continued population growth and inadequate land tenure systems, the current situation of the farmers will further deteriorate following increasing land sub-division and unemployment. As a result, the long-term viability of production systems is likely to be sacrificed to meet immediate needs. Therefore, in order to sustain an environmentally sustainable rural development, the small-scale subsistence producers must have the appropriate incentives to undertake those practices that serve to improve rather than attack their resource base.

The implication of inadequate farm income is the inability of the farmers to invest in their farms. This is a major bottleneck facing the agricultural sector, and in particular limits the adoption and maintenance of soil conservation measures. This handicap results in the inability of farmers to purchase essential farm implements; to afford farm labour to supplement the family labour; or to to use mineral fertiliser and chemicals to control pests and weeds, despite evidence that their application would raise crop yields. Furthermore, in this study it appeared that some small-scale subsistence farmers were often suspicious of innovations and untried methods of farming whose benefits are not easily foreseeable. In these study areas it was largely because the rural farmers cannot afford to take "risks" with labour, time and income, as has been found elsewhere (Kelly, 1983; Hawuck, 1985). Any risk of loss is intolerable. Thus, for effective soil conservation to be realised, the farmers' resource base must be large enough to provide a sufficient proportion for the initiation and maintenance of the soil conservation measures. For instance, about 83% of the farmers in the study had soil conservation measures which required immediate repair. This has been delayed by inadequate farm incomes.

Thus in order to make soil conservation attractive to the farmer, it is necessary to offer short- and long-term benefits in the form of increased or more assured income. This can be realised by targeting at production-oriented activities which will enable the farmer to produce more from the small farms. This will further allow the farmer to produce surplus for marketing purposes and also increase income, some of which could be saved for resource regeneration and soil conservation. This can be achieved by:

1. Subsidising the prices of farm inputs to a level that the rural poor can afford.
2. Supplying the farmers with free fertiliser on completion of a certain amount of conservation work or on the basis of well-maintained conservation work.
3. Increasing agricultural commodity prices. This would encourage farmers to produce more, to use more farm inputs and to adopt new technology. This may

be particularly important to the subsistence farmers, since, as the threat to survival is reduced and the constraints on resources eased, a switch to more sustainable agricultural practices may be possible.

4. Removal of food marketing restrictions as well as opening up of new marketing outlets so that the farmers are assured that their farm produce will reach the market.

Involvement of all land users in soil conservation projects is essential because land users make decisions particularly in response to the immediate social and economic pressures. The severity of soil erosion in an area is closely related to the physical characteristics of the area, the decisions made by the land users, and the nature and forces of the socio-economic pressures under which the decisions are made. There- fore, conservationists are likely to have greater success in promoting non-degrading systems by working within the frames of reference of the land: what the farmer has to do, wants to do, or can be persuaded to do with their land. This can be done by explaining to the farmer the objectives of the programmes and seeking ideas from the farmer which can be incorporated in the programme.

It is also important to increase the mobility of extension officers: it was observed that most of the field officers lack transport to visit farmers. The national government ideally should, through the Ministry of Agriculture and Livestock Development, ensure that the field extension officers are supplied with suitable transport to enable them to reach as many farmers as possible.

In the wider perspective, it is necessary to reduce population growth by increasing family planning; rapid population growth has exacerbated social and economic problems resulting in a population–resource imbalance. This has led to congestion and increasing pressure on land which is now manifest in farm fragmentation, landlessness, use of fragile marginal areas, overcultivation, overgrazing and defor- estation, all of which are contributing to soil erosion.

REFERENCES

Bernard, F.E. and Anzagi S.K. 1979. 'Population pressure in rural Kenya'. Unpublished report, Dept. of Geography, Ohio University Athens.

Ebdon, D. 1985. '*Statistics in Geography*'. Blackwell, Oxford.

Grigg, D. 1984. '*An Introduction to Agricultural Geography*'. Unwin, London.

Hawuck, F.W. 1985. 'Soil erosion and its control in developing countries'. In *Soil Erosion and Conservation* (eds S.A. El-Swaify, W.C. Moldenhauer and L.Q. Andrew). Soil Conservation Society of America, Iowa.

Hudson, N.W. 1981. 'Social, political and economic aspects of soil conservation'. In *Soil Conservation Problems and Prospects* (ed R.P.C. Morgan). Wiley, Chichester.

Kelly, H.W. 1983. '*Keeping the Land Alive: Soil Erosion, its Causes and Cures*'. FAO, Rome.

Mellor, J.W. and Brenam, R. 1986. 'The food crisis and environmental conservation in Africa'. *Food Policy*, 1.

Millington, A.C., Mutiso, S.K., Kirby, J. and O'Keefe, P. 1989. 'African soil erosion – nature undone and the limitations of technology'. *Land Degradation and Rehabilitation*, 1, 279–290.

Sombröek, W.G., Braun, H.M.H. and van der Pouw, B.J.A. 1982. '*Exploratory Soil Map and Agro-climatic Map of Kenya, 1980*'. Exploratory Soil Survey Report No. E1, Kenya Soil Survey, Nairobi.

Stocking, M.A. 1984. 'Rates of erosion and sediments yield in the African environment'. In *Challenges in African Hydrology and Water Resources* (ed. D.E. Walling). IAHS Publication No. 144.

Thomas, D.B. and Biamah, E.K. 1991. 'The origin, application, and design of the *fanya juu* terrace'. In *Development of Conservation Farming on Hill Slopes* (ed. W.C. Moldenhauer), pp. 185–194. Soil and Water Conservation Society, Iowa.

Wolde, F.T. and Thomas, D.B. 1986. 'The effects of narrow grass strips in reducing soil loss and runoff on a Kabete Nitizol, Kenya'. In *Soil and Water Conservation in Kenya* (ed. D.T. Thomas). Proceedings of the Third National Workshop, Kabete, Nairobi.

Wooldridge, R. 1984. *'Sedimentation in Reservoirs: Tana River Basin Kenya'*. Hydraulics Research Report No. 61, Wallingford.

CHAPTER 5

Phosphorus Transport from a Tropical Catchment and Implications for Sustainable Catchment Management: The Upper Tana, Kenya

NIC PACINI
Ministry of Environment, Rome, Italy

DAVID HARPER
Department of Biology, University of Leicester, UK

INTRODUCTION

More than any other nutrient, phosphorus is linked to the geology and history of the soil and is recognised as having the greatest influence upon the productivity of terrestrial agricultural and of aquatic ecosystems. Recent studies on phosphorus export from agricultural land in temperate regions have stressed the need for an understanding of the main factors implicated in mobilisation and transport of this element for an efficient management of nutrient fluxes in agroecosystems (Taylor et al., 1971; Sharpley and Syers, 1979; Kronvang, 1990; Sharpley and Smith, 1990; Sharpley et al., 1991). In tropical Africa the quantitative chemical relationships between land and water are as yet virtually unstudied (Viner, 1975). It is believed, however, that intensive weathering, under the sustained high temperature which characterises tropical climates, provides tropical soils with high abundance of amorphous iron, aluminium (Wolf et al., 1985) and metal oxide-coated clays (Froelich, 1988). These same compounds are responsible for the binding of free phosphate and for causing phosphorus deficiency in crops. Mineralogical research, fertiliser trials and greenhouse tests have shown that phosphorus is low in Kenyan soils and is commonly the primary limiting nutrient (Hinga, 1973; Nyandat, 1981).

In a study of the upper Tana river catchment (Harper, 1993; Pacini et al., 1993; Pacini, 1994; Pacini and Harper, 1996) we investigated catchment phosphorus

The Sustainable Management of Tropical Catchments. Edited by David Harper and Tony Brown.
© 1998 John Wiley & Sons Ltd.

sources, and identified the main forms and transport fluxes of the element with the purpose of providing guidelines for the sustainable management of phosphorus fluxes in tropical catchments.

LAND CHARACTER AND MANAGEMENT

The Tana river receives most of its drainage from the eastern slopes of the Nyandarua (Aberdare) Range (maximum altitude of Mt Kinangop 3,908 m) and from the Mt Kenya massif (maximum altitude 5,199 m) (Figure 5.1). The region, which is located between 1,200 and 2,200 m (the Kenya Highlands), consists of the most productive and most densely populated farming area in the country. Its landscape is dominated by a complex series of narrow valleys with steeply dissected ridges. Numerous small streams with moderate to steep gradients form a dendritic drainage. The upper catchment drains into the Masinga Reservoir, the upper of a cascade of five hydro-electric dams situated in the eastern part of the catchment at 1,056 m altitude (see Chapter 15). The surface area of the catchment is 7,335 km².

The Highlands in the western part of the catchment lie on Tertiary and Quaternary basalts and basaltic agglomerates. Phonolites, trachytes and olivine basalt interlain with ash deposits covered the area in the Miocene period and again in the Pleistocene

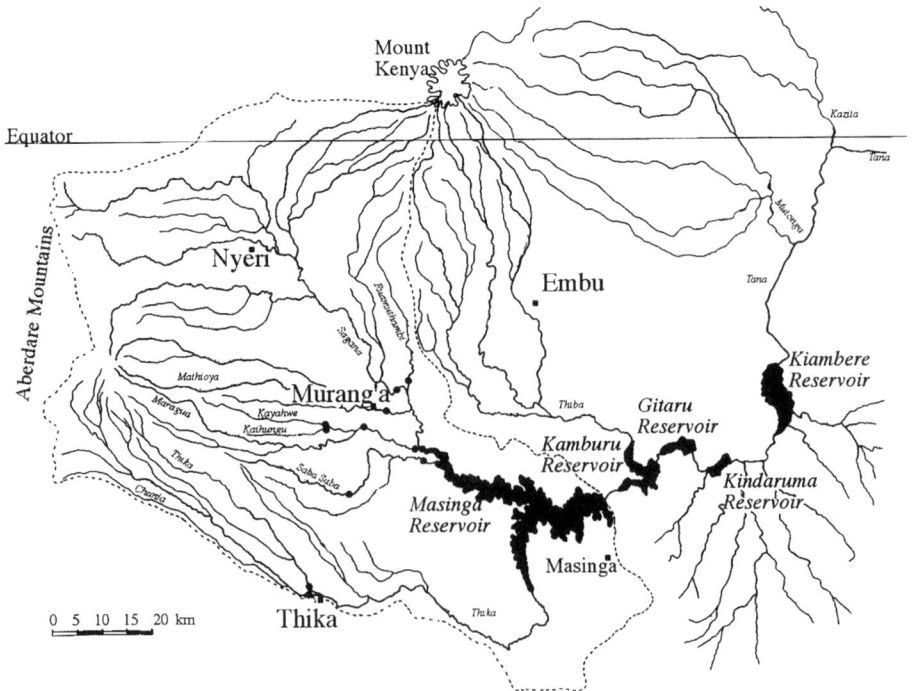

Figure 5.1 Location of the upper Tana catchment and its drainage streams. • = sampling stations

as products of the volcanism which gave rise to the Rift Valley and to Mt Kenya. The eastern part of the catchment, below 1,100 m, is characterised by a wide plain of irregular surface underlain by gneisses belonging to the Basement Complex, a Precambrian metamorphic formation which runs along the East African coastline from Somalia to Madagascar.

Within the slopes of the Nyandarua Range the soils are deeply weathered, rich in clay and uniformly red in colour. A continuous pedogenetic sequence extends from the high-altitude forest zone in the western part, characterised by andosols, histosols and lithosols, to nitosols in the mid slopes and ferralsols and vertisols in the eastern plains. In the lower slopes of the Nyandarua, soils exhibit a brown granular A horizon but lack a well-distinguished profile, characteristic of ferralitic soils in well-drained, high-relief regions of the humid tropics (Faniran and Areola, 1978). Among the main soil types humic and eutric nitosols are the most common. These friable clays, dominated by 1Si:1Al clay minerals, have a low erodibility and exhibit a moderate water retention capacity (Young, 1976). Samples of topsoil analysed by the National Agricultural Research Laboratories in Nairobi revealed moderate to strong acidity, low nutrient content and abundant extractable iron (Pacini, 1994). Nitosols represent an intermediate stage of ferralitic weathering on materials of fine texture and develop typically on well-drained, steep terrain. They are commonly threatened by erosion more because of the topographical situation in which they form than because of constitutional properties. When stripped of their vegetation cover they tend to dry out, harden and eventually erode, forming deep gullies along the prevailing slopes which are typically at 20–30°. The most remarkable manifestations of such erosion are seen along the numerous unsealed roads and footpaths which cut the hillsides. Such sites are common in the catchment and provide an unrestrained pathway to sediment and nutrient loss (see Chapter 2).

The rains are highly erosive and occur during two main rainy seasons (March–May, October–December). Mean rainfall in the western part of the catchment is between 1,400 and 2,000 mm yr^{-1}, with an average rainfall/evapotranspiration ratio of 65–80%. To the east rainfall averages only 650 mm yr^{-1}. Further details of the climate and geography of the area are reported by Atkins Land and Water Management (1984).

The area comprises four agroecological zones (Jaetzold and Schmidt, 1983) which range from the Forest Zone above 2,000 m to the Lower Coffee Zone at the foot of the volcanics (1,200 m). Tea and coffee are the main cash crops while maize and beans are part of the staple diet. Other vegetables and fruit trees are cultivated. Agroforestry is practised on some of the steeper slopes as a measure to reduce soil erosion. A small amount of artificial fertiliser is applied on cash crops (mainly on coffee) but also on subsistence crops such as maize and beans.

Population densities exceed 600 persons km^{-2} with an annual growth rate around 3%. The largest part of this population is well distributed on the farmland of the lower slopes. Very few settlements have any form of sewage treatment and effluent from the sewage treatment works of Thika (the largest town in the catchment) flows to the Athi river system to the south and therefore has no influence on the phosphorus dynamics of the Tana catchment. Thus effectively all the phosphorus is derived from diffuse sources.

METHODS

Water samples were collected in the mid channel from bridges by a polyethylene bottle attached to a weight that was slowly lowered through the water taking care not to cause sediment re-suspension. Samples were filtered in the field (pre-washed Sartorius Cellulose Acetate membrane filters, 0.45 µm) and analysed for pH, alkalinity, conductivity, silicate and soluble reactive phosphate (SRP) employing standard techniques (Mackereth et al., 1989) adapted to a portable spectrophotometer (Hach® Drel-2000, band width: 2 nm). Phosphate concentrations were often low and it was often necessary to include hexanol extraction to concentrate the phosphate before measurement. Samples for total phosphorus analysis were fixed immediately after filtration by addition of analytical grade chloroform to a concentration of 0.1%. Total phosphorus (TP) and total dissolved phosphorus (TDP) were digested in a muffle furnace at 500°C following Solórzano and Sharp (1980). Particulate phosphorus (PP) was estimated from the difference between total and total dissolved phosphorus. Suspended solids (SS) were measured by filtering samples in the laboratory through pre-weighed, oven-dried GF/C glass-fibre filters; then drying the filters at 105°C overnight and re-weighing. During the study 12 stations from 10 rivers (Figure 5.1) were sampled approximately bi-weekly, more often during the rainy seasons. In total some 45 samples from each station were processed for suspended solids while some 35 were analysed for phosphorus. Other chemical parameters were measured between 10 and 20 times at each station.

One sub-catchment, the Kaihungu, was chosen for an intensive study which allowed us to compute daily average concentrations of phosphorus and suspended solids, leading to a more accurate load estimate (Pacini and Harper, 1996).

CHEMISTRY OF RUNOFF

The general chemical composition of the rivers in the catchment (Table 5.1) reflects the volcanic nature of the bedrock which implies moderate porosity and low

Table 5.1   Stream water chemistry in the upper Tana (median values)

Chemical parameter	Chania	Thika	Kayahwe	Kaihungu	Mathioya	Sagana	Ruamu-thambi	Maragua	Tana	Saba Saba
SS	33	34	120	175	32	65	53	75	62	217
Cond.	54	67	87	148	77	88	82	72	80	240
TP	80	88	286	268	134	175	240	173	184	347
TDP	20	16	24	24	31	38	49	24	30	40
SRP	4	8	12	12	17	21	32	12	18	23
SiO_2	17	16	22	25	17	16	20	17	15	26
pH	7	7.0	6.9	7.1	7.3	7.0	6.5	7.0	6.9	7.3
Alk.	57	71	46	92	59	51	36	36	45	92

Units are $mg\,l^{-1}$ apart from phosphorus forms which are in $\mu g\,l^{-1}$ and conductivity (Cond.) which is expressed in $\mu S\,cm^{-1}$. Alkalinity (Alk.) is as $mg\,l^{-1}$ $CaCO_3$

resistance to weathering. The abundance of elements in solution reflects the selective leaching caused by the interaction of intense rains with mature, leached ("lateritic") soils under sustained high temperature.

The neutral pH and the alkalinities indicate moderately buffered conditions. The concentration of total dissolved solids, as indicated by conductivity, is low. In this geographical region the rainwater contribution is probably in the following order of importance: Na^+, Cl^-, Mg^{2+}, Ca^{2+}, SO_4^{2-} and K^+ (Kilham, 1971). The high leaching rate of Ca^{2+} and the selective supply of Cl^- and Na^+ by the rain determine the cationic dominance, which is $Na \geq Ca > Mg > K$. The concentration of other dissolved nutrients, such as SRP, TDP and dissolved oxidised nitrogen is low (Pacini, 1994) and is comparable to values reported by Wetzel (1983) as indicative of uncontaminated freshwaters.

FORMS OF PHOSPHORUS

Figures 5.2, 5.3 and 5.4 illustrate the variability in the concentration of phosphorus compounds. While total phosphorus values are dependent on episodic events of high transport of particulate material in suspension (Figure 5.5), soluble reactive and total dissolved phosphorus concentrations both show a relatively even distribution which reflects the buffering of phosphates by adsorption onto clay particles. The median values for soluble reactive phosphorus are comparable to average concentration of

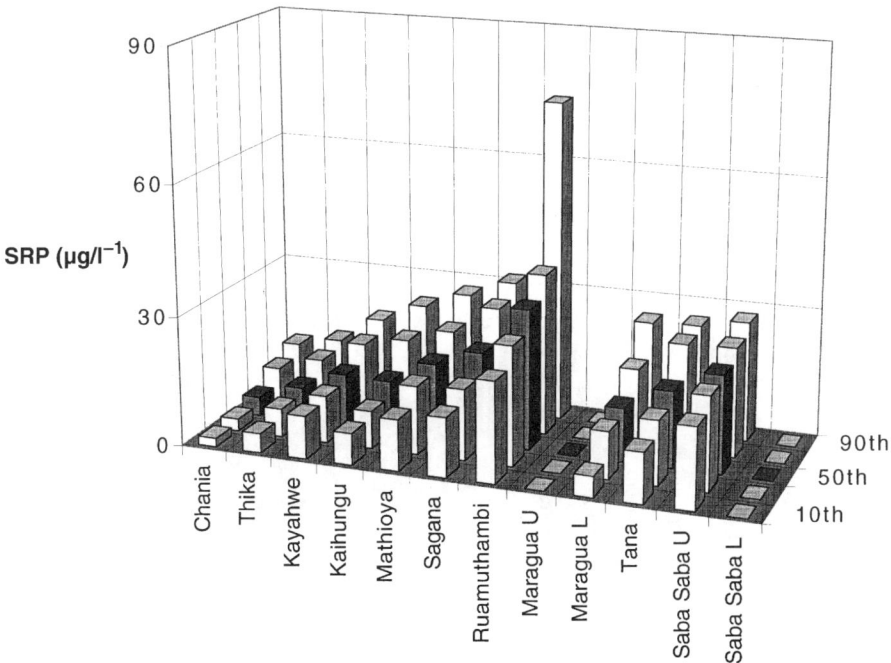

Figure 5.2   Concentration of soluble reactive phosphorus (SRP) in the upper Tana tributaries (expressed as percentiles)

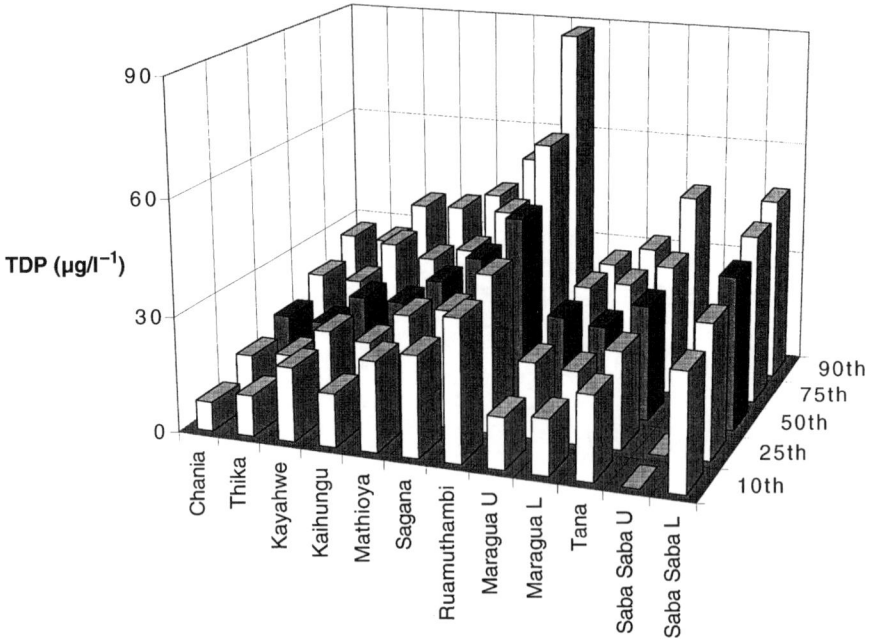

Figure 5.3   Concentration of total dissolved phosphorus (TDP) in the upper Tana tributaries (expressed as percentiles)

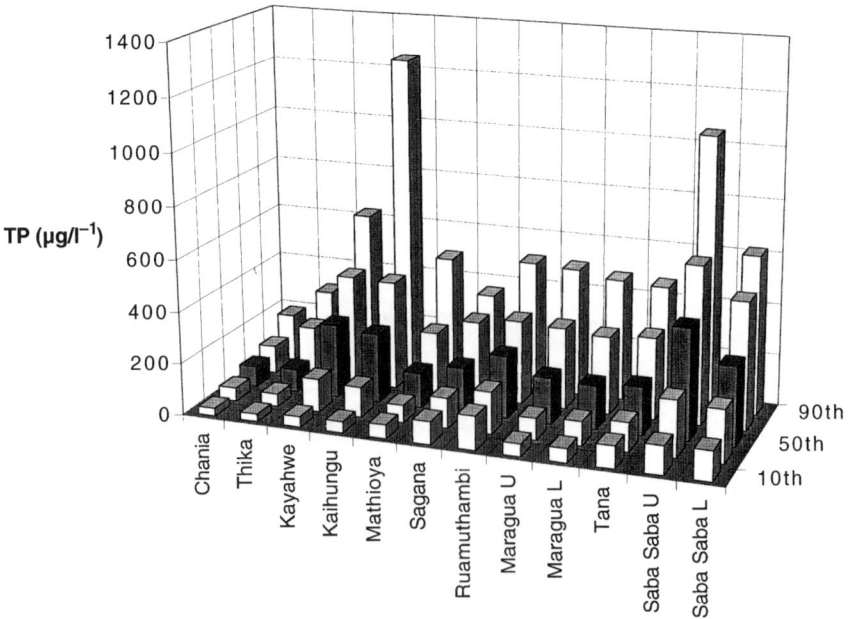

Figure 5.4   Concentration of total phosphorus (TP) in the upper Tana tributaries (expressed as percentiles)

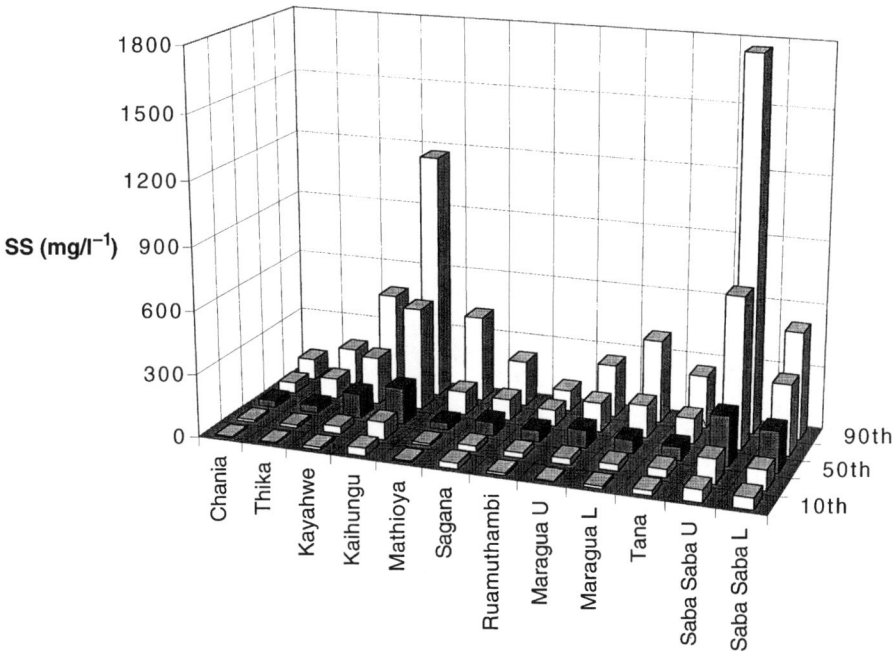

Figure 5.5   Concentration of suspended solids (SS) in the upper Tana tributaries (expressed as percentiles)

rivers in humid tropical regions estimated as 12 μg l^{-1} (Meybeck, 1981). The median concentration of total dissolved phosphorus shows that dissolved accounts for 10–25% of the concentration of total phosphorus (Figure 5.6). This proportion is under the influence of dilution and the effects of increased phosphorus sorption during high discharge periods. During baseflow, in most rivers, total dissolved phosphorus was often as much as 50% of total phosphorus, but such periods contribute little to the overall annual phosphorus export.

INFLUENCE OF VOLCANIC SPRINGS

While for most rivers nearly all the phoshorus was associated with the particulate matter, in the Ruamuthambi river, high concentrations of soluble reactive phosphorus (range: 23–90 μg l^{-1}) were common. At times, up to 95% of the total dissolved fraction was accounted for by soluble reactive phosphorus, indicating the presence of a source of inorganic phosphorus above the sampling station. This was identified as a groundwater inflow located at the discontinuity between the Mt Kenya olivine basalts and the impermeable biotite gneisses of the Basement Complex. This input of soluble phosphate is only partially readsorbed by the relatively low concentration of suspended solids carried by the Ruamuthambi river. Similarly to the Ruamuthambi, the Saba Saba river is under the influence of an alkaline spring. Here, as the riverbed loses its gradient and reaches the Basement Complex, the salts accumulate creating extensive pans of white concretions (*munyu*) in the fields along

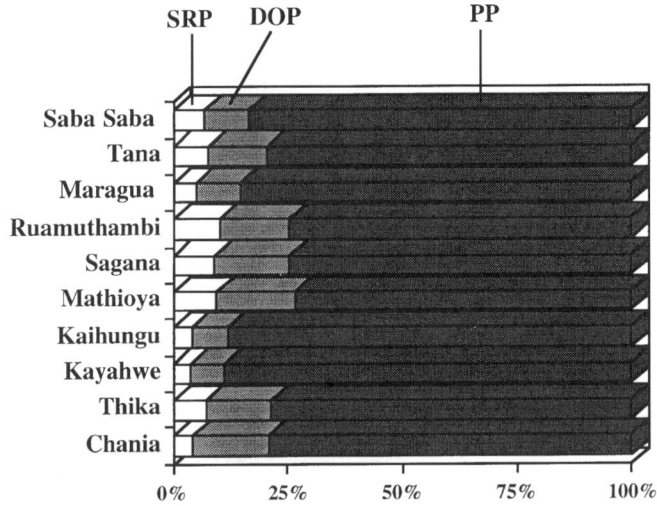

Figure 5.6   Mean percentage composition of phosphorus in the upper Tana tributaries.
DOP = dissolved organic phosphate

the riverbanks. The increase in soluble phosphate is buffered by the large amounts of
transported sediments (cf. Table 5.1).

Figure 5.7 illustrates the influence of springs in a cross-section of the catchment.
Two different scales are used for horizontal and vertical distances. Data on the
thickness of the volcanic layer and on the slope of the Basement Complex are from
Fairburn (1966).

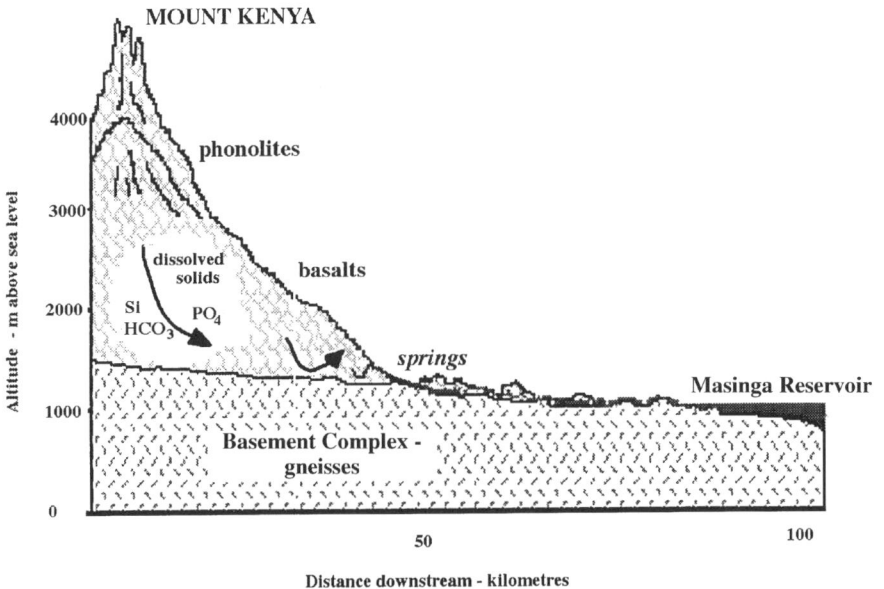

Figure 5.7   Conceptual model of groundwater discharge in the upper Tana catchment

ESTIMATES OF ANNUAL LOADS

Significant relationships between particulate phosphorus and suspended solids were established for each river to give an estimate of the mean phosphorus content of sediment (Table 5.2).

Phosphorus export coefficients (kg yr^{-1}) for each sub-catchment were calculated by multiplying estimates of suspended loads derived by Schneider (1993) by the estimated mean phosphorus content of the sediment of each river derived from the regression coefficient (b) (Table 5.2). Particulate phosphorus concentrations were estimated by subtracting a median total dissolved phopsphorus value from measured total phosphorus data. Export coefficients were converted to areal yields (kg km^{-2} yr^{-1}), presented in Table 5.3. Similarly, an estimate of phosphorus loading into Masinga Reservoir (180 kg km^{-2} yr^{-1}) was obtained by multiplying the average phosphorus content of sediment for the Tana river by the suspended solids load estimated to be 900–1,100 × 10^3 t yr^{-1} by Maingi (1991).

Table 5.2   Linear regression between particulate phosphorus (mg l^{-1}) and suspended solids (g l^{-1}). All correlation coefficients (r) are highly significant ($P < 0.001$). b is the regression coefficient which can be interpreted as representing the mean P content of suspended sediment.

River	b	r	n
Chania	1.1	0.91	35
Thika	1.1	0.91	35
Kayahwe	1.0	0.91	33
Kaihungu	0.2	0.53	136
Mathioya	1.2	0.75	34
Sagana	1.6	0.92	35
Ruamuthambi	2.8	0.77	32
Maragua	1.4	0.79	29
Tana	1.3	0.89	32
Saba Saba	0.6	0.77	29

Table 5.3   Estimates of suspended solids (SS) and total phosphorus (TP) yields for the sub-catchments

River	SS (t km^{-2} yr^{-1})	TP (kg km^{-2} yr^{-1})
Chania	100	110
Thika	73	80
Kaihungu	156	109
Mathioya	626	751
Sagana	67	107
Ruamuthambi	52	151
Maragua	266	346
Tana	139	180
Saba Saba	770	308

Besides statistical, analytical and sampling error, the significance of these estimates is subject to the high yearly variability of discharge from the Tana (Walling, 1985). The coefficient of variation of the 30-year monthly cumulated discharge for the Maragua river was shown to be 49% (Pacini, 1994). From historical records of discharge and suspended sediment concentrations Schneider was able to estimate a coefficient of variation in the annual suspended solids load in the same basin of 68% (Schneider, 1993).

DISCUSSION

PHOSPHORUS MOBILITY

With respect to their chemical composition, the rivers in the upper Tana are typical of Kilham's group of "common" African waters widespread around the continent (Kilham, 1971). In tropical regions river chemistry is primarily controlled by rock weathering which is here particularly intense and able to attenuate the effect of atmospheric deposition (Kilham, 1990). The nature and abundance of weathering products is relevant to the mobility of phosphorus forms in soils and sediments.

Conditions of intensive weathering are associated with the leaching of Ca^{2+} (Uriyo and Singh, 1978). The precipitation/dissolution equilibria of Ca-phosphate phases which commonly influence the occurrence of dissolved phosphate in temperate regions (Klotz, 1991) are less important in tropical regions (Golterman, 1988; Fardeau and Frossard, 1991) also due to the lesser availability of Ca^{2+} in tropical rivers. The high concentrations of soluble reactive silicate ($15–26\,mg\,l^{-1}$, from Table 5.1) found in the upper Tana (the world average is $13.1\,mg\,l^{-1}$ (Livingstone, 1963)) are related to the abundance of basalts and to the high annual mean temperature (Viner, 1975; Meybeck, 1981, 1986). The degradation of the clay structure and the release of large amounts of silica in runoff are well described processes leading to the formation of silica-poor clays such as kaolinite (Faniran and Areola, 1978; Meybeck; 1981). Kaolinite is the most widespread type of clay mineral found under tropical moist climates (Dixon, 1989, cited by Singh and Gilkes, 1992). Because of its specific chemical activity, kaolinite is believed to fix phosphorus by the "fast sorption step" (Borggaard, 1983, cited by Froelich, 1988), which is an expression of relatively strong chemical bonds and therefore concerns a phosphorus pool which is not readily available to crops (Parfitt, 1989). Besides the direct interaction with the clay surface, phosphorus adsorbs onto metal hydroxides which are themselves bound to the clay. Fertilisation trials on nitosols in the Embu district (the southern slopes of Mt Kenya) revealed that the application of superphosphate greatly increased the phosphorus pool occluded in Al and Fe (hydr)oxides showing good phosphorus retention capacity in these soils (Wapakala, 1976). Studies of the specific phosphorus sorption potential of soils of the upper Tana revealed moderate sorption indices (Hinga, 1973). This could be related to a low abundance of Al hydroxides which results in phosphorus sorption while clay is plentiful but does not limit the phosphorus sorption system (Hinga, 1973). Suspended sediments collected from rivers were largely composed of kaolinitic clays and had a phosphorus content close to the specific adsorption capacity of kaolinite (Table 5.2) estimated to $1.2-1.3\,\mathrm{mg\,P\,g}^{-1}$ by Viner (1987).

Besides the chemical composition of runoff, the temporal pattern of runoff determines phosphorus mobility. The highly erosive nature of rainfall in the humid tropical upper Tana makes such regions particularly susceptible to the effects of nutrient adsorption onto stream-borne particles. As a consequence, tropical rivers as a whole tend to carry a disproportionately large load of sorbed phosphorus which could constitute a substantial part of the global phosphorus flux to the oceans (Devol et al., 1989; Fox, 1990).

PHOSPHORUS EXPORT

Phosphorus export was in general higher than in most watersheds of igneous and sedimentary lithology cited by Dillon and Kirchner in North America (Dillon and Kirchner, 1974) with the exception of those under the direct influence of urban pollution. It compared instead with their estimate of mean phosphorus runoff from basins of volcanic origin in Washington State (Sylvester, 1961 and Emery, 1973 cited by Dillon and Kirchner, 1974). Examples of phosphorus runoff studies conducted in tropical basins are scarce. Grobler and Silberbauer (1985) related phosphorus export to runoff, grouping separately catchments of plutonic and sedimentary lithology in South Africa. Two sub-catchments in which rural land use was predominant were selected for comparison with the Kaihungu catchment, the most intensively studied in the upper Tana river catchment (Table 5.4). The regressions of total phosphorus versus runoff calculated in these catchments by Grobler and Silberbauer were applied to the higher rainfall conditions in the Kaihungu basin (Table 5.5).

Our estimate for the phosphorus yield from the Kaihungu falls in between the yields estimated from the linear models. The database from which these latter

Table 5.4   Comparison between the phosphorus (P) export from two sub-catchments within the basins of the Limpopo and Vaal rivers (South Africa) from Grobler and Silberbauer (1985) and the Kaihungu

Basin	Bedrock	Area (km^2)	P export ($kg\,km^{-2}\,yr^{-1}$)	Runoff (mm)	P/R
Limpopo	Granitic	1,171	0.8–3.8	12–45	0.07
Vaal	Sedimentary	16,153	1–68	0.2–12	5
Kaihungu	Volcanic	24	35–185	113–590	0.31

Table 5.5   Linear models between phosphorus (P) export and runoff (R) established by Grobler and Silberbauer (1985) in the Limpopo and Vaal catchments applied to runoff conditions in the Kaihungu catchment

Basin	Formula	P export ($kg\,km^{-2}\,yr^{-1}$)
Limpopo	$P = 0.17R - 1.71$	57
Vaal	$P = 1.31R - 5.14$	448

relationships were established did not contain data from volcanic basins. Smaling (1993) provided estimates of phosphorus runoff from a volcanic basin in Kisii, Western Kenya, employing empirical transfer functions and the Universal Soil Loss Equation (USLE, Wischmeier and Smith, 1978, cited by Smaling, 1993). Altitude, climate, topography, soils and population density in his study are comparable with the conditions found in the upper Tana. The soils were of the same type and had approximately the same phosphorus content as in our case study. With the USLE, Smaling estimated soil erosion and associated nutrient losses for the main land use units in the catchment. The contribution of erosion to phosphorus loss averaged $900\,kg\,P\,km^{-2}\,yr^{-1}$ for the entire Kisii district. This estimate is based on an off-plot soil erosion model in which phosphorus export is highly dependent upon differences in soil type, vegetation type and soil cover.

Being narrow and steep, the upper Tana catchments are not suitable for extensive cultivation as practised in Kisii, and are mostly occupied by subsistence farming which implies lower fertiliser inputs (see Chapter 4). In the highly populated slopes of the Nyandarua Range, the main sediment source for the upper Tana, off-plot erosion models are hardly relevant as the main sediment sources have been identified as footpaths, unsealed roads and bare ground surfaces (Chapter 2; Schneider, 1993; see also Dunne, 1979). Such sediment sources have a lower phosphorus content than agricultural topsoils. Reduced fertiliser inputs and alternative sediment sources could explain the difference between our measured estimates of phosphorus export $(80-750\,km^{-2}\,yr^{-1}$, Table 5.3) and those of Smaling (1993).

CONCLUSIONS

Phosphorus fluxes in the upper Tana are closely related to the mobility of soils and sediments within the catchment as particle-bound phosphorus constitutes more than 90% of the forms present. The main mechanism of phosphorus fixation is by sorption onto kaolinitic clays and their aluminium and iron coatings. Although the specific phosphorus sorption capacity of soils in the upper Tana is not high, the deep soil profiles and the abundance of transported river sediments ensure the sorption of large amounts of phosphorus from runoff. Free phosphate is very scarce in the rivers apart from where naturally phosphorus-enriched groundwater sources flow into the channel.

As stressed by some authors (Dillon and Kirchner, 1974; Grobler and Silberbauer, 1985), geology is a prime determining factor in explaining phosphorus export. In volcanic catchments it is often moderately high, due to the steep slopes which characterise volcanic terrain and to geochemical processes. Porous volcanic rocks allow the slow percolation of water at depth and the export of concentrated solutes through groundwater channels. Geochemical phosphorus sources are likely to be common in catchments such as the upper Tana, characterised by a relatively porous bedrock and high weathering rates (intensive tropical rain and high yearly temperature).

In the upper Tana these groundwater phosphorus sources may be relatively important in the overall export flux from the catchment as most other phosphorus sources, such as fertiliser-P and effluent-P produced by cattle and people living in the catchment, are more likely to remain bound within agricultural plots. River chemical composition shows that, in the upper Tana, inorganic phosphorus sources are well buffered by natural self-purification processes and have minimal impact on water quality parameters.

The tight relationship between phosphorus export and erosion rates allowed us to derive phosphorus yield figures directly from suspended yields. Erosion rates, however, cannot be established applying models which describe off-plot erosion as these are biased by various erosion protection structures widespread in the upper Tana and the increasing importance of alternative sediment pathways. Footpaths and other bare ground areas constitute an important sediment source that is not well taken into account by traditional erosion modelling. The derivation of phosphorus export yields from off-plot erosion models may imply overestimates because of a wrong identification of sediment delivery pathways. This study shows how phosphorus retention and pathways of sediment and phosphorus transport depend upon specific characteristics of tropical catchments which have to be investigated individually for the identification of precise strategies for the sustainable management of nutrient fluxes in these regions.

REFERENCES

Atkins Land and Water Management, 1984. '*Soil and Water Conservation Programme – Masinga Dam Catchment Areas*'. Final Report, Volume 1, W.S. Atkins, Cambridge.

Borggaard, O.K. 1983. 'The influence of iron oxides on phosphate adsorption by soil'. *Journal of Soil Science*, **34**, 333–341.

Devol, A.H., Richey, J.E. and Forsberg, B.R. 1991. 'Phosphorus in the Amazon river mainstem: concentrations, forms and transport to the ocean'. In *Proceedings of a Workshop on Phosphorus Cycling in Terrestrial and Aquatic Ecosystems, Vol. 3: South and Central America* (eds H. Tiessen, D. López-Hernández and I.H. Salcedo), pp. 108–128. SCOPE/UNEP, Paris.

Dillon, P.J. and Kirchner, W.B. 1974. 'The effects of geology and land use on the export of phosphorus from watersheds'. *Water Research*, **9**, 135–148.

Dillon, P.J. and Rigler, F.H. 1975. 'A simple method for predicting the capacity of a lake for development based on lake trophic status'. *Journal of the Fisheries Research Board of Canada*, **32**, 1519–1531.

Dunne, T. 1979. 'Sediment yield and land use in tropical catchments'. *Journal of Hydrology*, **42**, 281–300.

Fairburn, W.A. 1966. '*Geology of the Fort Hall Area*'. Report No. 73. Geological Survey of Kenya, Government of Kenya.

Faniran, A. and Areola O. 1978. '*Essentials of Soil Study (With Special Reference to Tropical Areas)*'. Heinemann, London.

Fardeau, J.C. and Frossard, E. 1991. 'Processus de transformations du phosphore dans les sols de l'Afrique de l'ouest semi-aride: application du phosphore assimilable'. In *Proceedings of the Workshop on Phosphorus Cycling in Terrestrial and Aquatic Ecosystems, Vol. 4: Africa* (eds H. Tiessen and H. Frossard), pp. 108–128. SCOPE/UNEP, Nairobi, Kenya.

Fox, L.E. 1990. 'Geochemistry of dissolved phosphate in the Sepik River and Estuary, Papua, New Guinea'. *Geochimica et Cosmochimica Acta*, **54**, 1019–1024.

Froelich, P.V. 1988. 'Kinetic control of dissolved phosphate in natural rivers and estuaries: A primer on the phosphate buffer mechanism'. *Limnology and Oceanography*, **33**, 649–668.

Golterman, H.L. 1988. 'The calcium and iron bound phosphate phase diagram'. *Hydrobiologia*, **159**, 149–151.

Greenland, D.J. 1973. 'Soil factors determining responses to phosphorus and nitrogen fertilizers used in tropical Africa'. *African Soils*, **17**, 99–108.

Grobler, D.C. and Silberbauer, M.J. 1985. 'The combined effect of geology, phosphate sources and runoff on phosphate export from drainage basins'. *Water Research*, **19**, 975–981.

Harper, D.M. 1993. Tana River Project. Progress Report to the European Community's STD2 programme, contract TS2-A-0256-UK.

Hinga, G. 1973. 'Phosphate sorption capacity in relation to properties of several types of Kenya soil'. *East African Agriculture and Forestry Journal*, **38**, 400–404.

Jaetzold, R. and Schmidt, H. 1983. '*Farm Management Handbook of Kenya: Natural Conditions and Farm Management Information. Volume II, Part B, Central Kenya*'. Ministry of Agriculture, Nairobi, Kenya.

Kilham, P. 1971. 'Biogeochemistry of African lakes and rivers'. PhD Thesis, Duke University, USA.

Kilham, P. 1990. 'Mechanisms controlling the chemical composition of lakes and rivers: Data from Africa'. *Limnology and Oceanography*, **35**, 80–83.

Klotz, R.L. 1991. 'Temporal relation between soluble reactive phosphorus and factors in stream water and sediments in Hoxie Gorge Creeek, New York'. *Canadian Journal of Fisheries and Aquatic Sciences*, **48**, 84–90.

Kronvang, B. 1990. 'Sediment-associated phosphorus transport from two intensively farmed catchment areas'. In *Soil Erosion on Agricultural Land* (eds I.D.L. Boardman, D.L. Foster and J.A. Dearing), pp. 313–330. Wiley, Chichester.

Livingstone, D. 1963. '*Chemical Composition of Rivers and Lakes*'. US Geological Survey Professional Paper No. 440G, USGS, Washington.

Mackereth, F.J.H., Heron, J. and Talling, J.F. 1989. '*Water Analysis: Some Revised Methods for Limnologists*'. FBA Publication No. 36, Freshwater Biological Association, Ambleside.

Maingi, J.K. 1991. 'Sedimentation in Masinga Reservoir'. MSc Dissertation, University of Nairobi.

Meybeck, M. 1981. 'Pathways of major elements from land to ocean through rivers'. In *River Inputs to Ocean Systems*, pp. 18–30. UNEP/UNESCO, Paris.

Meybeck, M. 1986. 'Composition chimique des ruisseaux non pollués de Franc'. *Bulletin de Sciences Géologiques*, **39**, 3–77.

Nyandat, N. 1981. 'The primary minerals in some Kenya top-soils and their significance to inherent soil fertility'. *East African Agriculture and Forestry Journal*, **46**, 71–76.

Pacini, N. 1994. 'Coupling of land and water; phosphorus fluxes in the upper Tana river catchment, Kenya'. PhD Thesis, University of Leicester.

Pacini, N. and Harper, D.M. 1996. 'Transport and fate of sediments and phosphorus in an equatorial catchment: the influence of geology, climate and land-use'. In *Sediment and Phosphorus* (eds B. Kronvang and L. Svendsen), pp. 104–105 Technical Report No. 178, NERI, Silkeborg, Denmark.

Pacini, N., Harper D.M. and Mavuti, K.M. 1993. 'A sediment-dominated tropical impoundment: Masinga Dam, Kenya'. *Verhandlungen Internationale Vereinigung Limnologie*, **25**, 1275–1279.

Parfitt, R.L. 1989. 'Phosphate reactions with natural allophane, ferrihydrite and goethite'. *Journal of Soil Science*, **40**, 359–369.

Schneider, H. 1993. Soil loss and sediment field in a tropical catchment: the upper Tana river basin, Kenya. Unpublished Ph.D. thesis, University of Leicester.

Sharpley, A.N. and Smith, S.J. 1990. 'Phosphorus transport in agricultural runoff: the role of soil erosion'. In *Soil Erosion on Agricultural Land* (eds I.D.L. Boardman, D.L. Foster and J.A. Dearing). Wiley, Chichester.

Sharpley, A.N. and Syers, N.N. 1979. 'Phosphorus inputs into a stream draining an agricultural watershed'. *Water, Air and Soil Pollution*, **11**, 417–428.

Sharpley, A.N., Troeger, W.W. and Smith, S.J., 1991. 'The measurement of bioavailable phosphorus in agricultural runoff'. *Journal of Environmental Quality*, **20**, 235–238.

Singh, B. and Gilkes, R.G. 1992. 'Properties of soil kaolinites from south-western Australia'. *Journal of Soil Sciences*, **43**, 645–667.

Smaling, E.M.A. 1993. 'An agroecological framework for integrated nutrient management with special reference to Kenya'. PhD Thesis, University of Wageningen, The Netherlands.

Solórzano. L. and Sharp, J.H. 1980. 'Determination of total dissolved phosphorus and particulate phosphorus in natural waters'. *Limnology and Oceanography*, **25**, 754–758.

Taylor, A.W., Edwards, W.M. and Simpson, E.C., 1971. 'Nutrients in streams draining woodland and farmland near Coshocton, Ohio'. *Water Resources Research*, **7**, 81–89.

Uriyo, A.P. and Singh, B.R. 1978. 'Studies on the distribution of inorganic forms of phosphorus and their relationship with the extent of weathering in some soil profiles in Tanzania'. *East African Agriculture and Forestry Journal*, **43**, 24–37.

Viner, A.B. 1975. 'The supply of minerals to tropical rivers and lakes (Uganda)'. In *Coupling of Land and Water Systems* (ed. A.D. Hasler). *Ecological Studies*, **10**, 227–261. Springer-Verlag, New York.

Viner, A.B. 1987. 'Nutrients transported on silt on rivers'. *Archivium Hydrobiologiae Beihefte (Ergebnisse der Limnologie)*, **28**, 63–71.

Walling, D.E. 1985. 'The sediment yield of African rivers'. *Proceedings of the Symposium on Challenges in Hydrology and Water Resources*, Harare, July 1984, pp. 265–283. IAHS Publication 144.

Wapakala, W.W. 1976. 'Changes in some chemical properties of a red clay soil: results of long term fertilizer and rotational trials in central Kenya'. *East African Agriculture and Forestry Journal*, **42**, 201–218.

Wetzel, R.G. 1983. '*Limnology*' (2nd edition). Saunders College Publishing, Philadelphia.

Wischmeier, W.H. and Smith, D.D. 1978. 'Predicting rainfall erosion losses – A guide to conservation planning'. *Agricultural Handbook No. 537*. US Department of Agriculture, Washington, DC.

Wolf, A.M., Baker, D.E., Pionke, H.B. and Kunishi, H.M. 1985. 'Soil tests for estimating labile, soluble, and algae-available phosphorus in agricultural soils'. *Journal of Environmental Quality*, **14**, 341–348.

Young, A. 1976. '*Tropical Soils and Soil Survey*'. Cambridge University Press, Cambridge.

CHAPTER 6

Agroecological Practices as Tools for the Sustainable Management of Catchments Susceptible to Erosion: Réunion Island

SYLVAIN PERRET
CIRAD, Montpellier, France

ROGER MICHELLON
CIRAD, Madagascar

JACQUES TASSIN
CIRAD, Ile de la Réunion

INTRODUCTION

On the small island of Réunion, in the Indian Ocean (Figure 6.1), population growth coupled with urban development on the coastal strip has resulted in intense pressure on the remaining agricultural systems which are fragile due to their situation on steep slopes experiencing high rainfall. On the island's western side (21°05′S, 55°20′E), deep volcanic ash soils range from 500 to 1500 m in elevation, according to climatic conditions. These andisols (Andepts) first show a dominant halloysitic character, then turn allophanic and gibbsitic (Zebrowski, 1975; Raunet, 1991).

Pelargonium (*Pelargonium* spp.) is the dominant crop, grown between 800 to 1200 m for the extraction of oil to be used in the perfume industry. Leaves and young stems are harvested and then water-distilled for essential oil extraction. At lower altitudes, sugar-cane dominates. Fallow lands, natural or exploited forests, and grasslands cover the higher areas. Mean annual temperatures range from 16 to 19°C. Mean annual rainfall ranges from 1200 to 1700 mm with 70% between January to April, mostly during tropical depressions or cyclones. These high-intensity rainfall events induce important runoff and soil losses on agricultural fields, enhanced by slope (averaging 15%), low conductivity and poor aggregation of degraded soils (Perret, 1993).

The Sustainable Management of Tropical Catchments. Edited by David Harper and Tony Brown.
© 1998 John Wiley & Sons Ltd.

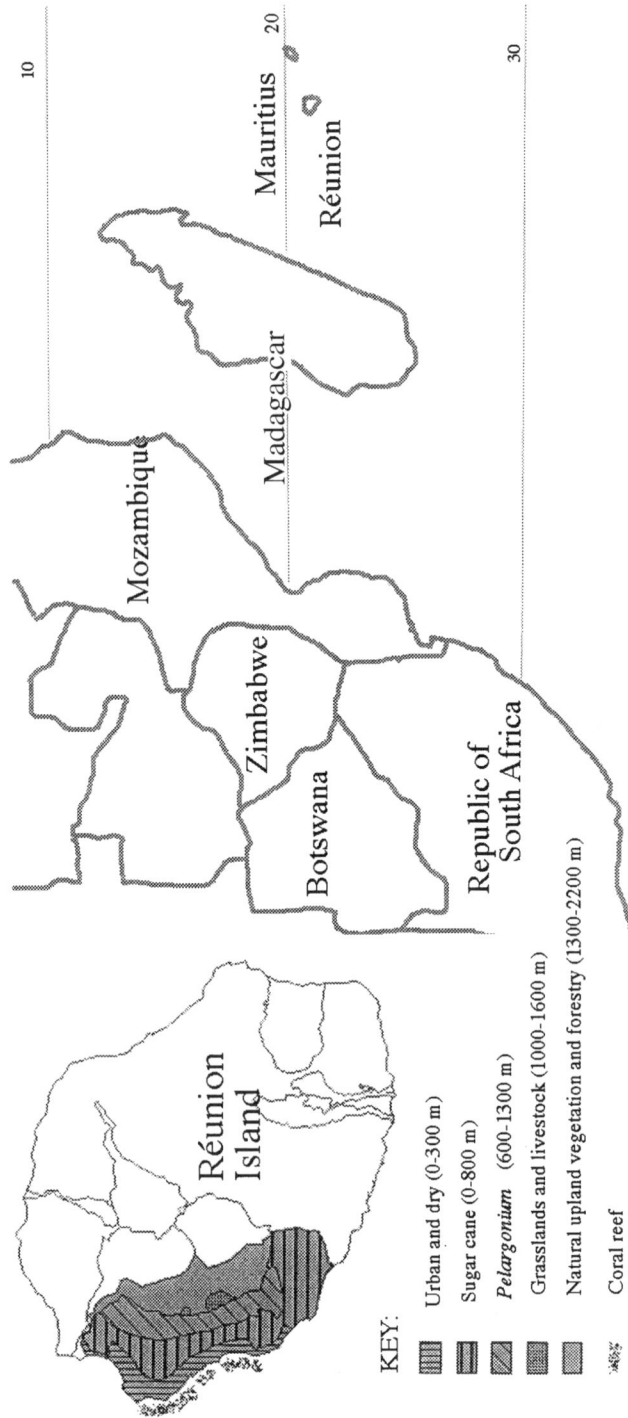

Figure 6.1   Location of Réunion Island, and illustration of vegetation zonation with altitude

The land husbandry also encourages erosion because it is mainly based on hoed crops (not only pelargonium, but also various food crops and vegetables (Garin, 1987)). Traditional systems were formerly based on shifting cultivation, alternating with long-term fallow which restored soil fertility, but with poor surface management to prevent erosion; this was mainly because farmers were not landowners.

At the present time, population settlement has forced farming systems to develop more intensive cropping practices, such as long-term monoculture without fallow, and weed control with hoeing. These practices increase soil degradation and lead to a decrease in fertility. For all crops, inputs have to increase (i.e. plant-health treatments, organic matter, mineral fertilisers) but still cannot overcome a decrease in yield and greater costs in both labour and money (Michellon and Garin, 1985).

Erosion is now so severe that few relict profiles remain. Nearly all tilled soils have lost the A horizon and now expose the Bt horizon, or even the deep-indurated horizon (volcanic tuff) or pyroclastic scree in the most degraded conditions (Raunet, 1991). Within the *Pelargonium* growing-zone, field erosion averages $20 \, t \, ha^{-1}$ during a usual year. It amounts to $50-200 \, t \, ha^{-1}$ during a rainy and cyclonic year (Bougère, 1988; measurements with standard erosion plots). It widely exceeds the tolerable soil-loss limit, usually accepted in the tropics as $5-15 \, t \, ha^{-1}$ (El-Swafy et al., 1982). In the coastal strip where drainage streams discharge this eroded silt, serious damage is being caused to coral reef ecosystems (Stieljes, 1993) and concern is being expressed about drinking water quality from aquifers at risk from pesticide and fertiliser influx.

Under such constraints, farmers progressively abandon the most degraded fields and *Pelargonium* culture. Currently, farming systems are evolving towards complex mixed-cropping systems, sometimes associated with goat or cattle breeding. But the persistence of hoeing on almost all of the cropped fields maintains erosion at very high rates. The steepness of the soils makes them unsuitable for mechanical rehabilitation (e.g. terracing and levelling, Nimlos 1992; Chapter 8) and so biological methods need to be investigated. "Agroecological" practices are increasingly being seen to fit small tropical farming systems (ICRAF, 1993), often using cover plants, legume intercropping and hedgerow planting (National Research Council, 1983; Lal et al., 1991).

TOWARDS A MANAGEMENT STRATEGY

Sustainable agriculture implies "successful management of resources for agriculture to satisfy changing human needs while maintaining or enhancing the quality of the environment and conserving natural resources" (Technical Advisory Committee, 1989, cited by Lal et al., 1991). The Réunion Island *Pelargonium*-growing area needs to become sustainable. In such conditions, soil loss is the primary factor to be taken into account, for establishing new soil conservation and productive cropping systems (Lal, 1989; Smith et al., 1992), and for developing the features of a more natural landscape (Thomas and Kevan, 1993). These systems are part of the global strategy towards sustainability, that includes diversification, integration and synthesis (Ikerd, 1993).

On severely eroded volcanic ash soils in Latin America, successful reclamation techniques have been based on mechanical methods such as terracing and levelling tillage (Nimlos, 1992; Chapter 8). The agroecological practices that are also most promising are the use of cover plants to restore soil and fertility (Lal, 1975; Lal et al., 1991), improving its different components such as aggregation and pore space, water infiltration, erosion and runoff resistance, nutrient cycling and soil organic matter. In Réunion a suitable plant for integration with the *Pelargonium*-growing system was *Calliandra*, using hedges as plot boundaries for increasing runoff control and soil restoration. *Calliandra* also provides wind protection and green fodder, as well as being resistant to insect attack. The aim of this work is to describe and quantify the components of soil degradation and restoration, and to assess the suggested agroecological changes.

METHODS

The CIRAD station of Trois-Bassins (ranging from 950 to 1020 m in elevation) is the main site for *Pelargonium* horticultural research in Réunion Island. Its mean annual rainfall is about 1500 mm with 70% between January and April. The andisols (Dystrandepts) cover volcanic ash, with a surface slope steepness averaging 15–20%. Water runoff and erosion processes were first studied on severely and moderately degraded bare soils under *Pelargonium* monoculture, and compared with *Pelargonium* cultivation associated with Kikuyu-grass (*Pennisetum*) and trefoil (*Lotus*) covered soils.

During July 1993, different plots on each *Pelargonium* harvesting system were analysed for soil structure and fauna, for aggregate size, stability and organic matter and for water transmission, taking into consideration the distance from *Calliandra* hedges. All plots had been severely degraded by long-term *Pelargonium* monoculture. The systems investigated were:

1. A long-term *Pelargonium* monoculture (25 years; plot 19).
2. A *Pelargonium* six-course rotation with food crops (six crop-cycles: first year–tobacco/maize (2 cycles/year), second year–potato/maize (2 cycles/year), third year–bean/maize (2 cycles/year), then 6 years in *Pelargonium* cropping, then potato/maize cropping cycles in 1993; plot 16).
3. *Pelargonium* association with kikuyu-grass (*Pennisetum clandestinum*, planted in 1989; plot 2).
4. *Pelargonium* association with greater bird's-foot-trefoil (*Lotus uliginosus*, planted in 1990; plot 1).

A sprinkler infiltrometer was used, developed by Asseline (1981) and Casenave (1982). It uses a deflector-nozzle fastened at the top of a 4 m high tower. The nozzle oscillates and continuously produces drops under 100 kPa pressure. The drop size, impact velocity, and kinetic energy are constant, and consistent with natural rainfall. The intensity varies as a function of the oscillation. The elementary plot surface was 1 m^2. Plot steepness ranged from 8 to 20%; rainfall intensity ranged from 36 to 83 mm h^{-1}.

On each situation, a soil profile was described, with particular reference to *in situ* structures, earthworm number and activity, and root system structure.

Aggregate size and stability were measured on a 30 g sample which was wetted, then separated into various size fractions by sieving through a nest of sieves under a gentle water flow to cause as little mechanical disruption of the aggregates as possible. Sieves with openings of 5.00, 2.00, 1.00, 0.50, 0.20 and 0.05 mm were used. The mean-weight diameter (*MWD*) of aggregates (Van Bavel, 1949) is equal to the sum of products of (a) the mean diameter of each size fraction d_i and (b) the proportion of the total sample dry weight w_i occurring in the corresponding size fraction, when the summation is carried out over all *n* size fractions, including the one that passes through the finest sieve:

$$MWD = \sum_{i=0.025}^{3.50} (d_i w_i) \qquad \text{thus, } 0.025 \text{ mm} \leq MWD \leq 3.50 \text{ mm} \qquad (1)$$

3.00 mm is the mean diameter of the remaining fraction on the sieve with an opening of 2.00 mm; 0.025 mm is the mean diameter of the fraction that passes through the sieve with an opening of 0.05 mm.

The same experiment was carried out with another 30 g sample, previously exposed to 20 kHz ultrasonic energy of 30 J mL^{-1} in water (e.g. 75 W × 120 seconds ~ 300 mL of water). This treatment causes both shattering and abrasion of the aggregates (North, 1976, 1979; Gregorich et al., 1988). Then, a new mean weight diameter MWD_{us} was calculated. A stability index was used:

$$S = MWD_{us}/MWD \qquad \text{thus, } 0 < S \leq 1 \qquad (2)$$

Previous experiments carried out on andisols have shown that *S* ranges from 0.3 (very poor stability) to 1 (full stability) (Perret, 1993).

In addition, a measurement of the total organic carbon (Anne method) and nitrogen (Kjeldahl method) were carried out on each sample.

Near each profile, tension infiltrometry was carried out in order to determine hydraulic conductivity K_i. The Triple Ring Infiltrometer at Multiple Suctions (TRIMS, Thony 1990) was developed from Clothier and White (1981) and Perroux and White (1988). It allows field-measurements of K_i from saturation to different negative matrix pressure *i*, to −200 mm.

RESULTS AND DISCUSSION

RUNOFF AND SOIL-LOSS PROCESSES

From 45 rainfall simulation experiments, runoff intensity (*R*), transported soil (*C*) and rainfall intensity (*I*) were related to soil degradation level and surface management. Figure 6.2 shows a rainfall-intensity threshold for runoff, starting at about 36 mm h^{-1}. After this, runoff increases with rainfall, according to a limit line calculated from the points that represent a maximum runoff intensity for a given rainfall. The equation is:

$$R = 1.23I - 46 \qquad (3)$$

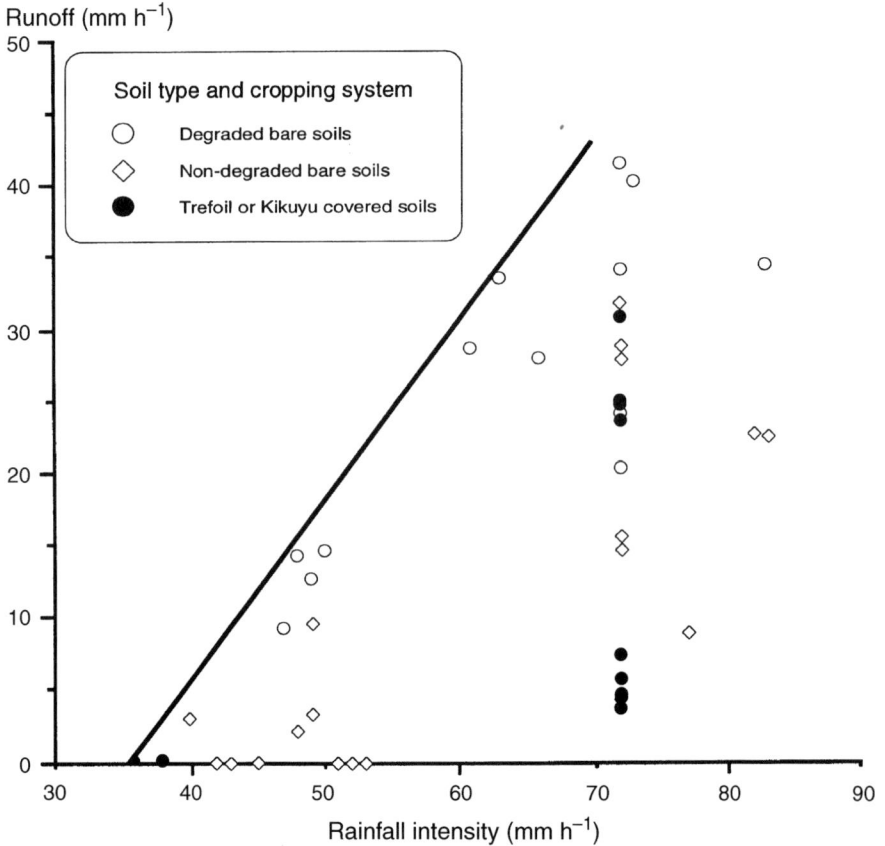

Figure 6.2   Variation in runoff with rainfall intensity under different soils and management

The points next to this limit represent exposed, severely degraded soils, with low roughness. Below this limit are moderately degraded soils and covered soils. The line indicates that infiltration, F (= rainfall less runoff), is not a constant, but decreases as a function of rainfall intensity, most probably due to slaking and surface structure degradation under strong rainfall. It is confirmed by the time-threshold for the commencement of runoff, which was 8 minutes on average under 50 mm h^{-1} rainfall, and only 2 minutes under 70 mm h^{-1} rainfall. Below 36 mm h^{-1}, there was no runoff. In addition, gradients of less than 20% had no effect on runoff intensity.

Live mulches cause little reduction of runoff, but kikuyu-grass and trefoil cover strongly reduce sediment transport (Figure 6.3). Their most important effect is to protect soil against raindrop impact, and to decrease velocity and carrying capacity of overland flow. Cover plants have an effect on structure and fauna to give better water infiltration, but this does not cause enough improvement to completely absorb runoff induced by strong rainfall. Table 6.1 summarises the effect of soil covering on runoff and erosion control, at different rainfall intensities. The cover plants' most important effect is to reduce erosion. Under strong rainfall, runoff remains important.

Soil carriage (kg h^{-1} ha^{-1})

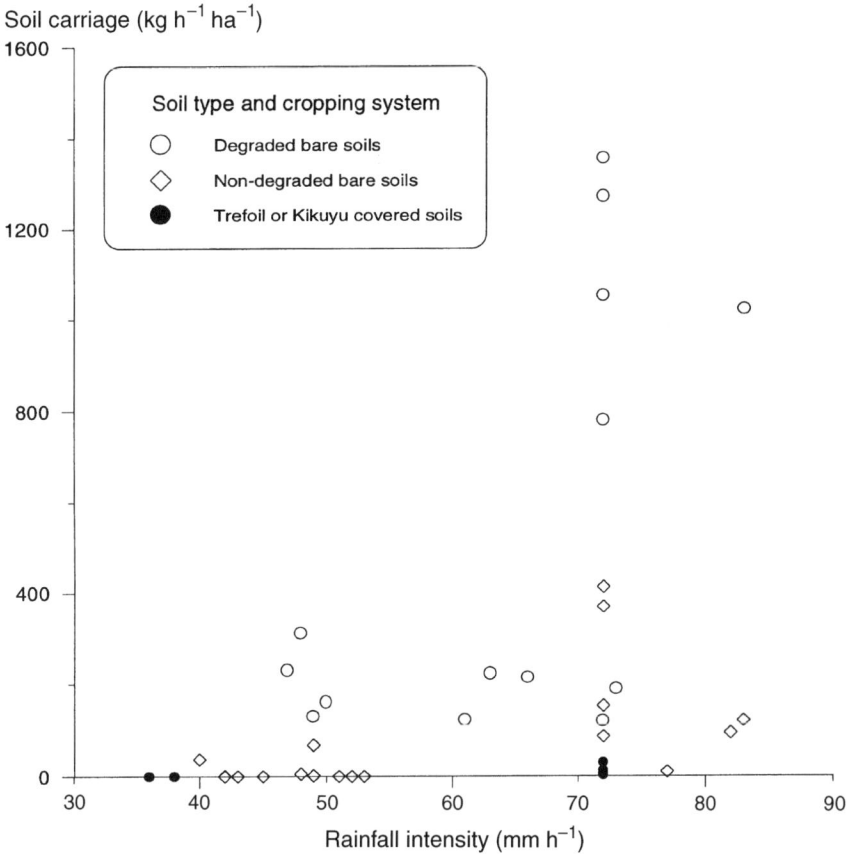

Figure 6.3   Variation in soil loss with rainfall intensity under different soils and management

Table 6.1   Rainfall simulation data for different rainfall intensities and soil managements

Soil type and management	n	Runoff (mm h^{-1})		Soil loss (kg h^{-1} ha^{-1})	
		$I = 45$	$I = 72$	$I = 45$	$I = 72$
Degraded bare soils	14	13	31	208	635
Non-degraded bare soils	19	2	22	10	201
Kikuyu/trefoil soils	12	0	15	0	7

Slope over 8–20%; soil loss expressed as oven-dried soil weight

Sediment transport measured during simulated rainfall on elementary 1 m^2 plots is very similar to that measured under natural rainfall, on 100 m^2 plots. These plots give results that are difficult to use for modelling erosion processes or to compare cropping systems and parameters. However, they do give information on sediment transport at a larger scale, including slope length effects, runoff accumulation, and long-term or successive rainfall influence.

The maximum sediment transport was approximately $3000 \, kg \, h^{-1} \, ha^{-1}$, over 3 days during a cyclonic rainfall (February 1988), on a degraded bare soil (with a 10% slope steepness), just after a harvest of potato (Bougère, 1988). That is the equivalent of soil erosion averaging 3 cm surface lowering during a single severe climatic event.

SOIL PROFILES: EARTHWORMS, STRUCTURE AND ROOT SYSTEMS

Figure 6.4 shows the changes in profiles from fallow to long-term *Pelargonium* monoculture, which leads to soil losses, degradation of structure, and loss of biology. This progressive degradation results in fertility loss, and strongly decreases yields. In addition, weed control becomes impossible to manage effectively (Michellon, 1986). Currently, most tilled soils are similar to the third profile and so there is an urgent need for new harvesting and soil husbandry systems.

New cropping systems have been developed to diversify production, to protect the soil and to restore fertility and yields (Michellon et al., 1991). Three examples of these are illustrated in Figure 6.5, which shows the recovery of profiles following rotation and cover-plant cropping systems, 3 years old, on a previously degraded bare soil.

A six-course rotation with foodcrops also supplies organic matter (manuring before planting) and protects soil structure and biology, increasing *Pelargonium* yields (Michellon, 1986). *Lotus uliginosus* and *Pennisetum clandestinum* provide living cover, which may be controlled with selective herbicides. Cover-plant roots restore structure and speed up the recolonisation by macrofauna. They do not compete with crops for water availability (Veillet, 1993). In addition, many associations increase yields and may allow low-input weed control (Burle, 1993). The other cover plants that have been tested are chiefly legumes (e.g. Kenya white-clover *Trifolium semi-pilosum*, perennial peanut *Arachis pintoï*, tick-clover *Desmodium* sp.).

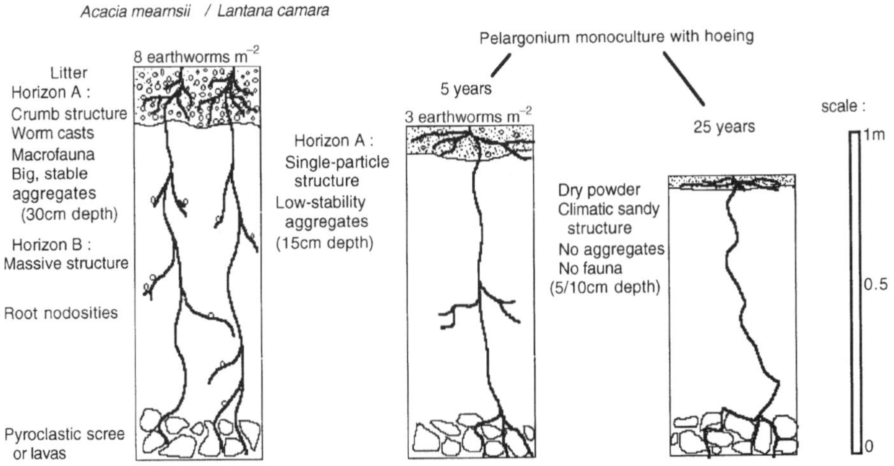

Figure 6.4   Changes of soil profile from bush-fallow caused by *Pelargonium* monoculture (plot 19)

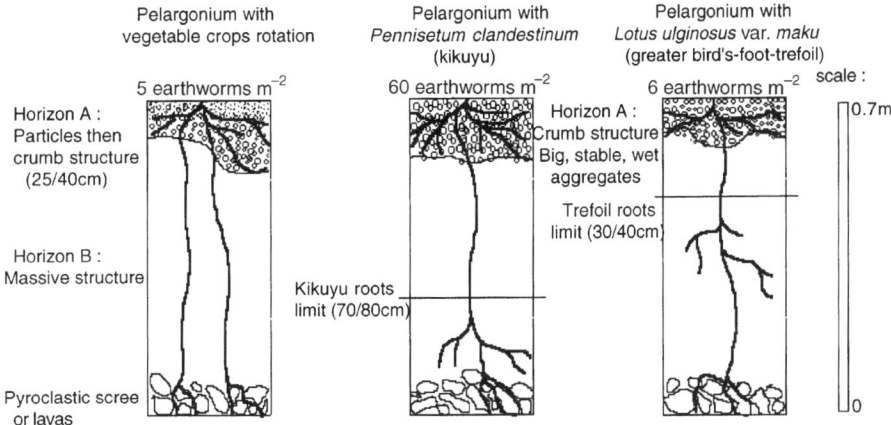

Figure 6.5   Restoration of soil structure by new cropping systems (plots 16, 2 and 1)

QUANTIFYING STRUCTURE, WATER TRANSMISSION AND ORGANIC MATTER EVOLUTION

Table 6.2 summarises the development of profiles under different cropping systems. Infiltration measured under rainfall simulations is very close to saturated conductivities measured with a disk permeameter (about $40\,\mathrm{mm\,h^{-1}}$ on degraded bare soils). These data show the increased permeability under covered conditions, and especially under the hedgerow, due to soil structural restoration. Soil fauna increases the turnover of organic matter and mineralisation beneath cover plants. Crop rotation alone cannot restore soil properties.

CALLIANDRA HEDGEROWS AS A TOOL FOR MANAGEMENT OF RUNOFF

In addition to different rotation and *Pelargonium* cultivation, the use of *Calliandra calothyrsus* hedgerows has been proposed, to provide forage during the dry season (to

Table 6.2   Restoration of soil properties under different cropping practices and surface management

Soil use and management	K_{sat} (mm h^{-1})	MWD (mm)	S	C (g 100 g^{-1})	N (g 1000 g^{-1})
Long-term fallow	250	2.50	0.92	17.2	13.5
Pelargonium monoculture on bare soil:					
– open field	40	1.10	0.52	7.1	6.2
– 1 m from the *Calliandra* hedgerow	70	1.60	0.60	8.0	7.9
– under the *Calliandra* hedgerow	225	2.36	0.84	8.4	6.7
Pelargonium/food crop rotation on bare soil	60	1.11	0.76	7.1	5.6
Pelargonium with kikuyu-grass cover	105	1.41	0.83	6.9	7.3
Pelargonium with trefoil cover	70	1.37	0.87	8.7	9.0

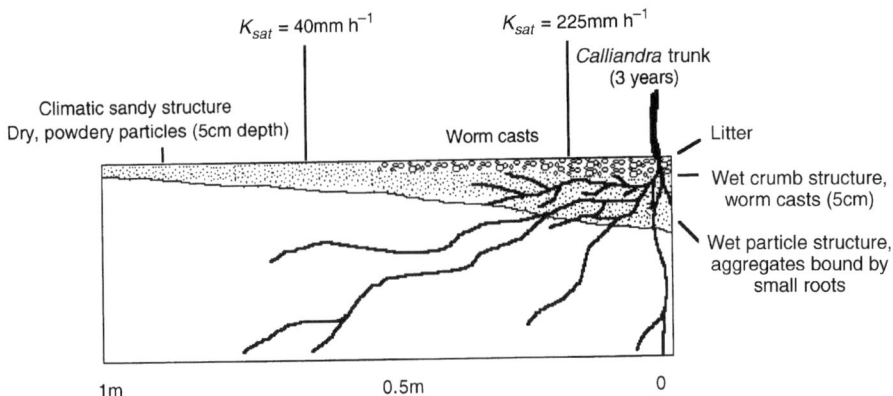

Figure 6.6   Restoration of a degraded soil profile by the protection of *Calliandra* hedgerow (plot 19)

feed animals that supply manure for vegetable crops), and to intensify soil protection and absorption of runoff. They also provide a wind protection for crops during cyclones (Maréchaux, 1993).

Figure 6.6 shows the spatial development of soil structure and fauna with distance from a hedgerow bottom. Under the hedge, a 1 m wide zone appears 2 years after plantation, and becomes significantly effective 3 years after plantation. *Calliandra* roots, the shade and microclimate under the hedge, and organic matter supplied by foliage drops, all improve earthworm numbers and activity. The earthworms then create a layer approximately 5 cm deep with their worm casts which develops into a stable organic crumb structure. Beneath this, the initially degraded structure becomes wetter and bound by fine roots. Disk permeameter measurements show an infiltration improvement. The soil below the hedgerows seems to be able to absorb runoff.

CONCLUSIONS

In Réunion Island, high-intensitiy rainfall induces important runoff in the high-altitude catchments, causing soil loss that directly damages agricultural systems and substructures. Downstream, eroded soil damages coral reef and lagoon ecosystems.

The solution to this problem lies with changes to agroecological practices. Live covering results in an immediate and important reduction of soil losses. Two years after planting, *Calliandra* hedgerows speed up soil fauna recolonisation, restore soil structure and increase local permeability. These associated and complementary cropping practices can quickly lead to complete erosion and runoff control. In addition, they can restore soil fertility of the whole field.

Because these biological management practices efficiently prevent erosion, they should compensate for the lack of any physical management techniques in Réunion's mountainous areas. Elsewhere, in similar circumstances, they could replace expensive mechanical operations, such as terracing, ridge and contour tillage. They also should improve husbandry systems to give a more productive agriculture, with low inputs and pesticide use. This agroecological approach can lead to sustainability in tropical catchments, improving or protecting all the components of the catchment: agricultural systems, soil and water resources, landscapes and ecosystems.

REFERENCES

Asseline, J. 1981. '*Construction d'un infiltromètre à aspersion*'. Note technique ORSTOM.

Bougère, J. 1988. 'Aperçu sur l'érodibilité des andosols cultivés à La Réunion'. In Les andosols de l'île de La Réunion. *CIRAD–CNRS–INRA–ORSTOM–Université*, Séminaire de Saint Denis, 24th May–1st June 1988, pp. 155–162.

Burle, D. 1993. 'Effets des couvertures végétales permanentes associées au géranium sur la fertilité des andosols de la Réunion'. *Mémoire ITA*, ENITA Bordeau/CIRAD CA.

Casenave, A. 1982. 'Le mini-simulateur de pluie: conditions d'utilisation et principes de l'interprétation des mesures'. *Cahiers ORSTOM, série Hydrologie*, **19**, 207–227.

Clothier, B.E. and White, I. 1981. 'Measurement of sorptivity and soil water diffusivity in the field'. *Soil Science Society of America Journal*, **45**, 241–245.

El-Swafy, S.A., Dangler, E.W. and Armstrong, C.L. 1982. '*Soil Erosion by Water in the Tropics*'. University of Hawaii, HITAHR, Research Extension Series 24.

Garin, P. 1987. 'Système de culture et itinéraires techniques dans les exploitations à base de géranium dans les Hauts de l'ouest de La Réunion'. *L'Agronomie Tropicale*, **42**, 289–356.

Gregorich, E.G., Kachanoski, R.G. and Voroney, R.P. 1988. 'Ultrasonic dispersion of aggregates: distribution of organic matter size fraction'. *Canadian Journal of Soil Science*, **68**, 395–403.

ICRAF, 1993. '*Integrated Management of Natural Resources on the East African and Central African Highlands*'. Consultative Workshop Report, Entebbe, Uganda, January 6–8, 1993.

Ikerd, J.E. 1993. 'The need for a system approach to sustainable agriculture'. *Agriculture Ecosystems and Environment*, **46**, 147–160.

Lal, R. 1975. '*Role of Mulching Techniques in Tropical Soil and Water Management*'. IITA Technical Bulletin, 1, International Institute for Tropical Agriculture, Ibadan, Nigeria.

Lal, R. 1989. 'Conservation tillage for sustainable agriculture: tropics versus temperate environments'. *Advances in Agronomy*, **42**, 85–197.

Lal, R., Regnier, E., Eckert, D.J., Edwards, W.M. and Hammond, R. 1991. 'Expectations of cover crops for sustainable agriculture'. In *Cover Crops for Clean Water* (ed. W.L. Hargrove), pp. 1–11. Soil and Water Conservation Society, New York.

Maréchaux, S. 1993. '*Les haies fourragères dans les Hauts de l'ouest à la Réunion: l'intégration du Calliandra calothyrsus pour une protection productive*'. Mémoire DESS, Paris XII/CIRAD Forêt.

Michellon, R. 1986. 'Stabilisation et maîtrise de systèmes de production à base de géranium'. *Rapport Annuel IRAT–CIRAD, Réunion*, 155–161.

Michellon, R. and Garin, P. 1985. 'Recherche-Système dans les Hauts de l'ouest; Synthèse des résultats obtenus: le géranium, le haricot, la pomme de terre, le tabac, le maïs, les fruitiers tempérés'. *Bilan de la Recherche-Système dans les Hauts de l'ouest. Journées du 25–27/11/1985*, CIRAD Réunion, 129–201.

Michellon, R., Perret, S. and Roederer, Y. 1991. 'Conservation et gestion des sols et des cultures'. *Rapport Annuel CIRAD, Réunion*, 81–84.

National Research Council, 1983. 'Calliandra: *A Versatile Small Tree for the Humid Tropics*'. National Academy Press, Washington, DC.

Nimlos, T.J. 1992. 'Reclamation of indurated, volcanic-ash materials in Latin America'. *Soil Restoration, Advances in Soil Science*, **17**, 153–170.

North, P.F. 1976. 'Towards an absolute measurement of soil structure stability using ultrasound'. *Journal of Soil Science*, **27**, 451–459.

North, P.F. 1979. 'Assessment of the ultrasonic method of determining soil structural stability in relation to soil management properties'. *Journal of Soil Science*, **30**, 463–472.

Perret, S. 1993. 'Propriétés physiques, hydriques et mécaniques de sols andiques de la Réunion. Facteurs d'évolution des horizons culturaux, implications agronomiques et écologiques'. Thèse de doctorat ENSAM/CIRAD SAR 01/93.

Perroux, K.M. and White, I. 1988. 'Designs for disk permeameters'. *Soil Science Society of America Journal*, **52**, 1205–1215.

Raunet, M. 1991. '*Le milieu physique et les sols de l'île de La Réunion. Conséquences pour la mise en valeur agricole*'. CIRAD-IRAT/Région Réunion.

Smith, G.D., Coughlan, K.J., Yule, D.F., Laryea, K.B., Srivastava, K.L., Thomas, N.P. and Cogle, A.L. 1992. 'Soil management options to reduce runoff and erosion on a hardsetting Alfisol in the semi-arid tropics'. *Soil and Tillage Research*, **25**, 195–215.

Stieljes, L. (ed.) 1993. '*Atlas des risques majeurs à la Réunion*'. Conseil général/Région Réunion/ Ministère de l'Environnement, BRGMR 37747.

Thomas, V.G. and Kevan, P.G. 1993. 'Basic principles of agroecology and sustainable agriculture'. *Journal of Agricultural and Environmental Ethics*, **1**, 1–19.

Thony, J.L. 1990. '*Infiltromètre à succion contrôlée sorptivimètres. Notice d'utilisation provisoire*'. Institut Mécanique Grenoble.

Van Bavel, C.H.M. 1949. 'The mean-weight diameter of soil aggregates as a statistical index of aggregation'. *Soil Science Society of America Proceedings*, **14**, 20–23.

Veillet, S. 1993. '*Etude de l'évolution de l'état hydrique d'un andosol selon différents systèmes de culture*'. Mémoire DAA, ENSAM/CIRAD.

Zebrowski, C. 1975. 'Etude d'une climatoséquence dans l'île de La Réunion'. *Les Cahiers de l'ORSTOM, Série Pédologie*, **13**, 255–278.

CHAPTER 7

The Importance of Geopedology in Sustainable Use of Tropical Catchments: Sodic Soils and Land Use Scenarios in Northern Amazonia

CARLOS SCHAEFER
Departamento de Solos, Universidade Federal de Viçosa-Minas Gerais, Brazil
JOHN DALRYMPLE
Department of Soil Science, University of Reading, UK

INTRODUCTION

Since the failure of the African groundnut scheme in the 1950s, it has become normal practice to make an appraisal of land resources wherever new land is developed or changes of land use proposed. Such general purpose land evaluations should ideally include both field and laboratory based surveys of all relevant environmental conditions such as climate (including nature and reliability of weather patterns); land surface properties (including slope percentages, directions and lengths); soils (including field mapping, descriptions and laboratory analyses); hydrology (including nature and depth of water tables); and vegetation (including contemporary land use). Such data are best presented in terms of their three-dimensional spatial relationships, catchment-based and illustrated in the form of catenas and block diagrams.

Such land evaluation surveys, which have been assisted by aerial photography and more recently by satellite remote sensing techniques, have regrettably rarely achieved the integration upon which they depend for success. An essential part of this integration is a full consideration of the socio-economic as well as the physical environmental constraints, but only recently have the necessary socio-economics been considered. One example of general purpose land evaluation which has been successfully implemented is the Canada Land Inventory. This comprehensive survey of land capability, which has been designed and used for agriculture, forestry,

The Sustainable Management of Tropical Catchments. Edited by David Harper and Tony Brown.
© 1998 John Wiley & Sons Ltd.

recreation and wildlife, was approved in 1963 and completed by the early 1970s (Canada Land Inventory, 1970; Jackson and Marwell, 1971; Rees, 1977). This system of land evaluation was based on the land capability classification system originally formulated in the USA in the late 1930s and modified in the 1960s (Klingbiel and Montgomery, 1961). The scheme is based on permanent limitations for agricultural use, of which there are eight land classes. Within each class the limitations might be any one or more of the following: erosion hazard, water excess, stoniness, salinity, shallow profile, or climatic. In essence, the system relies on previously mapped soil boundaries, since the land capability boundaries are identified by clustering the soil mapping units (usually soil series or associations).

In the late 1970s and 1980s land capability became replaced by land suitability. Specific guidelines have been published for rain-fed crops (FAO, 1983); for forestry (FAO, 1984); for irrigated agriculture (FAO, 1985); and for extensive grazing (FAO, 1991). The essential difference between capability and suitability is that in the former case each class covers a broad range of crops in moderately high levels of management, whilst suitability specifically relates to each individual crop, use or management practice. At all scales from an individual field to farm, catenary sequence catchment and region, four levels of suitability ("non-suitable" and three levels of "suitable") may be established and mapped. At each scale, suitability may vary with the given, or projected management practice.

At the global scale, use has been made of this system by FAO to establish its Agroecological Zones Programme. Using existing information and, wherever possible, GIS as a database for storage, retrieval and presentation, this project has constructed in map form a first approximation of the present and potential use of the world's land resources. The eventual aim is to build up individual country resource databases (Kassam et al., 1991; FAO, 1993).

An alternative approach developed during the 1980s has been to concentrate on the soil pedon and its properties, rather than on soils as a component of land. This is the concept of "agrotechnology" transfer initiated by USAID (Soil Management Support Services) and developed in the Benchmark Soil Project (Silva, 1985). With suitable modifications this method also forms the basis of the Network System of the International Board for Soil Research and Management (IBSRAM, 1985). The principle here is that, if a soil pedon in one part of the world can be classified (at family level in Soil Taxonomy (Soil Survey Staff, 1994) or at an equivalent level in the FAO/UNESCO system (FAO, 1988), and its management and crop yield is known, then under a similar management regime in, for example, another continent similar yields can be expected.

In all cases however, "ground truth" is essential. This is as true of the "top-down" approach such as the FAO framework (Smyth and Dumanski, 1993) as it is of the "bottom-up" approach such as the European Union (EU) individual holistic case studies (Cheverry and Stoops, 1992). Thus before mapping new land or evaluating an alternative land use strategy for land already in some form of use, it is necessary to clearly establish the future use objectives and to identify and map only those properties of soil and land that are necessary to establish the suitability for that objective. Tables of class groups of soil and land properties (both for specific crops and for other land uses) are now readily available (Landon, 1991), but obtaining "ground truth" information, especially in the tropics, is time-consuming and thus expensive. Not surprisingly, most of the successful soil/land evaluation studies have been at detailed scale, and where there is both local/indigenous community support and responsibility for implementation combined with the survey. The land allocation study of the direct action programme for the Kabupaten Sangau SSDP sites, West Kalimantan, Indonesia (Tricon Jaya PT and Dale International Ltd, 1990) is a good example. Inherent in all studies of suitability must be the ability to recognise the negative as well as the positive outcome, and not hide negatives in uncertainty. It is in this context that a case study of sustainability in a catchment in northern Amazonia is presented.

An immediate obstacle to the promotion and management of sustainable agriculture in Amazonia is the general lack of knowledge of the soils in those areas where indigenous as well as introduced colonisation of the land has taken place. As a consequence, policy-makers are often tempted to establish development programmes for homogeneous settled environments that are based on generalisations from the few existing surveys available, whereas, in fact, a complex pattern of variability of both soils and land uses exists.

Whilst some scientists, including Sanchez et al. (1982), consider at least 45% of the soils in Amazonia to be potentially suitable for grazing and agriculture, others, including the authors, hold serious doubts as to the feasibility of overcoming the severe constraints of the natural, and especially the geopedological, environment. However, the pressure on the land is increasing at an alarming rate and, although increased information about short-term impacts on forest areas is now becoming available, savanna areas and their soils are being neglected, and are regarded as marginal or even unimportant, in the Amazonian scenario.

The state of Roraima, with an area over 230,000 km^2, a little smaller than Great Britain, is the most environmentally complex and diverse in the Brazilian Amazonia. In Roraima, the area comprising the Surumu/Parimé catchment, covering approximately 255,000 ha, has been recommended by two different surveys as the most suitable for the development of both intensive agriculture and improved pasture in Roraima. These technical surveys were based mainly on chemical data, which showed apparent high fertility (Brasil, 1975, 1980; Figure 7.1). Due to the relative flatness of the local relief, specific emphasis was placed on intensive, irrigated rice production with differing levels of management. This development programme was proposed prior to, and regardless of, more detailed studies in this area, and the known domination of sodic soils (planosols, solodic planosols and solodised solonetz; Brasil, 1975).

In the human context, this catchment forms part of a larger, non-demarcated indigenous reserve (Raposa–Serra do Sol), where sparsely settled *colonos* (settlers) live alongside Macuxis and Uapixana people. Roraima currently has one of the fastest-growing economies in Amazonia, and so domestic public concern is raised with reference to all planning issues and proposed development programmes. As a consequence, Brazilian researchers are urged to identify alternative land uses and devise policies and research programmes that require a multidisciplinary approach to provide a comprehensive view. In addition, the government seeks precise and detailed information concerning the region's natural resources (MacMillan and Furley, 1994).

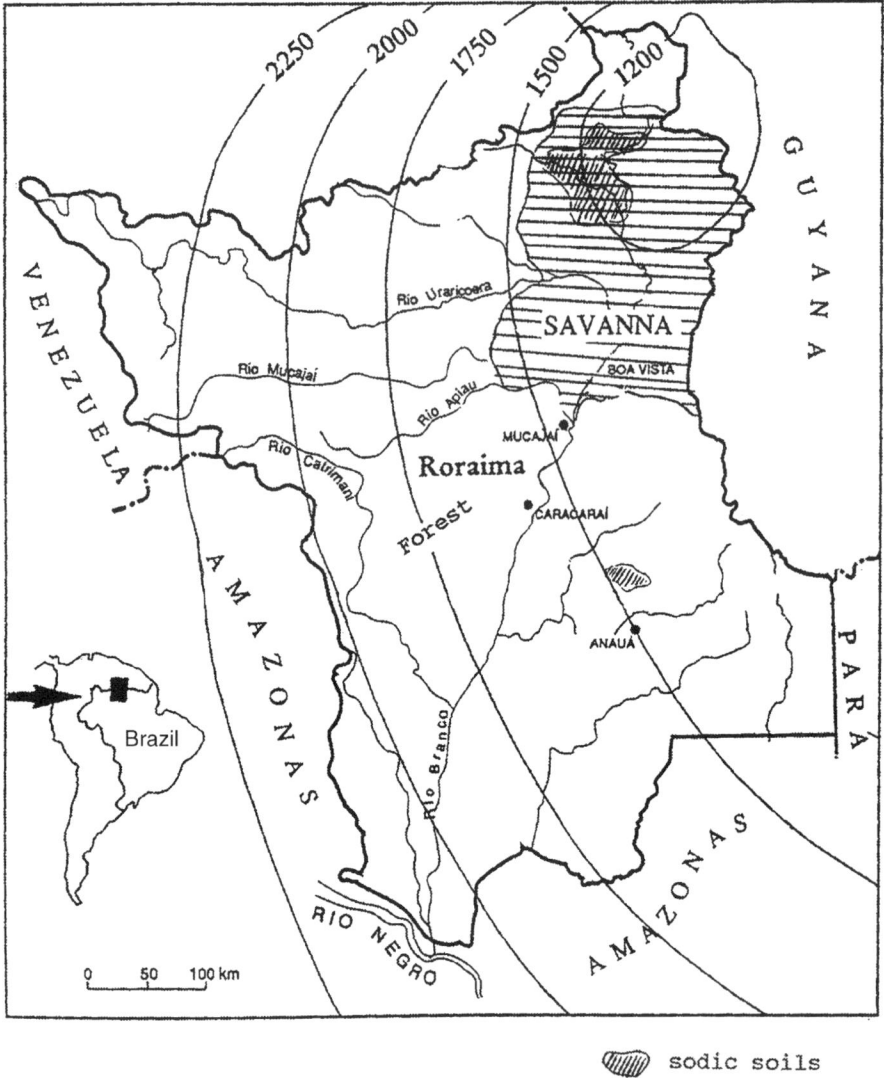

sodic soils

Figure 7.1 Location of Roraima State in Amazonia, with isohyets (mm yr^{-1}), vegetation and occurrence of sodium-affected soils (after Schaefer and Dalrymple, 1996)

As very little information is at present available with reference to sodic soils in Roraima, the purpose of the present work is twofold: (a) a geopedological and ecological study of a selected toposequence across a savanna–forest boundary, and, against this background (b) to discuss the implications to the land use and regional policies, based upon the assessment of different land use scenarios.

The presence of semi-arid-like soils, normally sodic, in north Amazonia was first noted in Brasil (1975). It describes their occurrence under savanna and, rarely, under tropical forest, but it provided no detailed characterisation of their properties nor discussion as to the likely environmental and ecological processes involved in their genesis. More recent studies have described such soils in greater detail (Schaefer et al., 1993; Schaefer and Dalrymple, 1996) and illustrated their widespread, spatial importance in the north Amazonia landscape (Schaefer and Dalrymple, 1995). It is in north-eastern Roraima that the most extensive areas of sodium-affected soils occur within the Amazon basin. The apparent "unfitness" of the geomorphology, vegetation and the present climatic regime with these soils is such as to suggest the presence of these sodium-affected soils to be the result of climatic and environmental changes in the late Quaternary (Figure 7.2; Schaefer, 1991; Schaefer and Dalrymple, 1995). Evidence for Quaternary climatic changes in north Amazonia have been recognised and studied by many workers from geological, geomorphological, climatological, ecological and soil viewpoints (Wijmstra and van der Hammen, 1966; Eden, 1974; Brasil, 1975; Ab'Saber, 1982; Tricard, 1985; Schubert, 1988; Schaefer and Dalrymple, 1995).

FIELD MATERIALS AND METHODS

In the Surumu/Parimé basin (Figure 7.2), the Paricarana catchment is both the central and the largest area of sodic soils in north-eastern Roraima State, occurring under tropical conditions with a rainfall higher than $1,200\,\mathrm{mm\,yr^{-1}}$ and strong

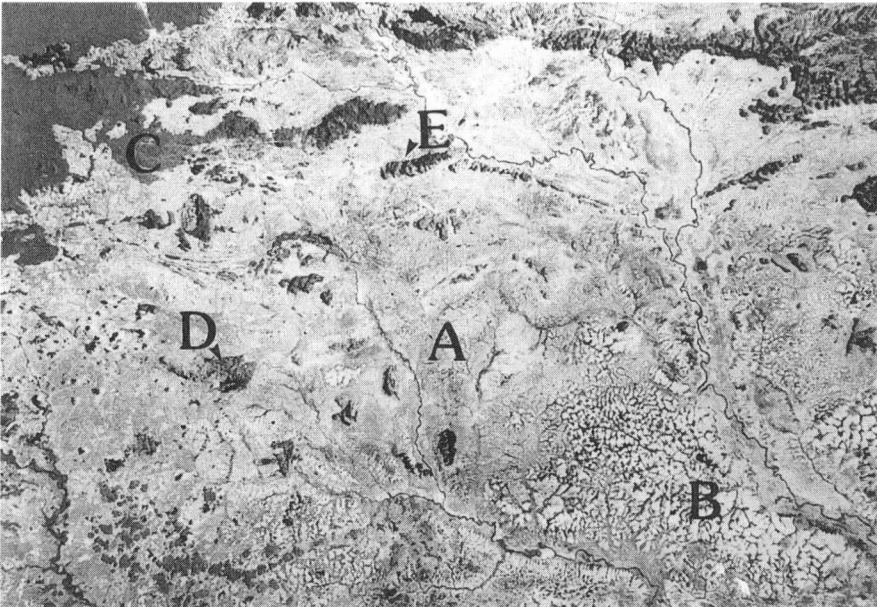

Figure 7.2   Landsat image (March 1991, end of dry season) of the Surumu/Parimé catchment, showing the central area of sodic soils (A), bordered by the dissected white-sand plateau (B) and the forested slopes of the Pacaraima mountains to the left. Extensive indigenous burnings (D) and the inselbergs associated with deciduous forest (E) can be seen (Schaefer and Dalrymple, 1995)

seasonality. The associated vegetation is "Xerophytic Savanna" and Savanna Woodland, dominated by Lixeira (*Curatella americana*) and Muricis (*Byrsonima* spp.) trees with a grass cover of *Trachopygon* and *Andropogon* spp. Many inselbergs occur in association with rock outcrops and boulders (Figure 7.3), and they are covered by a dry forest vegetation with species adapted to withstand seasonal drought, like xeromorphic orchids and Cactaceas (Schaefer, 1991; Schaefer and Dalrymple, 1995). The change between the two vegetation types is usually abrupt.

The landscape is dominated by an extensive flattened surface – the Rio Branco pediplain. It is typified by abundant forested inselbergs with boulders, whose genesis is possibly related to long dry phases in the late Quaternary (Barbosa and Ramos, 1959; Schaefer and Dalrymple, 1995). Geologically, this landscape consists predominantly of acid and intermediary volcanic rocks, including rhyolites, rhyo-dacites and andesites – the Surumu Formation (Brasil, 1975) – and granitic rocks, forming inselbergs. Although these rocks are rich in quartz, Na-plagioclase and K-feldspar, large tracts of the weathered bedrock are often overlain by Quaternary sediments that include colluvial and alluvial deposits and, frequently, buried weathered profiles and paleosols. The area is currently used for extensive cattle grazing both by indians and *colonos*.

The fieldwork was undertaken during the dry season, 1991 (Schaefer and Dalrymple, 1996). Three soils, comprising a toposequence and ranging from 125–118–105 m in altitude, were identified, described and sampled in Paricarana catchment (part of the Surumu/Parimé basin; Figure 7.4). With respect to soil structure, columnar peds

Figure 7.3   Aerial view of the sodium-affected soil area, near Rio Paricarana. Inselbergs of granitic/granodioritic rocks are prominent on the rhyolitic/dacitic floor below. Inselbergs are covered in deciduous forest that contains straggling low bushes, cacti and xeromorphic orchids. Sharp boundaries occur between the different rocks and the deciduous forest/xerophytic savanna. Indigenous burning can be seen in the background. (Photo by Carlos Schaefer, 1991)

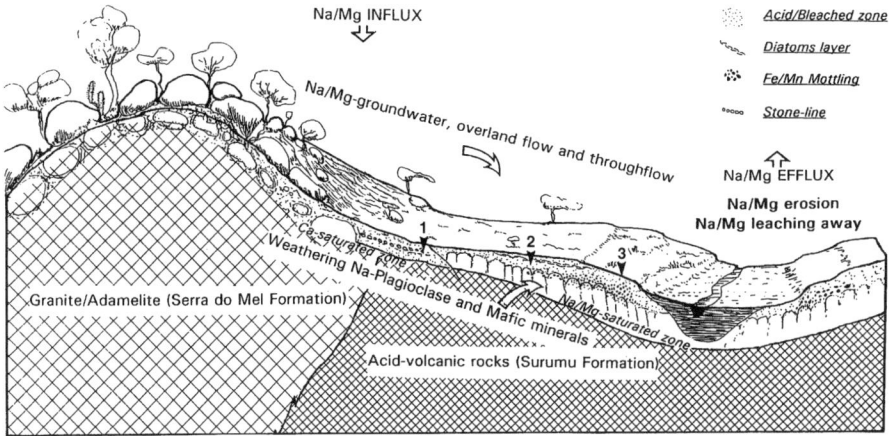

Figure 7.4 Catenary sequence of Na/Mg-affected soils, with columnar structure, in the Paricarana catchment. (1) shows a semi-arid soil, dominated by smectite and saturated with calcium in the exchange complex. The lower position (2) presents columnar structure, associated with high levels of magnesium and sodium, sandy and silty pre-weathered sediments, dominated by kaolinite and illite. Sedimentary layering and mottling is concentrated in the lowest part (3). The main source of Na and Mg is the relative richness of Na-plagioclase and Mg-bearing minerals, associated with the extensive flattened surface and strong seasonality (Schaefer and Dalrymple, 1996)

are only present at middle and lower slope positions (positions 2 and 3), where the soils and their associated columnar peds have developed under savanna from rhyodacite volcanic rocks. The parent material at the upper slope position (position 1), under dry forest, is a granite/adamelite. Here, the pattern of weathering is different and prismatic and blocky peds characterise this soil.

Sampling for elemental chemical, physical and mineralogical analyses were made at 20 cm intervals to 1.20 m at each site. Full details of the analytical techniques used are given in Schaefer and Dalrymple (1996). Thin sections of the soils were also made and described micromorphologically, whilst additional quantitative chemical data were obtained using Scanning Electron Microscopy (SEM) and Energy Dispersive X-ray (EDX).

Whenever possible, the land use scenarios approach followed the recommendations of Davidson (1992) and Dalal-Clayton and Dent (1993). The integrated assessment of financial input and social relevance were based mainly on information abstracted from Roraima's censuses (IBGE, 1985), whilst the grades of social relevance were partially based on the work of Fearnside (1983) on the ecological evaluation of development alternatives in Amazonia.

RESULTS

The soil analyses are shown in Tables 7.1–7.3. The results presented and discussed are considered in more detail in a study of soils in northern Amazonia presented by Shaefer and Dalrymple (1996).

Table 7.1 Exchangeable cations, available phosphorus, pH, organic carbon and micronutrients for different soil profile depths, at slope positions 1, 2 and 3 on the toposequence. Average of three replicates, except organic carbon. (Schaefer and Dalrymple, 1996)

Slope position	Soil profile	Exchangeable cations (cmol$_c$ kg^{-1})							Micronutrients (ppm)			%		pH	% Org. C
		Ca	Mg	K	Na	S	Al	P	Fe	Cu	Zn	Na/S+Al	Mg/S+Al		
1	A$_1$	1.2	0.1	0.26	0.15	1.71	0.0	0.45	32	0.4	2.1	–	–	5.5	1.3
	1Bt	4.9	1.7	0.15	0.13	6.88	0.05	0.10	38	0.1	1.1	–	–	5.0	0.2
	2Ab	4.0	2.1	0.20	0.21	6.51	0.05	0.10	88	0.4	0.9	–	–	4.5	0.4
	2Bt$_{n21}$	4.5	0.1	0.13	0.26	4.99	0.0	0.00	90	0.3	2.4	5.2	2.0	4.8	0.1
	2Bt$_{n22}$	4.6	0.3	0.17	0.54	5.61	0.0	0.00	3	0.4	0.9	9.6	5.3	6.0	0.1
2	A$_1$	0.4	0.2	0.18	0.12	0.90	0.25	0.25	79	0.1	2.8	–	–	4.5	0.9
	E	0.4	0.2	0.11	0.20	0.91	0.18	0.10	72	0.2	1.4	–	–	4.5	0.1
	E/B	2.0	3.8	0.15	0.38	6.33	0.2	0.00	12	0.1	0.5	6.0	60.0	4.8	0.03
	Bt$_{n21}$	2.9	6.5	0.12	1.11	10.63	0.0	0.00	15	0.2	0.9	10.4	61.1	6.0	0.03
3	A$_1$	0.6	0.3	0.10	0.14	1.14	0.0	0.35	89	0.4	0.8	–	–	5.4	0.7
	E	0.5	0.3	0.08	0.14	1.02	0.0	0.00	87	0.2	1.5	–	–	4.9	0.15
	2Ab	0.4	0.2	0.06	0.12	0.78	0.18	0.15	82	0.2	0.7	–	–	4.6	0.08
	2C	0.4	0.1	0.08	0.16	0.74	0.38	0.00	190	0.3	1.1	21.6	13.5	3.7	0.12
	3Bt$_{n21}$	0.5	0.3	0.10	0.12	1.02	0.24	0.00	185	0.8	0.7	11.8	29.4	3.7	0.10
	3Bt$_{n22}$	4.2	7.6	0.28	0.36	12.44	0.0	0.00	23	0.2	1.1	2.9	61.0	5.5	0.03

Table 7.2 Physical properties and chemical composition of the clay fraction for different soil profile depths, at slope positions 1, 2 and 3 on the toposequence. (Schaefer and Dalrymple, 1996)

Slope position	Soil profile	Colour (Munsell)	Depth (cm)	Structure	Bulk density (g cm^{-3})	Clay composition (%)					Texture (%)			
						SiO$_1$	Al$_2$O$_3$	Fe$_2$O$_3$	TiO$_1$	MnO	CS	FS	Silt	Clay
1	A$_1$	10YR 6/2	0–15	–	1.15	47.6	26.6	18.4	7.0	0.4	3	17	74	6
	Bt$_1$	10YR 6/1	15–40	WM SB	–	47.6	25.8	15.6	10.4	0.6	4	20	66	10
	2AB	10YR 6/3	40–60	–	1.08*	39.8	31.1	21.6	6.9	0.6	14	15	63	8
	2Bt$_{21}$	10YR 6/2	60–100	WM SB	–	49.9	29.4	14.5	5.9	0.3	1	14	74	11
	2Bt$_{22}$	10YR 7/2	100–150	WM SB	1.46	47.0	31.9	16.4	4.6	0.1	2	8	59	31
2	A$_1$	10YR 7/2	0–10	–	1.38	66.1	23.6	7.1	2.6	0.6	2	44	51	3
	E	10YR 8/2	10–20	–	–	63.5	23.5	12.4	0.0	0.6	1	35	62	4
	E/B	10YR 8/2	20–65	VW CO	–	59.3	31.1	7.5	1.8	0.3	2	38	44	16
	Bt$_{n21}$	10YR 8/1	65–150	WM CO	1.55	56.2	31.2	10.0	2.4	0.2	3	27	46	24
3	A$_1$	10YR 6/3	0–10	–	1.40	47.8	31.5	11.3	7.6	1.8	51	33	13	3
	E	10YR 7/3	10–30	–	–	67.6	21.9	8.0	1.8	0.7	41	40	13	6
	2Ab	10YR 6/2	30–50	–	142*	44.4	23.1	28.9	2.7	0.9	65	27	6	2
	2C	10YR 8/2	50–100	VW CO	–	47.4	39.0	8.4	4.5	0.7	25	31	39	5
	3Bt$_{n21}$	10YR 8/1	100–150	VW CO	1.52	51.1	36.3	9.9	2.3	0.4	1	5	85	9
	3Bt$_{nm22}$	10YR 8/1	150–180	WM CO	1.65**	50.0	40.5	8.7	0.5	0.3	41	20	16	23

WM SB = weak/moderate sub-angular blocky
VW CO = very weak columnar
WM CO = weak/moderate columnar
CS = coarse sand; FS = fine sand
*=taken from the mid-horizon
**=taken from the inner column

Table 7.3   Clay mineralogy (glass-mounted, oriented, DCB treated, and heated at 350 and 500°C) and mineralogy of the whole soil (fine earth, cavity mount) for different soil profiles, at slope positions 1, 2 and 3 on the toposequence. (Schaefer and Dalrymple, 1996)

Slope position	Soil profile	Clay mineralogy	Mineralogy whole soil (< 2 mm, powder)
1	A_1	kaolinite, illite, smectite	quartz, K-feldspar, Na-plagioclase, kaolinite, mica, ilmenite (tr.), anatase (tr.), *hematite(tr.)*
	Bt_1	smectite, illite, kaolinite	quartz, K-feldspar, Na-plagioclase, mica, kaolinite, ilmenite (tr.), smectite (tr.)
	2Ab	kaolinite, illite	quartz, K-feldspar, Na-plagioclase, mica, kaolinite, ilmenite (tr.)
	$2Bt_{21}$	smectite, illite, kaolinite	quartz, K-feldspar, Na-plagioclase, mica, smectite, kaolinite, ilmenite (tr.), anatase (tr.)
	$2Bt_{22}$	smectite, illite, kaolinite	quartz, K-feldspar, Na-plagioclase, mica, smectite, kaolinite, ilmenite (tr.), anatase (tr.)
2	A_1	kaolinite, illite, smectite (tr.)	quartz, K-feldspar, Na-plagioclase, traces of mica, kaolinite, ilmenite (tr.)
	E	kaolinite, illite, smectite (tr.)	quartz, K-feldspar, Na-plagioclase, traces of mica, kaolinite, ilmenite (tr.)
	E/B	illite, smectite, kaolinite	quartz, K-feldspar, Na-plagioclase, mica, kaolinite, interstrat. smectite/mica (tr.)
	Bt_{n21}	illite, smectite, kaolinite	quartz, K-feldspar, Na-plagioclase, mica, kaolinite, smectite (tr.)
3	A_1	kaolinite, illite, interstrat. smectite/illite	quartz, K-feldspar, Na-plagioclase, mica (tr.), kaolinite, ilmenite, anatase
	E	kaolinite, illite, goethite, interstrat. smectite/illite	quartz, K-feldspar, Na-plagioclase, mica, kaolinite, interstrat. smectite/mica (tr.)
	2Ab	illite, interstrat., smectite/illite, kaolinite	quartz, K-feldspar, Na-plagioclase, mica, kaolinite, interstrat. (tr.), ilmenite
	2C	illite, interstrat., smectite/illite, kaolinite	quartz, K-feldspar, Na-plagioclase, mica, kaolinite, smectite (tr.), interstrat. (tr.), ilmenite
	$2Bt_{n21}$	smectite, illite, kaolinite	quartz, K-feldspar, Na-plagioclase, mica, kaolinite, smectite (tr.), ilmenite
	$2Bt_{n22}$	smectite, illite, kaolinite	quartz, K-feldspar, Na-plagioclase, mica, kaolinite, smectite (tr.), ilmenite

interstrat. = interstratified 2:1 clays, with broad peak areas; tr. = trace

Perhaps the most significant findings are the lower sodium saturation values of the horizons identified in the field as Bt_{n21}, $3Bt_{n21}$ and $3Bt_{nm22}$, which are not as high as their well-developed columnar structural peds would suggest. At the upper slope position 1 on the toposequence, the exchange complex is dominated by calcium throughout the soil profile (Table 7.1 and Figure 7.5). The sodium saturation of the

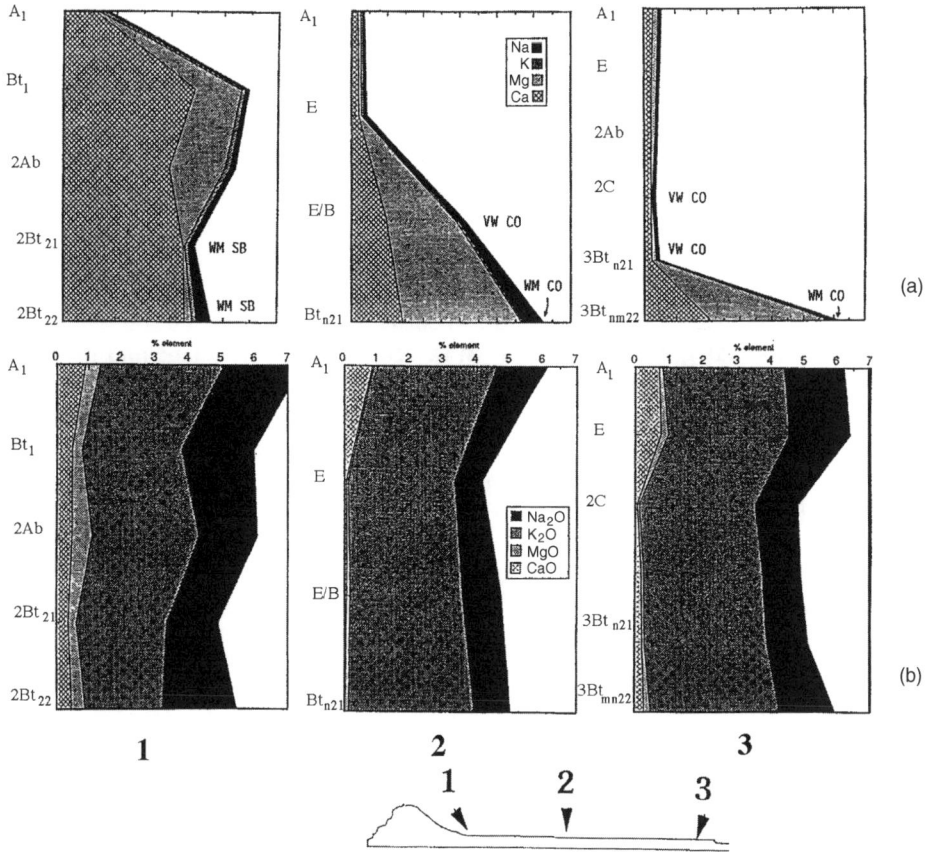

Figure 7.5   Exchangeable Ca, Mg, K and Na (a); distribution of total Na_2O, K_2O, MgO and CaO (b); both for different soil depths at slope positions 1, 2, 3 as shown on the toposequence. Abbreviations for structure are as used on Table 7.2

diagnostic horizons ranges from 5.2% to 9.6% of the total CEC. Even using the sum of exchangeable Na plus Mg, the values of percentage saturation are still lower than 15% (the limit of classification as a solonetz). Thus this soil is solodic rather than sodic. This is reflected in the field by the presence of an argillic Bt horizon and not the natric Btn horizon and well-developed columnar peds that characterise the soils in the middle and lower slope positions 2 and 3. The eutrophic character of the soil pedon at slope position 1 and the absence of any clear E (albic) horizon are further evidence that this higher-lying soil under a deciduous forest has a different genesis from the other two soils. It is considered that this is more related to its position in the landscape than to its specific and contrasting parent material, since even the combined availability of Na and Mg have not resulted in the formation of columnar peds.

A complication to interpretation of the vertical distribution of the values of the chemical elements in the soil at position 1 is the presence of a paleosol. This buried 2Ab horizon (0.4% organic carbon) has higher values of exchangeable Mg and K, as compared with its overlying and underlying horizons. This confirms that this soil has been covered and buried by younger colluvial material of different composition and accounts for the observed chemical variations within the total profile. The pattern of distribution of total Na_2O, K_2O, CaO and MgO down the profile, however, suggests a richness in Na-bearing minerals (Figure 7.5). The potassium content is high (ranging from 3.7% in the A1 to 2.3% in the $2Bt_{21}$ horizon) but, in comparison with these total values, the amounts of exchangeable potassium are low, ranging from 0.26 to 0.13 meq $100\,g^{-1}$ respectively). This soil, therefore, has a large potassium and sodium reserve which stems from its richness in K-feldspar, illite/mica, and Na-plagioclase (XRD; Table 7.3). Thus, on mineral weathering of this soil, there is the potential for the release of large amounts of sodium into the soil solution. However, despite the limited degree of weathering in this soil, its good drainage and its upper slope position in the landscape have resulted in considerable amounts of Na leaching downslope with its consequent accumulation in the two lower slope positions by throughflow. In contrast, most of the total Ca and Mg occurs in exchangeable form and thus is not so easily leached or removed under the prevailing pedogeomorphological conditions. This is despite the "Typic Tropustic" soil moisture regime with its strong seasonality and high rainfall (van Wanbeke, 1981). A further severe constraint to any agricultural use of this soil in upper slope position 1 is the presence of stone fragments and boulders, both on the soil surface and within the A and B horizons.

In contrast, there are marked differences in the soil at position 2 in the toposequence. Here, the exchangeable Na only comprises 10.4% of the total CEC. The exchange complex of the diagnostic Bt_{n21} horizon is dominated by Mg^{2+}, whilst the sum of Mg^{2+} plus Na^+ reaches 66% in the E/B and 71.5% in the Bt_{n21} horizon. Thus, on the basis of Na^+, the soil at position 2 can be classified as solodic, or natric (if Na^+ plus Mg^{2+} values are used). Again this soil contains K-feldspar and Na-plagioclase, together with smectite, kaolinite and illite (Table 7.3). In the lower position 3 in the toposequence, although the main source of sodium derives from the parent material, which is Na-plagioclase rich, the relatively high concentrations of Mg and Na in this soil are unlikely to result just from the greater intensity of leaching (low pH values) and poor drainage. Colluvial additions from upslope by throughflow are thought to be significant and this is also reflected in the high subsurface values for the sum of the bases. The presence of ilmenite and anatase is thought to reflect that down valley, alluvial additions of pre-weathered sediments have occurred (Table 7.2) and this is confirmed by the presence of a buried paleosol 2Ab horizon.

Interestingly, with respect to the soils in slope positions 2 and 3, other previously described sodic soils from north Amazonia (Brasil, 1975; EMBRAPA, 1982, 1984) also contain high Mg as well as Na values. Whether there is a direct connection between these soils and past climatic changes, as originally suggested by Schaefer (1991), is still unclear, but there is convincing evidence for drier paleoclimates having occurred in this particular area of Roraima during the Quaternary (Schaefer and Dalrymple, 1995).

Soil pH and bases' sum behaviour are closely related. There is a progressive trend in all three soils from the top to the bottom of the profiles, of reducing acidity, accompanied by a marked increase in the concentrations of Ca^{2+} plus Mg^{2+} in the exchange complex. However, there is evidence of solodisation having occurred. This is deduced from the low pH values in the E/B horizon (position 2) and the 2C and 3Btn21 horizons (positions 3) and, more particularly, the presence of columnar peds in the diagnostic subsurface horizons in both these slope positions on the toposequence. This suggests that these columnar, structural peds are relict features. Furthermore, since they remain even after vertical and lateral leaching has taken place, such structural, three-dimensional patterns of columnar peds can be maintained despite considerable changes in the composition of the soil (Figure 7.6).

With respect to the availability of micronutrients, only Fe is within the recommended range (Sanchez, 1976; CFSEMG, 1989). The availability of Zn, and especially Cu, is very low but this is to be expected given the absence of primary minerals containing these microelements. Also, the sandy/silty texture and low

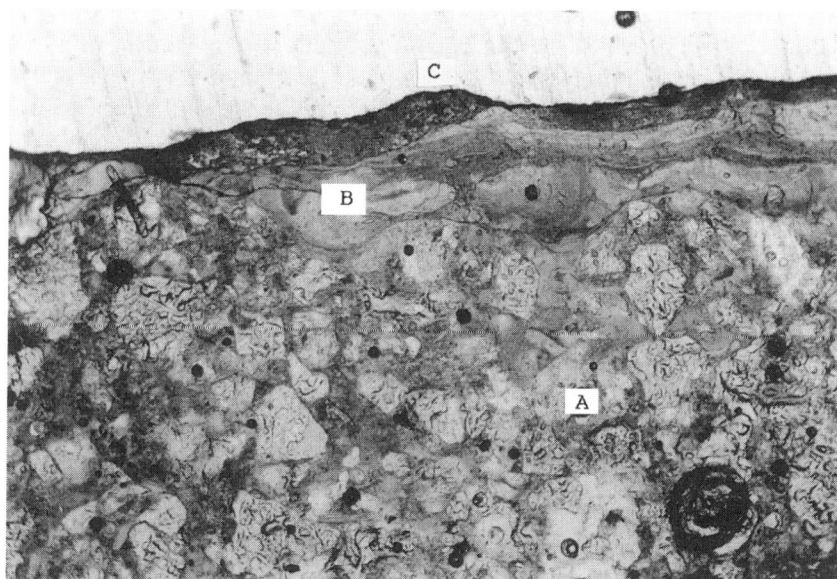

	A	B	C
SiO_2	51.84	47.38	18.65
Al_2O_3	14.88	28.25	7.53
FeO	0.04	6.12	3.35
MnO	0.10	0.28	4.10
K_2O	14.29	0.13	0.19
Na_2O	0.24	0.31	0.16

1mm

Figure 7.6   Microphotograph of the indurated top of a columnar ped in the $3Bt_{nm22}$ at slope position 3. The Energy Dispersive X-ray data is shown below, as % weight. (A) = feldspathic groundmass, with high silica and potassium content, due to the dominance of quartz and K-feldspar. (B) = intermediate crusting layer, formed by indurated clay, mainly smectite, with higher iron content and low manganese. (C) = external silcrete coating, manganese-rich, which forms a seal enclosing the column. The relatively low values of SiO_2 and Al_2O_3 indicate the presence of organic matter, and suggest that most of the silica is amorphous

exchange capacity of the clay fraction of the A horizons in these soils results in low pH values despite low levels of exchangeable Al. However, their B horizons do have higher pH values resulting from the presence of 2:1 minerals and, in the case of lower slope positions 2 and 3, exchange complexes dominated by Mg and Na. Thus, although eutrophic, the imbalance in Ca:Mg ratios and the nature and spatial pattern of the columnar peds are such as to counter any beneficial effects. Again, these soils occurring in the lower slope positions of the toposequence are also unsuitable for most agricultural purposes.

CHEMICAL AND ECOLOGICAL CONSEQUENCES

The presence of Mg/Na-affected soils in the lower positions coincide with and mark the abrupt ecologic boundary. The Ca-saturated zone with well-aggregated soils is covered by dry forest, whilst the Mg/Na-affected soils with columnar structure are associated with xerophytic savanna. The whole soil toposequence is currently becoming progressively leached and acidified, from upper slope position 1 to a maximum at slope position 3, at the bottom of the toposequence. In spite of the high amounts of Na_2O in all three soils on the toposequence, the exchange level of Na is concentrated in the lower slope positions, and, particularly, in their subsurface Btn and Btnm horizons, thus creating sharp and abrupt changes between horizons with respect both to their levels of Na- or Na + Mg-saturation and the domination of 2:1 clay minerals. It is thought that the resulting structural patterns of columnar peds reflect the behaviour of Mg plus Na more than the presence of Na itself. There is unpublished evidence from an experimental pedological study to support this (Schaefer, 1994). Given the lack of variability in the total amounts of chemical elements and high variability in their exchange levels, it would seem that differential redistribution of soluble products of weathering has occurred by throughflow and overland-flow and that they are responsible for the relative Na accumulation in the lower part of the landscape in this catchment. Groundwater fluctuations are thought to be relatively insignificant. The maintenance of Na levels as well as Ca, Mg and K in such climatic conditions can be explained by the Tropustic nature of the climatic conditions (high seasonality, with up to 5 months of evapotranspiration excess) and the presence of a broadly distributed flat surface – the Rio Branco Pediplain. Thus the present climatic regime ought not to favour the presence of sodium-affected soils.

LAND USE SCENARIOS AND IMPLICATIONS

In Table 7.4, the various options for potential land use for the sodic soils in Roraima are illustrated. It is considered that, contrary to the recommendations in previous soil and land evaluation surveys (Brasil, 1975, 1980), suitable alternatives for intensive land use are very limited. Severe constraints for most land use scenarios exist and, *inter alia*, indicate the extreme unsuitability of these soils for agricultural purposes.

The worst possible scenario is irrigated rice, because it requires a very high financial input in association with very high to severe soil constraints. As a result of controlled

Table 7.4   Assessment of pedo-ecological and economic factors under different land use scenarios for the sodic soils in Roraima

Land use scenarios	Financial input	Constraints					Social relevance	Ecological impacts
		Exchangeable bases	Micronutrients	Structural (physical)	Use of fire	Erosion hazard		
Indians' extensive cattle grazing	negligible	moderate	high	severe	frequent	very high	desirable	long-term high
Improved pastures	moderate	high	very high	very severe	rare	high	intermediate	moderate
Rice (irrigated farming)	very high	very high	very high	very severe	none	moderate	unknown	very high
Annual crops	high	very high	very high	very severe	rare	very high	intermediate	very high
Perennial (citrus/ cashew)	moderate/high	moderate	moderate	moderate	none	moderate	unknown	moderate

flooding between April and September, the stability of aggregates would greatly decrease. This would favour further physical and chemical degradation of the soil structure and changing morphological patterns in the columnar peds in association with the promotion of higher sodium and magnesium accumulations with higher levels of groundwater.

In respect to other land use scenarios, the soils constraints are generally very high. The main exception is perennial citrus and cashew, which are well adapted to the particular set of physical and chemical features which characterise these soils, particularly for cashew which is a native fruit of this part of Amazonia. However, cashews as a cash crop are unlikely to generate a significant economic return, given the great distance from the markets in southern Brazil, a problem commonly associated with cash crops in Roraima (Barrow and Paterson, 1994). In addition, the social relevance of the establishment of this cash crop cannot be easily estimated and depends on the kind of agricultural system adopted. Two possibilities exist: smallholdings organised in co-operatives, with a more equitable income distribution, or large landownership/agricultural companies, usually directed at distant and international markets, with uneven equitability. It is clear that no single development option considered in Table 7.4 is likely to attain a sustainable mode, and the potential ecological impacts range from moderate to very high. Even in the case of indigenous cattle grazing, which is usually associated with cassava planting, the erosion hazard is very high as a result of the frequent use of fire during the dry seasons.

Although only moderate financial input would be required for establishing improved pastures, the constraints to overcome soil deficiencies on a long-term sustainable basis are expected to be to high. The prospects of reducing use of fire could lead to reducing erosion, yet it is unlikely to have a significant effect in controling the overall degradation process of soils and land in this catchment, as everywhere in Roraima persisting gullies and sheetwash erosion are common on improved pastures of Coloniâo (*Panicum maximum*) and especially, Quicuio da Amazônia (*Brachiaria humidicola*). Furthermore, these introduced species are less resistant to recurrent fire than the indigenous *Trachypogon* from the savannas of Roraima.

In the long term, the cost of weed control could be too high, although in many adjacent areas the ranchers do invest in herbicides, whilst older pastures are normally annually burned to renew the grass cover. The social relevance of such improvement in pastures would certainly be considerable, in view of the steady decline in income experienced by many small and medium size ranchers, which has been associated with the degradation of older pastures caused by accelerated erosion with high intensity of grazing.

In summary, the chemical and physical properties of the soils in the lower slope positions impose serious limitations on land use and soil management. With low humus content and high sand/silt content, little or no A horizon structural aggregation occurs. These soils are hard-setting and prone to superficial crusting, even under natural environmental conditions. This could explain the adaptation of a very poor xerophytic savanna to this particular set of characteristics. Under cultivation, one can expect increasing erosion and hard-setting, rendering the amelioration of the soils' structural pattern extremely difficult to attain. From the chemical point of view, the acidic upper horizons combined with sodic/solodic lower horizons create a soil environment of serious constraints that is very difficult to overcome. For example, the use of gypsum application to remove sodicity in subsurface horizons would be prevented by the relative acidity of the upper horizons, and so be a severe constraint to any proposed land use.

The erosion hazard associated with these soils under indigenous/*colonos* cattle grazing requires attention, as natural gullies are widespread today in the plains, especially near the inselbergs; this feature has been reported elsewhere in the Roraima savannas (Eden and McGregor, 1994) (Figure 7.7). The resulting, newly exposed, natric Bt horizon renders the eroded surface a hard cemented medium in which the establishment of natural vegetation is almost impossible, and may result in desertification (Figure 7.8). The present land use in the indigenous reserve is chiefly extensive cattle ranching and small cultivations of cassava, promoted by Macuxis and Uapixanas indians and pioneer settlers (Schaefer et al., 1993). Thus, it is clear that any agricultural improvement based on an entrepreneurship model is a challenging undertaking. The indigenous adaptation to this adverse environmental setting seems to be, ultimately, a reasonable appraisal of the cost/benefit involved.

In view of the severe soil and land constraints, the permanent demarcation of the Raposa–Serra do Sol indigenous reserve is ecologically as well as socially desirable. Any trial of modern agricultural settlement in this unique environment would eventually lead to its rapid deterioration, thus wasting one of the least known environments in Amazonia: the xerophytic savannas of the Surumu/Parimé catchment.

Figure 7.7   Naturally developing gully erosion near Paricarana river (photo C. Schaefer)

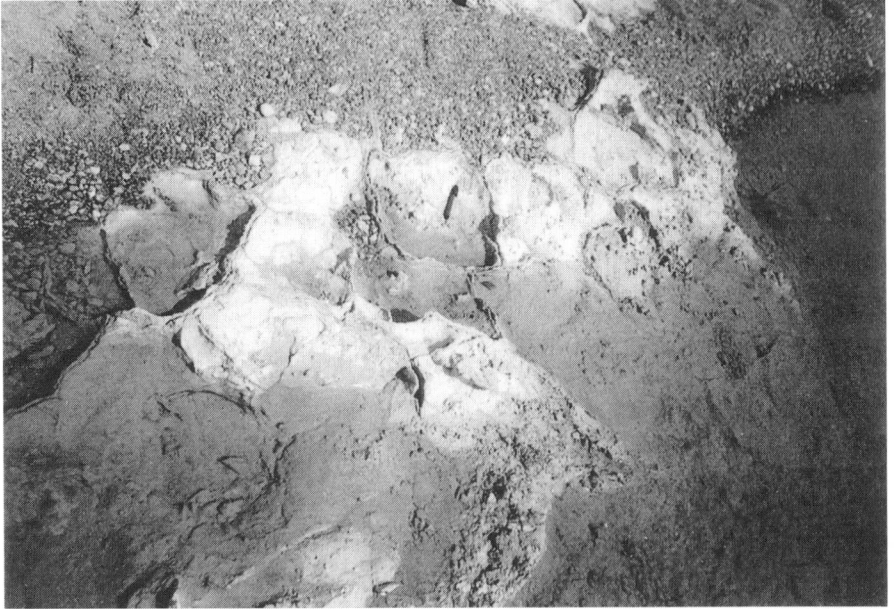

Figure 7.8 Natural gully erosion exposing the indurated top of the natric horizon, with columnar structure. The cracks between columns are filled up with iron/manganese oxides, which are resistant to erosion (photo C. Schaefer)

REFERENCES

Ab'Saber, A.N. 1982. 'The paleoclimate and paleoecology of Brazilian Amazonia'. In *Biological Diversification in the Tropics* (ed. G.T. Prance), pp. 41–59. Columbia University Press, New York.

Barbosa, O. and Ramos, J.R.A. 1959. 'Território do Rio Branco: Aspectos principais da Geomorfologia, da Geologia e das possibilidades minerais de sua zona setentrional'. *Boletim Divisão Geologia e Mineralogia*, **196**, 46. Rio de Janeiro.

Barrow, C. and Paterson, A. 1994. 'Agricultural diversification. The contribution of rice and horticultural producers'. In *The Forest Frontier: Settlement and Change in Brazilian Roraima* (ed. P.A. Furley), pp. 153–181. Routledge, London.

Brasil, 1975. '*Projeto Radambrasil*'. Folha NA 20 Boa Vista e parte das folhas NA 21 Tumucumaque, NB 20 Roraima e NB 21. Ministério das Minas e Energia, Rio de Janeiro.

Brasil, 1980. '*Aptidao Agriíola das Terras de Roraima*', 15. Ministério da Agricultura Brasilia.

Canada Land Inventory, 1970. '*Objectives, Scope and Organisation, Report 1*', Lands Directorate, Ottawa.

CFSEMG (Comissão de Fertilidade de Solo no Estado de Minas Gerais), 1989. '*Recomendacões para o uso de corretivos fertilizantes em Minas Gerais*'. Belo Horizonte.

Cheverry, C. and Stoops, G. 1992. '*New Challenges for Soil Research in Developing Countries: An Holistic Approach*'. Proceedings of the Workshop EU Life Sciences and Technologies for Developing Countries (STD 3), Rennes, France.

Dalal-Clayton, B. and Dent, D. 1993. '*A Review of Land Resources Information and its Use in Developing Countries*'. Environmental Planning Issues. No. 2. Environment Planning Group, IIED, London.

Davidson, D. 1992. '*The Evaluation of Land Resources*'. Longman, London.

Eden, M.J. 1974. 'Paleoclimatic influences and development of savanna in Southern Venezuela'. *Journal of Biogeography*, **1**, 95–101.

Eden, M.J. and McGregor, D. 1994. 'Deforestation and the environment'. In *The Forest Frontier: Settlement and Change in Brazilian Roraima* (ed. P.A. Furley), pp. 86–109. Routledge, London.

EMBRAPA (Empresa Brasileira de Pesquisa Agropecuaria), 1982. 'Levantamento de reconhecimento de alta intensidade dos solos e da aptidão agrícola, no munícipio de Caracaraí. RR'. *Boletim Técnico*, **79**, 1–173.

EMBRAPA (Empresa Brasileira de Posquisa Agropecuaria), 1984. 'Caracterização das frações argila e silte em solos de área do pólo Roraima'. *Boletim Pesquisa*, **25**, 1–59.

FAO, 1983. '*Guidelines: Land Evaluation for Rainfed Agriculture*'. Soils Bulletin 52, FAO, Rome.

FAO, 1984. '*Land Evaluation for Forestry.*' Forestry Paper No. 48, FAO, Rome.

FAO, 1985. '*Guidelines: Land Evaluation for Irrigated Agriculture*'. Soils Bulletin 55, FAO, Rome.

FAO, 1988. '*FAO–UNESCO Soil Map of the World*', revised legend. World Soil Resources Report 60, FAO, Rome.

FAO, 1991. '*Guidelines: Land Evaluation for Extensive Grazing*'. Soils Bulletin 58, FAO, Rome.

FAO, 1993. '*Agroecological Assessments for National Planning: the Example of Kenya*'. Soils Bulletin 67, FAO, Rome.

Fearnside, P.M. 1983. 'Development alternatives in the Brazilian Amazon: An ecological evaluation'. *Interciência*, **8**, 65–76.

IBGE, 1985. Censo Econômico e Agropewário de Roraima p. 229. Rio de Janeiro.

IBSRAM, 1985. '*IBSRAM Highlights*', 1985. Funny Publishing Limited Partnership, Bangkok.

Jackson, C.I. and Maxwell, J.W. 1971. '*Landowners and Land Use in the Tantramar Area, New Brunswick*'. Report 9, CLI, Department of Regional Economic Expansion, Ottawa.

Kassam, A.H., van Velthuizen, H.T., Fischer, G.N. and Shah, M.H. 1991. '*Agro-ecological Assessments for National Planning: the Example of Kenya. Resource Data Base and Land Productivity*'. World Soil Resources Report No. 71, FAO, Rome.

Klingebiel, A.A. and Montgomery, P.H. 1961. '*Land Capability Classification.*' USDA Soil Conservation Service Agricultural Handbook 210.

Landon, J.R. 1991. '*Booker Tropical Soil Manual*'. Longman, London.

MacMillan, G. and Furley, P.A. 1994. 'Land use pressures and resource exploitation in the 1990s'. In *The Forest Frontier: Settlement and Change in Brazilian Roraima* (ed. P.A. Furley), pp. 185–205. Routledge, London.

Rees, W.E. 1977. '*The Canada Land Inventory in Perspective*'. Report 12, Lands Directive, Ottawa.

Sanchez, P.A. 1976. '*Properties and Management of Soils in the Tropics*'. Wiley, New York.

Sanchez, P.A., Bandy, D.E., Villachica, J.H. and Nicholaides, J.J. 1982. 'Amazon Basin soils: management for continuous crop production'. *Science*, **216**, 821–827.

Schaefer, C.E.G.R. 1991. 'Environments in northeastern Roraima: Soils, palynology and paleoclimatic implications'. MSc Thesis, Universidade Federal de Viçosa, in Portuguese.

Schaefer, C.E.G.R. 1994. 'Soils from northeastern Roraima, Amazonia. Geomorphology, genesis and landscape evolution,' Unpublished PhD Thesis, University of Reading.

Schaefer, C.E.G.R. and Dalrymple, J.B. 1995. 'Landscape evolution in Roraima, North Amazonia: Planation, paleosols and paleoclimates'. *Zeitschrift für Geomorphologie*, **39**, 1–28.

Schaefer, C.E.G.R. and Dalrymple, J.B. 1996. 'Pedogenesis and relict properties of soils with columnar structure from Roraima, north Amazonia'. *Geoderma*, **71**, 1–17.

Schaefer, C.E.G.R., Rezende, S.B., Correa, G.F. and Lani, J.L. 1993. 'Chemical characteristics and pedogenesis of sodium-affected soils from Roraima, north Amazonia'. *Revista Brasileira de Ciência de Solos*, **17**, 471–478.

Schubert, C. 1988. 'Climatic changes during the last glacial maximum in northern South America and the Caribbean: A review'. *Interciência*, **13**, 128–137.

Silva, J.A. 1985. '*Soil-based agrotechnology transfer*'. Benchmark Soils Project, Department of Agronomy and Soils, Hawaii Institute of Tropical Agriculture and Human Resources, University of Hawaii.

Smyth, A.J. and Dumanski, J. 1993. '*An International Framework for Evaluating Sustainable Land Management*'. World Soil Resources Report 73, FAO, Rome.

Soil Survey Staff, 1994. '*Keys to Soil Taxonomy (6th edition)*'. US Department of Agriculture SMSS Technical Monograph, 19.

Tricart, J. 1985. 'Evidence of Upper Pleistocene dry climates in northern south America'. In *Environmental Change and Tropical Geomorphology* (eds I. Douglas and T. Spencer). Manchester University Press, Manchester.

Tricon Jaya PT and Dale International Ltd, 1990. *Second Stage Development Programme – Feasibility Studies and Direct Action Programme – Kanpaten Sanggau SSDP Sites, West Kalimantan for Republic of Indonesia*, Ministry of Transmigration. Volume II – Annex A – Land Allocation Study.

van Wanbeke, A. 1981. '*Soils of the Tropics; Properties and Appraisal*'. McGraw-Hill, New York.

Wijmstra, T.A. and van der Hammen, T.V. 1966. 'Palynological data on the history of tropical savanna in northern South America'. *Leidse Geologische Mededelingen*, **38**, 71–90.

CHAPTER 8

Soil Restoration and Conservation: The "Tepetates" – Indurated Volcanic Soils – in Mexico

PAUL QUANTIN
Dijon, France

CHRISTIAN PRAT
ORSTOM, Montpellier, France

CLAUDE ZEBROWSKI*
ORSTOM, Bondy, France

INTRODUCTION

The soils of the high volcanic plateaus of Mexico frequently show indurated horizons, locally called "tepetates" (which means stone matting in Nahuatl). They occur mainly after erosion of the superficial soil on the piedmonts of the volcanic massifs and the plateaus, thus denuding quasi-sterile areas. In the Mexico valley at the foot of the Sierra Nevada and in Tlaxcala State, they account for between 30 and 40% of the soils dedicated to food crops and they affect most of the rural population. The shortage of agricultural soils is a serious problem in this overpopulated area close to Mexico City (Figure 8.1). Therefore, considerable effort has been made for about 20 years, both locally and with international assistance, to try to reclaim tepetates into fertile soils and to control erosion.

In order to convert these soils back to productive agriculture, a cultivation sequence is used:

1. "Roturation" (deep subsoiling, deep ploughing, and disc harrowing) to split up and loosen the tepetate in order to allow water and air to infiltrate.
2. Terracing, generally at reduced slope and following contours.

---

*Unfortunately died in an airplane crash at Bogota, 20th April 1998.

The Sustainable Management of Tropical Catchments. Edited by David Harper and Tony Brown.
© 1998 John Wiley & Sons Ltd.

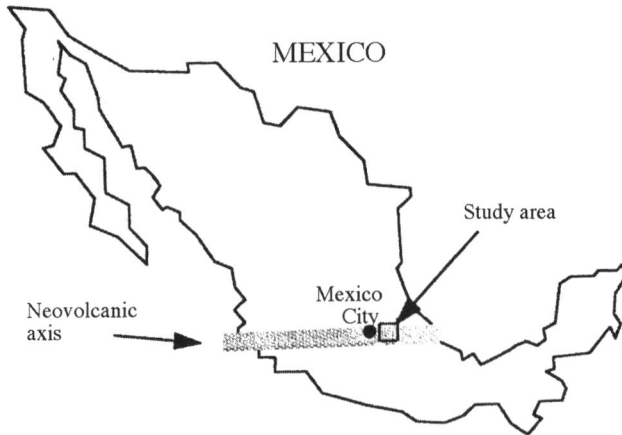

Figure 8.1   Location of the tepetate soils in Mexico

3. An organic and/or mineral fertilisation to replace the original lack of nitrogen, phosphorus and organic manure (humus, microorganisms).
4. A crop rotation adapted to the farmers' experience with wheat or barley first, then with maize and beans. Ridging must be employed with maize; this is a complex operation over three periods. In May, a furrow is made by a mouldboard plough and the maize is sown. In July manual weeding is carried out after emergence and the maize begins to be earthed-up; in August this is completed and ridges and depressions characterise the crop field.

Production approaches average for the region after 3 or 5 years of cultivation (Marquez et al., 1992; Quantin, 1992). Initial studies conducted by the European Community (CCE-STD2) between 1989 and 1992 were devoted to understanding tepetates (features, origin and properties); to the measurement of erosion in small plots; and finally to the agronomic and socio-economic consequences of agricultural reclamation. A subsequent study has tested at small-plot scale the regeneration and conservation of the indurated volcanic soils elsewhere in South America (mainly in Ecuador and Mexico). The major results are given in a synthesis elsewhere (Quantin, 1992). This chapter highlights the results of the first programme measuring erosion in small plots in the Mexico valley and Tlaxcala State as well as the success of agricultural production in the first years of recultivation. It also reports the initial results of the water balance and erosion studies (Prat et al., 1993).

Erosion was studied in small experimental plots, most of them being 22×2 m (Wischmeier standards) and with a mean slope ranging from 8 to 9%. The tepetates tested were of two types, t_2 and t_3 (Quantin, 1992), with a fragipan consistency (hard when dry and friable when moist). They contain 30–40% clay but are weakly cemented, probably by silica. Their total porosity ranges from 40 to 55%, but their macroporosity is low, generally lower than 5% and sometimes even zero, so that hydraulic conductivity is lower than 1 mm h^{-1} and air porosity is very limited. Thus, they outcrop after soil erosion and remain sterile.

Measurements were made in four stations (San Miguel Tlaixpan, El Carmen, Matlalohcan and Tlalpan) at similar altitude (2,500 to 2,600 m), slope (8 to 9%) and climate (average air temperature 13°C, rainfall 700–800 mm). They enabled local variations in the rainfall regime and intensity to be tested. A cultivated agricultural soil (bare or with maize ridging) was compared with a tepetate denuded by erosion, and with a bare *"roturé"* tepetate or one covered with a cultivated plant (flat cultivation of wheat, or maize ridging). Rainfall (rainfall intensity pattern), the volume of runoff water and the weight of eroded soil were measured. Periodic measurements of soil moisture, density–porosity and particle size were also made and the evolution of the surface features – the formation of crusts, the decrease in the size of aggregates and the infiltration rate – were observed.

RAINFALL EROSIVITY

The hygrothermic climatic regime is of "ustic–isomesic" type (mean temperature 13°C, mean rainfall 700–800 mm). The rainy season from May to October during the summer period alternates with a six month dry and somewhat colder season. The analysis of rainfall erosivity was carried out on the 1991 rainfall patterns recorded in the four stations. The EI.30 Wischmeier index (1958) was converted into American units in order to be comparable with other data from the literature. Calculations were also made in international units of I.30 $(mm\,h^{-1})$ and EI.30 $(MJ\,ha^{-1} \times mm\,h^{-1})$. The AIm erosivity index (Lal, 1976) was also calculated (Baumann, 1992).

Rainfall and the cumulative index of erosivity at the four stations are shown in Table 8.1. Rainfall was about 100 mm higher on the Tlaxcala slope (El Carmen, Tlalpan and Matlalohcan) facing the Texcoco slope in the Mexico valley (San Miguel Tlaixpan). Thus, there is a slope effect from the eastern to western sides of the Sierra Nevada. Moreover, given that there is the same total amount of rainfall on the Tlaxcala slope, there is variation in number of rainy days and particularly their intensity and erosivity index. EI.30 ranged from 234 to 429. The more irregular the rainfall regime (at Matlalohcan), the more erosive it appeared to be. This effect is confirmed by the runoff and erosion recorded (see below).

The calculated values of the erosivity index in 1991 ranged from 200 to 429, considered low to medium by Roose (1981) for a "tropical dry" area. The frequency of highly erosive rainfall events is actually very limited with reference to the calculated

Table 8.1 Rainfall and rainfall erosivity in 1991

| Station | Rainfall | | Erosivity | |
	H (mm)	N (days)	EI.30[a]	AIm[b]
San Miguel T.	669	99	200	
El Carmen	779	120	234	261
Tlalpan	803	112	357	330
Matlalohcan	775	96	429	418

[a] Wischmeier index, American units. [b] Lal index

Table 8.2   Rainfall frequency at different levels of intensity and erosivity

I.30[a]	PE 25–30		ME 30–50	FE >50	
EI.30[b]	100–200	200–400	400–1000	>1000	Total
San Miguel T.	4	–	–	–	4
El Carmen	–	3	3	–	6
Tlalpan	–	3	3	2	8
Matlalohcan	–	9	3	1	13

[a] I.30: mm h^{-1}. [b] EI.30: MJ ha^{-1} × mm h^{-1}. PE: weakly erosive, ME: moderately erosive, FE: highly erosive

values I.30 and EI.30 and to the results of soil loss observed after each rainfall. Table 8.2 shows these exceptional rainfall events and classifies them by order of erosive intensity for each station.

Given a similar rainfall regime, each station is therefore distinguished by the amount and particularly the intensity of erosive rainfall events. Thus there are few erosive events each year, but they are responsible for 60–80% of the total erosion.

The values of the AIm index proposed by Lal (1976) for tropical areas are close to those of EI.30. The calculation of EI.30 for short rainstorms may be less accurate; it affects the annual cumulated value and it is not precise enough to establish a correlation with the intensity of runoff and erosion. It would be necessary to make a more precise analysis of the rainfall intensity at a time step of 10 minutes, or even of 5 minutes, for a better understanding of the effect of high rainfall intensities during a short period of time.

RUNOFF AND EROSION

Measurements were made over 2 years following tepetate "roturation". Runoff and erosion data obtained from four stations in a year when the same cultivated plant (maize ridging) was used in all the stations are shown in Table 8.3. In the same year rainfall erosivity, observation of the soil water regime and of the evolution of the surface features during the rainy season were measured.

The tepetates t_2 and t_3 denuded by erosion show a similar behaviour. The average annual runoff rate is close to 70%. It is not total, for initial rainfall and small rainfall events infiltrate through shrinkage cracks. But runoff reaches 80–90% when there are heavy rainstorms. Erosion is normally low, ranging from 5 to 10 t ha^{-1} yr^{-1} (except in t_3 at Tlalpan) due to the compactness and stability of the original tepetate.

A single deep "subsoiling" of the denuded tepetate decreased runoff by nearly half, 40%, and erosion by 2 t ha^{-1} yr^{-1}. There was also an improvement in the infiltration and surface roughness, without any destabilisation of aggregates. The effect of a complete "roturation" of the denuded tepetate (deep subsoiling followed by several deep ploughings and disc harrowings in order to get an optimum aggregate in the first year followed by a deep ploughing in the second year) varied with the rainfall erosivity. In the less erosive station (San Miguel Tlaixpan), the runoff rate ranged

Table 8.3 Measurement of soil loss and runoff from tepetates

Station	Treatment	Plot length (m)	Erosion (t ha^{-1})	Runoff (%)
San Miguel Tlaixpan	Bare tepetate t_3	22	5.05	88[a]
	Ploughed, bare T	22	21.89	44[a]
	T, L + B, bare	22	1.04	12[a]
	T, L + B, maize	22	1.24	5[a]
	T, L + B, maize	10	1.18	5[a]
	Soil, L + B, bare	22	1.1	12[a]
	Soil, L + B, maize	22	1.79	11[a]
El Carmen	t_3 Bare tepetate	3	8.0	70[b]
	t_2 Bare tepetate	3	6.3	67[b]
	t_2 T, L bare	22	78.0	34[b]
	t_2 T, L + B, maize	22	23.0	11[b]
Tlalpan	t_3 Bare tepetate	3	41.0	68[b]
	t_2 Bare tepetate	3	7.5	65[b]
	t_3 T, L bare	22	128.0	43[b]
	t_3 T, L + B, maize	22	26.0	21[b]
Matlalohcan	t_3 Bare tepetate	6	8.8	75[b]
	t_3 T, R + shrubs	6	26.0	61[b]
	Soil, bare	22	47.0	34[b]
	Soil, savanna	22	0.3	10[b]

T: Tepetate (t_2 or t_3), L: subsoiled and ploughed, R: subsoiled, B: ridged.
[a] Average estimate of the three most erosive rainfalls. [b] % of the total rainfall

from 10 to 20% and erosion amounted to 21 t ha^{-1} yr^{-1} In the Tlaxcala area, characterised by heavier rainstorms, runoff ranged from 30 to 40%. In the station characterised by moderately erosive rainstorms (El Carmen) erosion increased rapidly to 72–78 t ha^{-1} yr^{-1} and in the station characterised by two highly erosive rainstorms (Tlalpan) to 128 t ha^{-1} yr^{-1}). Without any soil-conserving practices or vegetative cover, the "*roturé*" and ploughed tepetate is therefore highly erodible. Fine aggregates in these soils are unstable.

The effect of ridging, whether the cultivated tepetate was devoid of vegetation or planted with maize, was spectacular in the less erosive station (San Miguel Tlaixpan): runoff was reduced to 5% under maize and to 12% on bare soil, and erosion was reduced to 1 t ha^{-1} yr^{-1} in both cases. But given the rainfall intensity, even under maize, runoff increased to 10% in the moderately erosive station (El Carmen) and to 20% in the most erosive station (Tlalpan). Erosion increased dramatically to 23 t ha^{-1} yr^{-1} in tepetate t_2 at El Carmen and to 26 t ha^{-1} yr^{-1} in tepetate t_3 at Tlalpan. Above a certain threshold therefore, runoff may no longer be controlled, ridges may break and erosion may be strong. Moreover, maize whose growth is too late has a less obvious effect on erosion in the weakly erosive stations (San Miguel Tlaixpan), but the comparison with a bare ridged soil was not made in the most erosive stations.

In original soil (before any accelerated erosion), under natural shrubby savanna, there was only a low runoff rate of around 10% and almost no erosion, even though this was at the most erosive station (Matlalohcan). At that station, however, the same

Table 8.4   Measurements of erosivity and runoff on bare tepetate in 1992

Rainfalls		EI.30	Runoff
number	%	$(MJ\ ha^{-1} \times mm\ h^{-1})$	(%)
40	59	< 7	< 40
15	22	7–25	40–50
10	15	25–80	50–70
3	4	> 80	> 70

soil devoid of its vegetation cover showed runoff increased to 34% and erosion to $47\,t\,ha^{-1}\,yr^{-1}$. The effect of ridging was effective in the weakly erosive station on the bare soil as well as in the soil cultivated with maize; it restrained runoff to 11–12% and erosion to $1–2\,t\,ha^{-1}\,yr^{-1}$. This result was similar to that observed in the cultivated tepetate, although the soil was less stable than the tepetate. It would undoubtedly be different in more erosive rainstorms, but measurements have not been made. No observation has been made either on the cultivated soil and on the bare unridged soil, in the weakly erosive station.

In a subsequent year, new measurements were made in the natural tepetate (t_3, at San Miguel Tlaixpan), but in a large plot of $1,800\,m^2$ (instead of $44\,m^2$) and with a slope of 8–10% (Prat et al., 1993) in order to control for any size effects. The rainfall regime was similar to that previously observed (90 rainy days, three to four erosive rainstorms only). Table 8.4 shows the erosivity and runoff values as related to the amount of rainfall during the observation period (78 rainfall events).

During the observation period, the total soil loss amounted to about $10\,t\,ha^{-1}\,yr^{-1}$, about double the amount measured in the small plot at San Miguel Tlaixpan, but the same order of magnitude as in the whole four stations (5 to $10\,t\,ha^{-1}\,yr^{-1}$). The runoff rate of erosive rainstorms (EI.30 > 80) was higher than 70%. Thus, the larger plot confirmed the original results.

The results for tepetates can be compared to those observed in the cangahua in Ecuador (Custode et al., 1992) during a 5-year period (1987–91) for a mean rainfall of 660 mm, similar to the less erosive San Miguel Tlaixpan station, but with a steeper slope of 22% (Table 8.5).

The runoff rates measured on original soil can be compared to those reported by

Table 8.5   Runoff and erosion on cangahua in Ecuador

Treatment	Runoff (%)		Erosion (t ha^{-1})
	Mean	Max.	
Bare cangahua, surface cultivated	20–30	60–90	96.6
Traditionally cultivated cangahua	7–14	30–55	18.9
Improved cultivated cangahua	2–9	9–55	4.5

Roose (1981) in a "tropical dry" area on "ferruginous tropical" soils with an annual mean rainfall of 850 mm: these ranged from 10 to 15% under savanna, from 35 to 43% on bare soil, and from 10 to 40% on cultivated soil. Erosion was also very low under savanna, ranging from 0.2 to 0.7 t ha^{-1} yr^{-1}; moderate on cultivated soil ranging from 1 to 14 t ha^{-1} yr^{-1}; but somewhat less high on bare soil ranging from 10 to 35 t ha^{-1} yr^{-1}. As a whole, the ranges are similar.

EVOLUTION OF SURFACE FEATURES

Surface features were recorded according to Casenave and Valentin's method (1989) in plots of "*roturé*" tepetate and soil cultivated with ridged maize and on a bare "*roturé*" tepetate (Jerome, 1992). Six observations were made, at the beginning and at the end of the three cultivation periods (from sowing to the first weeding; up to the second weeding and ridging of maize up to harvesting). Features recorded included crust formation (structural and depositional), erosion patterns, aggregate slaking, modifications in the surface relief as well as measurements of porosity, infiltration rate and soil moisture. The observations followed the soil processes which lead to runoff and erosion.

In the case of the soil and tepetate cultivated with ridged maize, modifications occurred due to the formation of surface and structural crusts on ridges and of depositional crusts in depressions. In the soil, this was rapid, for the total slaking of aggregates was rapid and considerable (50–80% by the end of cultivation). In the cultivated tepetate, the changes were slower and more gradual, for the aggregates of more than 2 mm were more stable and their slaking was only partial (decrease in size). During the first cultivation period (up to the first weeding), the changes were slow, aggregates were stable and infiltration remained rapid. But during the second period and above all the third period, the process became more rapid: the aggregate slaking increased; the crust formation became more intense and general; porosity decreased close to the original conditions prevailing before "roturation"; the infiltration rate slowed down considerably to low levels (Table 8.6). Thus, runoff and sensitivity to erosion became higher. In fact, the aggregates smaller than 2 mm slaked completely

Table 8.6   Development of porosity and infiltration rate

Cultivation period	1		2		3	
	Early	Late	Early	Late	Early	Late
Tepetate porosity	58.1	53.2	57.7	54.3	51.1	50.6
Infiltration rate on ridge (mm h^{-1})	96	45	48	28	21	10
Infiltration rate on furrow (mm h^{-1})	52	16	24	12	8	2
Soil porosity	56.4	50.9	55.3	51.5	50	48.9
Infiltration rate on ridge (mm h^{-1})	68	32	16	8	10	4
Infiltration rate on furrow (mm h^{-1})	52	0	12	2.4	5	0

Initial tepetate porosity = 44%. Infiltration rate = 0.3–0.5 mm h^{-1}

and gave rise to the formation of crusts and soil loss. A laboratory experiment had showed that an "optimum" aggregate should be no smaller than about 3 mm. Thus, excessive fragmentation of tepetate (harrowing, weeding) into a fraction smaller than 2 mm gives rise to encrusting, runoff and erosion. Contrary to previous belief, the ploughed tepetate is not stable, even under maize; aggregates decreased in size, those smaller than 2 mm slaked, the material becomes packed and encrusted. The clear implications are that cultivation techniques must limit fragmentation by avoiding pulverisation; other devices and methods need to be tested. The Mexican method of maize ridging, divided into three successive operations of ploughing, weeding and earthing up, intensifies fragmentation and is therefore dangerous. It would be advisable to ridge in one operation and to sow maize in the ridge. Moreover, ridges should be partitioned so that they could not be suddenly breached by high rainstorms.

In the case of the "*roturé*" bare tepetate, the originally cloddy surface was gradually covered with a structural crust through the slaking of fine aggregates; then a depositional crust was formed in small depressions. Runoff increased and the surface became covered with rill-erosion which grew deeper when rainstorms were heavy and erosion became stronger. Thus, without any cultivation, the cultivated tepetate became easily erodible by heavy rainstorms.

AGRONOMIC IMPLICATIONS

Analyses of tepetates in laboratory and greenhouse experiments showed that this material displayed certain constraints which are responsible for its sterility: low macroporosity and hydraulic conductivity, lack of organic matter, of nitrogen and phosphorus and of symbiotic microorganisms specific to cultivated plants (e.g. maize and bean). But the material also showed properties suitable for fertile soil regeneration: 30–40% clay which permits (after structural improvement) a good retention of water and provision of micronutrients, a high microporosity, good structural stability, neutral or weakly alkaline pH, and an adequate content of Ca, Mg, K, and micro-elements. Thus, the reclamation of an agricultural soil can be rapid, profitable and sustainable, providing a good structure has been restored and through moderate fertilisation with nitrogen and phosphorus (according to crop requirements) and if possible manuring.

Manual agricultural reclamation of tepetates had already started before Spanish colonisation, with organic enrichment and using soil-conserving terraces ("metepantles"). These practices fell into disuse but presently are being reintroduced, using mechanisation and, of course, changed crops and fertilisation techniques. An agronomic, productivity and economic assessment of the experimental plots (Marquez et al., 1992) as well as socio-economic observations of the farming community (Zahonero, 1992) indicate that sustainable use is achievable, under adequate support conditions.

Fertility and productivity increased steadily for 3 to 5 years after the start of cultivation. Not all crops were initially successful; maize and beans did not yield well even on a well-loosened and fertilised tepetate soil, while wheat and vetch achieved

Table 8.7 Crop yields on tepetates (t ha^{-1})

	Maize	Bean	Broad bean	Wheat	Vetch
First year[a]	0–0.2	0.1–0.2	0.3–0.8	2–4[c]	2–4[d]
Third year[a]	2.2–2.5	–	–	1.5	–
Fifth year[a]	2.5–3.1				
Mean[b]	1.8	0.75	1.33	2	

[a] Year of cultivation since "roturation"; with mineral fertilisation without organic manure
[b] Mean of the yields observed in the area
[c] Wheat in experimental plots and fine texture; in other plots 1–2 t ha^{-1} (mean 15)
[d] Vetch in experimental plots, dry matter yield

normal yields (Marquez et al., 1992, Navarro and Zebrowski 1992). Table 8.7 shows an experimental comparison of four traditional plants (wheat, maize, bean and broad bean) and vetch (a green manure).

The cultivated tepetates can yield wheat, vetch and, less successfully, broad beans from their first year of cultivation, but not maize and beans. It is necessary to obtain a rather fine texture (< 2 mm), but this carries the considerable risk of increased erodibility referred to above. Without manure, a fine soil and a moderate mineral fertilisation of N/P 120/60 can yield 4 t ha^{-1} of wheat, 4 t ha^{-1} of vetch and 0.4 t ha^{-1} of broad bean. With manure and 60 units of P without addition of nitrogen, wheat can yield 6 t ha^{-1}, vetch 6 t ha^{-1} and broad bean 0.8 t ha^{-1}. In this case, nitrogen fertilisation is of no value, decreasing the wheat yield. Failure of maize and bean cultivation in the first year seems to be due to a deficiency of symbiotic microorganisms which are necessary for good plant nutrition (Alvarez-Solis et al., 1992; Ferrera, pers. comm.). In order to provide organic fertilisation of "*roturé*" tepetates in the absence of manure or compost, vetch or a fodder grass such as oats need to be used as green manure. The soil enrichment of specific symbionts then allows a satisfactory yield of maize and bean.

After 2 to 5 years of cultivation, maize and bean yields improved. In the second year, maize yielded from 1.2 to 1.7 t ha^{-1} in experimental plots without manure but with mineral fertilisers (N/P 120/80). In the third year, maize yields ranged from 2.2 to 2.5 t ha^{-1}, and in the fifth year, from 2.5 to 3.1 t ha^{-1}, which was high (mean for the region 1.8 t ha^{-1}).

Reclaimed tepetates are a significant, even necessary, agricultural soil resource for small farms of < 20 ha per family (Zahonero, 1992). The terrace works are also good protection against erosion. Crop rotation has been adapted to the year of recultivation: wheat in the first year, then maize associated with broad bean or bean. By the third year of cultivation, productivity reaches levels similar to normal soils in the region. By about 8 years, the initial costs of reclamation can be repaid (farmers receive State assistance) through the cash crops (wheat, barley). Subsistence crops (maize, broad bean, bean) then prevail gradually over cash crops. Profitability for farms of less than 15 ha, which are marginally self-sufficient, is less certain because the financial yield of cash crops does not allow repayment: here State assistance would be necessary to enable works to be carried out at a lower cost.

CONCLUSIONS

Overall, tepetates of fragipan type, once they have been correctly used, are a sustainable agricultural resource with good productivity. This resource is a necessity particularly for farms of < 20 ha per family. Erosive rainstorms in the areas of Mexico and Tlaxcala are few, so "*roturé*" tepetate is rather stable and not very erodible provided too fine a fragmentation (*c.* 2 mm) is avoided. Ridging associated with maize cultivation is very effective in low-erosive situations; but it is inadequate in areas experiencing highly erosive rainstorms (I.30 > 50 mm ha^{-1}). Without any ridging or vegetative cover, cultivated tepetates rapidly become unstable and highly erodible. Repeated cultivation techniques (ploughings + two weedings and earthing up for maize in Mexico) make slaking of aggregates more rapid which leads to crust formation, runoff and erosion. Therefore, it is necessary to limit the fragmentation of tepetates by reducing tillage frequency. Moreover, partitioned ridges must be constructed to reduce the risks of erosion by high rainstorms.

A "*roturé*" tepetate improved by a moderate mineral fertilisation (N 60–120, P 60) or an organic one (manure + P 60) can be productive from the first year for plants such as wheat and vetch and less productive for broad bean. It is not productive for maize and bean, due to deficiency in symbiotic microorganisms. However, maize yield is satisfactory by the third year and becomes optimum within 5 years. Organic improvement (manure or green manure) and the insemination of symbionts could speed up the process. The operation becomes profitable, leading to recovery of initial costs within 8 years, but additional State assistance (works at a lower cost, loan at reduced interest) is required for small farms of 15 ha or below.

REFERENCES

Alvarez-Solis, J.D., Ferrera-Cerrato, R. and Zebrowski, C. 1992. 'Analisis de la microflora asociada al manejo agro-ecológico en la recuperación de tepetates'. *Terra, Suelos Volcánicos Endurecidos* **10**, 419–424.

Baumann, J. 1992. '*Investigaciones Sobre la Erodibilidad y el Regimen Hídrico de los Duripanes tepetates Rehabilitados para los Cultivos, en el Bloque de Tlaxcala*'. Final Report CEE TS2-A212C, European Commission, Brussels.

Casenave, A. and Valentin, C. 1989. '*Les Etats de Surface de la Zone Sahélienne; Influence sur l'Infiltration*'. ORSTOM, Col. Didactiques, Paris.

Custode, E., De Noni, G., Trujillo, G. and Viennot, M. 1992. 'La cangahua en el Ecuador: caracterisación morfoedafológica y comportamiento frente à la erosión'. *Terra, Suelos Volcánicos Endurecidos*, **10**, 332–346.

Jerome, G. 1992. '*Etude des réorganisations superficielles sous pluies naturelles sur un sol volcanique induré, le tepetate, dans la vallée de Texcoco, Mexique; comparaison avec un sol non induré*'. Memoire Fin d'Etudes Inst. Sup. Tech. O. M., Juin 1992.

Lal, R. 1976. 'Soil erosion in Alfisols in Western Nigeria'. 3: Effects of rainfall characteristics. *Geoderma*, **16**, 389–401.

Marquez, A., Zebrowski, C. and Navarro, H. 1992. 'Alternativas agronómicas para la recuperación de tepetates'. *Terra, Suelos Volcánicos Endurecidos*, **10**, 465–473.

Navarro, H. and Zebrowski, C. 1992. '*Etude des Sols Volcaniques Indurés "tepetates" des Bassins de Mexico et Tlaxcala, Mexique; leur Production Agricole*'. Final Report CEC TS2A 212C, European Commission, Brussels.

Prat, C., Oropeza, J.L. and Janeau, J.L. 1993. 'Resultados del primer año de investigación del programa ORSTOM-CP sobre la rehabilitación de los tepetates de Mexico'. *Xll Congr. Int. Ciencia del Suelo, Salamanca*, oral communication.

Quantin, P. 1992. '*Etude des Sols Volcaniques Indurés "tepetates" des Bassins de Mexico et de Tlaxcala, en Vue de leur Réhabilitation Agricole*'. Final Report, CEC TS2A 212C, European Commission, Brussels.

Roose, E. 1981. '*Dynamique Actuelle des Sols Ferralitiques et Ferrugineux Tropicaux d'Afrique Occidentale. Etude Expérimentale de Transferts Hydrologiques et Biologiques de Matières sous Végétations Naturelles ou Cultivées.*' Trav. Doc. ORSTOM No. 130.

Wischmeier, W. H. 1958. 'Rainfall energy and its relationships to soil loss'. *Transacions of the American Geographical Union,* **39**, 285–292.

Zahonero, P. 1992. '*Des Lits de Pierre sur l'Altiplano. Contribution à l'Analyse de la Mise en Valeur des "tepetates", Sols Indurés d'Origine Volcanique, dans la Région de Tlaxcala*'. Final Report, CPSL/ISTOM, Paris.

CHAPTER 9

Environmental Management for Sustainable Selective Logging in Tropical Rainforests

IAN DOUGLAS

Department of Geography, University of Manchester, UK

INTRODUCTION

The concept of managing tropical forests for sustainable yield is old. In 1950, Unstead wrote: "The earlier practice of simply cutting down the trees ... is being replaced by a careful utilisation of the forest – a coordinated system of felling selected trees and arranging for the better growth or the replanting of others" (Unstead, 1950, p. 196). The key to such selective logging practices is the ability of the forest to recover from the logging operation and for the vegetation to regrow and retain a high proportion of its original plant and animal diversity. The degree of success depends in part on the severity of the impact of the logging and the damage done to the forest canopy and floor. Analysing changes in stream catchment ouputs is one way of gauging the severity of that impact and the rate at which the forest reverts towards a pre-logging state.

Sustainable-yield logging operations require the rotation of annual logging coupes or cycles, ranging from 25 to 60 years, depending upon the regrowth period of the main stock species and the silvicultural system being used to manage the forest. Common to all management methods is a period of severe disturbance as access roads are constructed and logging operations take place, followed by a period of recovery as the stand regenerates prior to the next intended harvest (Greer et al., 1995).

Selective logging in tropical forests typically removes only "merchantable species", representing from perhaps 20 to 60% of the standing volume. Normally only the trees of commercially valuable species of over 60 cm breast height diameter (sometimes over only 30 cm, depending on local regulations) are removed with the small ones being left to grow larger for the next logging cycle. In Indonesia, for example, concession management is based on a 35-year rotation, which assumes that there will be commercial volumes of merchantable species 35 years after the original felling. The

The Sustainable Management of Tropical Catchments. Edited by David Harper and Tony Brown.
© 1998 John Wiley & Sons Ltd.

actual amount of the canopy of the forest removed in selective logging depends on the density of "merchantable species" of appropriate size. Forests in eastern Sabah, Malaysian Borneo, for example, are extremely rich in large trees of the family Dipterocarpaceae. Sabah alone has at least 180 species in this family, but, as many have similar timber properties, they can be marketed under just a few commercial timber names (Whitmore, 1984). This relative abundance of "merchantable species", coupled with the strong export demand for logs in the 1980s, encouraged extraction levels in Sabah which were among the highest in the world (Sundberg, 1983). In some concessions, the volume of timber removed was as high as $120 \, m^3 \, ha^{-1}$, although typical figures would be somewhat less. During the 1980s Yayasan Sabah, for example, cut an average of about $70 \, m^3 \, ha^{-1}$ from its concession (Marsh and Greer, 1992).

Unless great care is taken, such high levels of extraction are inevitably associated with heavy damage (Fox, 1970; Nicholson, 1979). Among the many doubts expressed about the selective logging system (Ewing and Chalk, 1988) are:

1. That without proper monitoring, operators may remove the smaller trees.
2. That even if they are not removed illegally, young trees are frequently damaged during initial logging operations and do not recover.
3. That the character of the forest is so changed by the removal of much of the forest cover that there is no assurance that the younger trees will survive.

LOGGING OPERATIONS AND THEIR EFFECTS ON THE CANOPY AND GROUND SURFACE: CREATING A NEW SURFACE HYDROLOGY

Normal selective logging operations do not usually expose a high proportion of the total ground surface, even though they may damage as much as 70% of the canopy. Serious damage to trees of no timber value can occur if the vines and creepers which hang from one tree to another are not cut before logging. The fall of one tree can bring several others down. In logged-over forest, the high rate of damage to tree crowns is all too apparent, save where particular care is taken.

The key to understanding the hydrologic changes caused by logging is to recognise the extension of the surface drainage network by the creation of a new system of roads and tracks. To remove logs, a hierarchy of log drag-paths (snig tracks or skid trails), temporary access roads, log landing areas, trucking yards and main log lorry haulage routes is required. The cost of constructing and maintaining roads is a major part of the total cost of a selective logging operation. Usually the access roads are opened up in the dry period some 9 to 12 months before logging is due to occur, but as soon as such earth-moving operations begin, changes to the hydrology and earth surface process systems of the forest occur.

When logging commences, much more of the forest floor is disturbed. In the timber concessions of Sabah and Sarawak, for example, logging is usually by tractor yarding and/or high lead lines. Tractor yarding involves dragging logs up snig tracks by tracked caterpillar tractors with the whole log dragged along the ground, while high lead logging involves setting up a jib on a spur or high point and dragging logs upslope by cable, so that only the end of the log touches the ground. While tractor yarding creates wide, compacted snig tracks, high lead logging disturbs a large area in the vicinity of the jib (Douglas et al., 1995). The main access roads, or primary routes, usually tend to follow the contour and have well-constructed stream crossings, carry pneumatic-tyred vehicles during logging and have a gravel cover, making them more stable and less readily eroded than the snig tracks. However, once logging ceases the primary routes cease to be maintained and may start to erode.

Each track and road produces a potential pathway for water movement. Good practice requires adequate drainage to divert flows down tracks into the adjacent forest where it is slowed and impeded by vegetation, usually infiltrating and so moving to streams slowly. Poor practice sees the flows along roads creating roadside gullies running directly into channels, thereby extending the ephemeral, storm period, integrated drainage network.

NATURAL DISTURBANCE AND THE CONSEQUENCES OF LOGGING

In a natural forest, about 20% of the rainfall is intercepted and lost back to the atmosphere and most of the remaining rainfall reaches the ground, penetrates through the litter layer and enters the surface horizon of the soil. The litter, albeit thin in places, protects the forest floor against erosion. However, disturbance by animals, such as foraging by pigs, removal of soil by termites, building of soil pillars by cicadas and the making of tracks by deer, creates local bare patches which erode rapidly during heavy rain. Falling trees create natural gaps in the canopy and sometimes uproot soil. However, save where in steep country natural landslides provide large sediment sources, all these natural disturbances are highly localised and rarely provide more than a few metres of concentrated downslope runoff. Snig tracks and log landing areas operate at a different scale. The mechanised logging techniques used in selective systems result in dense track and haulage road networks and large log landings (log assembly, debarking and loading areas) occupying from 16% of Malaysian logging sites (Burgess, 1973) to 25% of sites in Australia (Gilmour et al., 1982).

COMPACTION AND INFILTRATION

Infiltration rates are usually high under vegetation and litter cover in undisturbed tropical forest as a consequence of soil surface protection and high macroporosity (Anderson and Spencer, 1991). Thus overland flow rarely occurs except where the forest soil is temporarily exposed. Rainsplash erosion may be seen in the natural forest where the litter layer is thin, with individual leaves sometimes capping a soil pillar 2 to 3 cm in height, but there is little downslope movement of the detached material. Thus surface soil removal in natural forest is small or even negligible. When the soil surface is exposed, raindrop impact destroys surface aggregate structure, loosens fine soil particles and blocks macropores and even some micropores. Any

Table 9.1   Effect of canopy cover on runoff and soil erosion after Ruangpanit (1985). Data for 41 storms totalling 1128 mm rainfall

% canopy cover	Surface runoff (m^3 ha^{-1})	Soil erosion (t km^{-2})
20–30	194.2	65.28
40–50	177.8	51.23
50–60	183.4	45.69
60–70	113.2	37.26
70–80	121.3	29.80
80–90	102.3	28.51

reduction in the forest canopy thus leads to increased soil erosion (Table 9.1) with loss of the soil litter layer the most critical factor.

Bare areas, including roads and tracks, become heavily compacted as the energy of raindrops compacts the surface layer of the soil and overland flow carries the fine particles away. The result is a slight increase in soil density over time. Average soil moisture content beneath the road also decreases as infiltration is reduced (Helvey and Korchenderfer, 1990).

Average values for infiltration rates into surface soils in forest at the Danum Valley Field Centre, Sabah, Malaysian Borneo (Figure 9.1), were 88 mm h^{-1} for undisturbed forest, 73 for forest regenerating after logging and 15 on logging tracks abandoned 12 years previously (Van der Plas and Bruijnzeel, 1993). A comparison of 15-minute rainfall intensities with average saturated infiltrabilities and infiltration curves suggests that on old logging tracks the frequency and volume of overland flow will increase substantially. Such tracks therefore play a key role in the post-logging dynamics of the forest.

SURFACE RUNOFF AND SEDIMENT TRANSPORT DURING AND AFTER LOGGING

While the recovery of forest after selective logging has been assessed by a variety of indicators, especially comparison of the abundance and biodiversity of mammals, birds and insects in undisturbed forests and areas recovering after logging (Johns, 1992; Lambert, 1992), few of the studies of the hydrologic changes caused by logging extend far into the post-logging recovery period.

Erosion following logging, causing high suspended sediment loads in rivers, is a great problem throughout the humid tropics. Along the coasts of many tropical countries, especially islands and peninsulas of South-east Asia, plumes of sediment project seaward from river mouths, much of the sediment coming from areas of logging or land clearance. For example, in Palawan, the most south-westerly island of the Philippines, the marine resources of the coastal zone of Bacuit Bay (Dixon, 1990), particularly those exploited by fishing industries and tourism, are threatened by increased offshore sedimentation. While a reduction in marine resources has no direct impact on timber production, logging activities and associated soil erosion are having a major negative effect on the marine-dependent resource sectors of this island

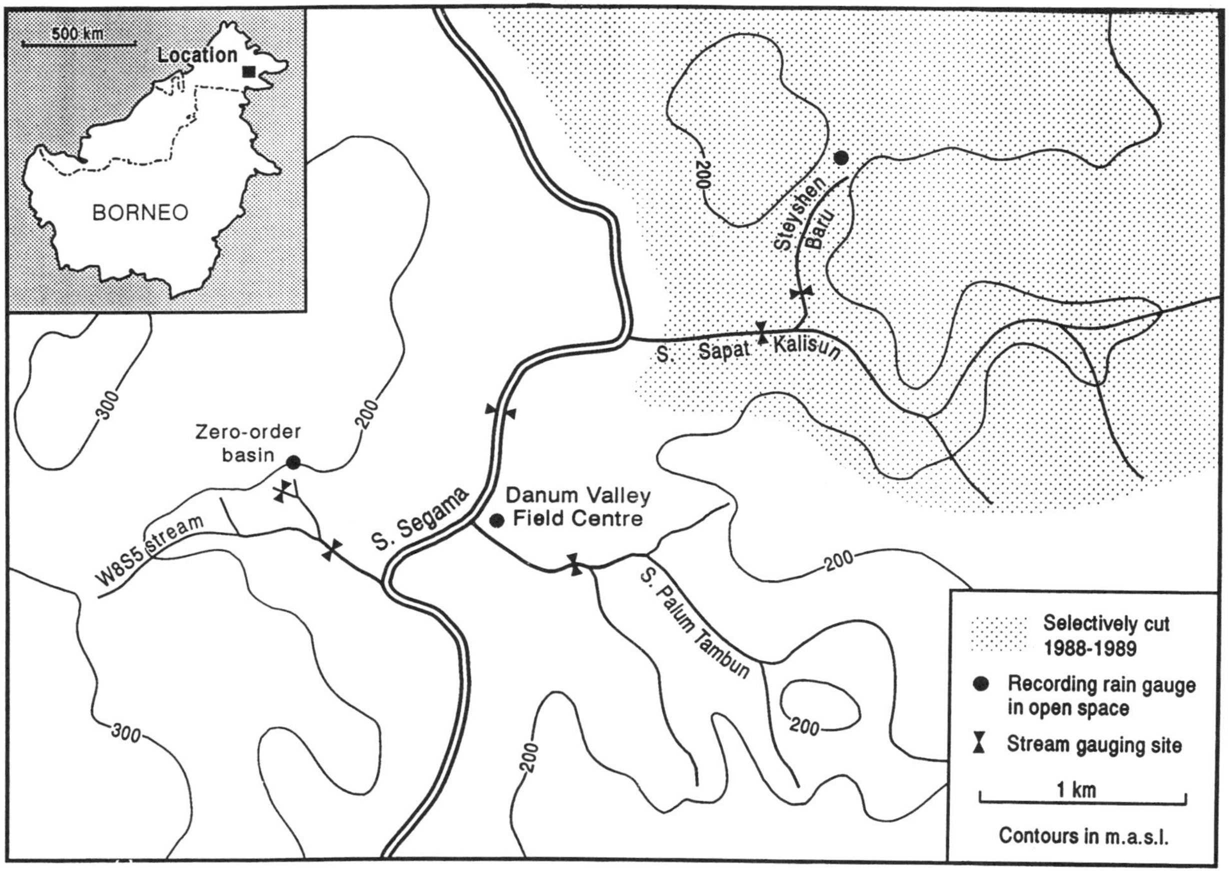

Figure 9.1   Location of study area

which aims to develop tourism that maintains the integrity of the environment (Luce, 1995).

The erodibility of tropical soils varies greatly with the nature of the parent materials. Rates of erosion are also strongly influenced by rainfall regimes and storm intensity. Cyclone-affected areas suffer some of the most serious erosion, but high rainfalls are also associated with high relief. Erosion hazards are particularly high in tropical steeplands, not only areas of high elevation, but also the rugged terrain associated with areas of folded sedimentary rocks. High intensity rainfalls and, in many places, deeply weathered soils pose particular hazards once the protective forest cover is disturbed. Since 1980 several studies of the effects of forest practices in the humid tropics on sediment yields have emerged (Roche, 1981; Cassels et al., 1982; Fritsch 1983, 1993; Douglas et al., 1990; Malmer 1990; Nik and Harding, 1990). They used paired investigations of catchments ranging from a few hectares to several square kilometres in size on varying lithologies and terrain, but all clearly showed that roading, logging practices and the relationship of the timing of forestry operations to extreme weather conditions are major factors in the increase of sediment yield during and after logging.

The results of most studies of erosion and sediment yield caused by logging or land conversion should be seen as indications of orders of magnitude rather than as reliable quantitative estimates of such impacts (Bruijnzeel, 1993). Relatively few good experimental basin studies, where land use is altered in one catchment, while the other is kept as a control, have been conducted in the humid tropics. While a major study has been done in French Guyana, some of the best work has been in Peninsular Malaysia (Table 9.2).

In the Bukit Berembun study in Negeri Sembilan, peninsular Malaysia, one catchment was logged according to normal practices, a second logged under the constraints of protective buffer strips and careful log extraction, while a third was kept undisturbed. Both logged catchments showed an increase in water yield after logging, but the catchment logged without control had much the greater increase. However, the higher yields were not matched by increased storm runoff. Much more sediment was discharged from the uncontrolled logging area than from the other two catchments (Nik and Harding, 1990).

Although good data are available on changes in volumes of water leaving logged catchment areas, less is known about the changes in the outputs of dissolved and solid mineral matter. A major effort to tackle that deficiency was made in the Bukit Berembun study (Figure 9.2). In the uncontrolled logged catchment annual sus-pended yields changed from being 1.5–3 times those of the undisturbed control catchment to being 4–20 times greater. However, in the catchment logged with conservation practices, the increase in sediment yield was only a doubling (Zulkifli Yusop et al., 1990). Five years after logging, the sediment yield from the conserva-tion-logged catchment was lower than that of the undisturbed catchment, but from the uncontrolled logged catchment it remained almost four times greater. The implication is that careful logging reduces the hydrologic and erosional impact of forestry activity, but that, in the normal type of logging operation, not only is the change in runoff and sediment yield large, but it persists for at least 5 years after logging has ceased.

Table 9.2 Comparison of sediment yields of forested and selectively logged catchments in Malaysia

Catchment name	Catchment area (km^{-2})	Suspended sediment load (t km^{-2} yr^{-1})	Bedload (t km^{-2} yr^{-1})	Study period	Land use in study period	Source
Sg. Batangsi	19.8	2826	1264	Mar. 87–Apr. 89	Logging continuous	Lai et al., 1995
Sg. Chongkak (a)	12.7	2476	619	Mar. 87–Jun. 88	Logging to Apr. 87	Lai et al., 1995
Sg. Chongkak (b)	12.7	1335	334	Jul. 88–Oct. 89	Post-logging recovery	Lai et al., 1995
Sg. Lui	68.1	90	22	Mar. 87–Jun. 89	Rural, 80% forest	Lai et al., 1995
Sg. Lawing (a)	4.7	54	125	Aug. 88–Jul. 89	Undisturbed	Lai et al., 1995
Sg. Lawing (b)	4.7	1129	414	Jan. 93–Dec. 93	Logging from Jan 93	Lai et al., 1995
Berembun C1	0.129	27	–	–	Unsupervised logging	Baharuddin, 1988
Berembun C2	0.042	19	–	–	Forested	Baharuddin, 1988
Berembun C2	0.297	12	–	–	Supervised logging	Baharuddin, 1988
Stn. Baru	0.49	1632	–	–	Unsupervised logging	Douglas et al., 1993
Sg. W8S5	1.7	118	–	–	Forested	Douglas et al., 1993

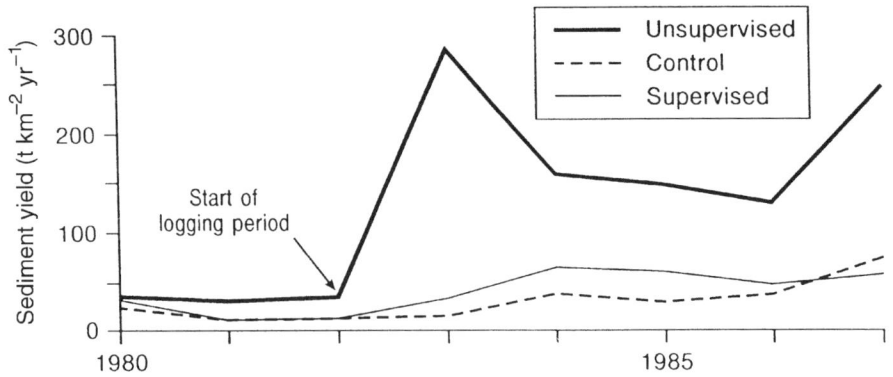

Figure 9.2 Sediment yield as a function of different logging practices in the Bukit Berembun experimental catchments, Peninsular Malaya

In Sabah, detailed work in the Sipitang area has thrown light on many of the hydrologic impacts of logging, afforestation and fire (Malmer, 1993). Infiltration rates of $154 \, mm \, h^{-1}$ in the forest fall to $0.28 \, mm \, h^{-1}$ on new roads and $0.63 \, mm \, h^{-1}$ on old roads (Malmer and Grip, 1990). Cutting and burning led to an increase of water output of 1008 mm over the first 32.5 months during treatment and plantation establishment, while clear-felling, tractor logging and burning led to an increase of 1190 mm (Malmer, 1992). Suspended sediment yields from logged, manually logged and tractor-extracted catchments of 3 to 18 ha were 90, 210 and $390 \, t \, km^{-2}$ respectively over the first 18.5 months during and after extraction (Malmer, 1990). However, this small catchment work has not been able to address the problem of the relationship between headwater processes and events 0.5 to 1 km downstream.

The catchment work at Danum Valley in Sabah (Douglas et al., 1992a, 1992b) has begun to address many of these issues, and is almost unique in the humid tropics in combining detailed work on hillslope hydrology and zero-order basin dynamics with processes in larger catchments. It also has some of the best records of sediment loads and hydrochemistry for anywhere in the tropics. Here logging has an immediate impact on sediment yields, with quite moderate storms, such as those of December 1988, carrying over $10 \, t \, km^{-2} \, day^{-1}$ from the affected areas of the selectively logged Steyshen Baru catchment. In relatively small rain events, sediment transport in newly logged areas may be discharge-limited, with much channel aggradation (Figure 9.3). Soon after initial road building in 1988, the gravel and boulder bed of the Baru stream became covered with a layer of fine silt which persisted and grew during the logging process. Some patches of thick clay remained on the channel bed at least until late 1994, but by that time most of the fine silt deposits had been removed and the channel had returned to its original gravel and boulder state.

The Baru catchment, logged in 1989, had a total sediment yield of $1632 \, t \, km^{-2}$ for that year, $1017 \, t \, km^{-2}$ the following year and only $96 \, t \, km^{-2}$ in the first 7 months of 1991, while W8S5, the undisturbed natural rain forest control catchment, had comparative yields of 118, 117 and $36 \, t \, km^{-2}$ respectively Longer after logging, sediment yields again became supply-limited, requiring higher discharges to move the quantities that had been carried by moderate flows immediately after disturbance.

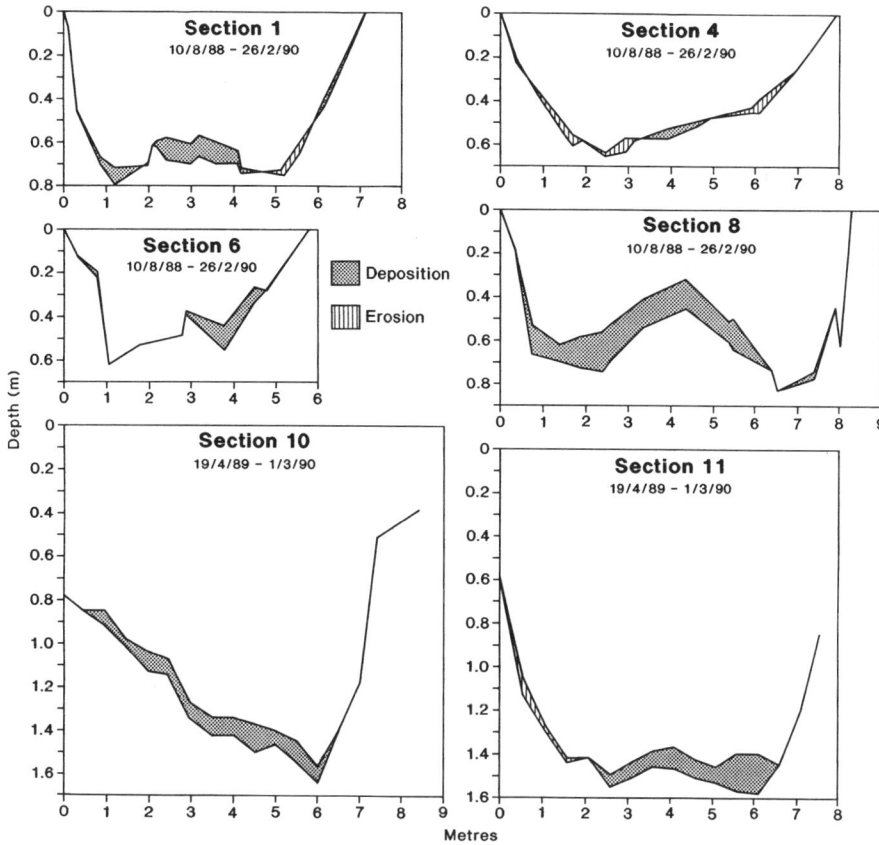

Figure 9.3   Patterns of aggradation and scour during and after logging at various cross-sections of the Sg. Steyshen Baru catchment, Ulu Segama, Sabah, Malaysia

THE IMPACT OF ROADS AND LOG HAULING TRACKS

The impact of roads on the post-logging sediment budget is clearly evident (Figure 9.4). Where roads are cut into hillsides, conditions are not always stable and landslides may occur, resulting in a further road cut back into the slope, so enlarging the total disturbed area. Also, cutting slopes at excessively steep angles can lead to debris slumping and changes in stability of the slope, resulting in debris blocking roadside drains and increased downslope sediment supply.

In a northern Palawan case study, estimates of erosion indicated that, although roads accounted for only 3% of the total area, they were responsible for over 80% of the total erosion (Table 9.3) (Dixon, 1990). The extent to which road construction caused greatly increased sediment production in Huay Ma Feung stream in Thailand was investigated in the rainy season of 1983. Storms produced simultaneous peaked suspended sediment concentrations and discharge hydrographs with maximum values at 28,000 mg l^{-1} and 1.46 m^3 s^{-1} respectively. Bedload accounted for more than 80%

Figure 9.4   Photograph showing a steep, badly eroded logging road descending to a stream crossing (where the people are) in a steep part of the Ulu Segama Forest Reserve near Danum Valley, Sabah (Photo I. Douglas)

of the total load of $10,494 \, t \, km^{-2} \, yr^{-1}$. Sedimentological evidence and prevailing soil hydraulic properties indicate that the bulk of such sediment originates from roadside gullies, slumps and landslides (Henderson and Witthawatchutikil, 1985).

Plot studies on logging tracks at Danum Valley show that, on compacted abandoned logging track surfaces, infiltration is reduced so much that 52% of the rainfall becomes overland flow, compared to only 5% under natural forest.

Table 9.3 Erosion in the Bacuit Bay logging concession area, Palawan, Philippines (after Dixon, 1990)

Land cover	Area (km^2)	Erosion (t km^{-2} yr^{-1})	Total (t × 10^9)	% of total
Roads	1.23	14 130	17.4	84
Cut forest	4.80	260	1.2	6
Uncut forest	37.0	60	2.1	10
Total	43.03		20.7	100

Sediment yield from this unprotected track was 19,050 t km^{-2} yr^{-1} some 9 months after abandonment but with the growth of vegetation this had dropped to 1,050 t km^{-2} yr^{-1} a year later. Barriers across tracks reduced sediment yield considerably and demonstrated that simple earth barriers successfullly held back water and sediment. Subsequent adoption of the practice by the logging contractor helped to reduce erosion.

NUTRIENT LOSSES DUE TO SELECTIVE LOGGING

The key issues of the selective logging nutrient budget are the loss of nutrients in harvested wood and any additional export of nutrients in fluvial solute, sediment and organic loads. As the nutrient content of tropical soils is extremely variable, depending on parent material, it is unwise to make overall generalisations.

Relatively little attention has been paid to the actual removal of nutrients from logged tropical rainforest catchments in the harvested wood. At Mendolong, Sipitang, Sabah, amounts of some elements, especially calcium, are low in the soil (103 kg ha^{-1} exchangeable Ca; 727 kg ha^{-1} total Ca, in the top 50 cm of the soil). Loss of calcium through log removal after clear-felling represented about 19% of the total amount in the ecosystem, suggesting that plantation forestry (clear-felling and replanting) in this area may not be possible without replacing the calcium removed. The sandstones and mudstones in this part of Sabah are particularly deficient in nutrients. In other areas, the removal of nutrients in logs by clear-felling may not be a problem.

Selective logging removes fewer saw logs than clear-felling and a large part of the biomass of felled trees remains in the catchment area. Nutrient removal may not be as serious a problem as accelerated loss of nutrients to streams by erosional processes. Conventionally, nutrient ouput assessments are restricted to a study of the chemical elements in solution in rivers. However, there are other nutrient losses associated with the removal of chemical elements in the solid mineral load of suspended sediment and bed material and in the floating organic load, from leaves and twigs to logs and other coarse woody debris.

Conventional studies of solute load at Danum Valley, Sabah, showed that as runoff increased immediately after logging activity began, exports of solutes increased. However, the magnitude of the increase was relatively small, of the order of 20–80% more than the solute load under natural forest. When water discharge reverted to the levels in the undisturbed forest, some 12 months after selective logging ceased, solute concentrations returned (similar to those in the control catchment). Loss of nutrients in solution is thus important but not great, and is short-lived.

A visible increase in organic load follows the disturbance caused by logging. Part of this increase is immediate, with the supply of plant material ejected directly into streams and also entrained in overland flow which carries it downslope to rivers. Another part is delayed, as broken plant material decays and either falls or is carried into stream channels. More debris dams develop in streams affected by logging activity (for example six every 100 m in the selectively logged Ulu Jauh catchment compared to around three every 100 m in the W8S5 control catchment at Danum Valley, Sabah). Collapse of these debris dams releases a pulse of downstream nutrient transport in floating organic matter and coarse woody debris.

Nevertheless, the 10–20-fold increase in sediment yield following logging is probably the main source of nutrient export by fluvial systems. At Danum Valley, sediment yields rose by this magnitude for 2 years after logging. Much of the disturbed sediment on slopes and stored in channels had been evacuated by the end of the 2 years and small to moderate rain events (storms of 10 to 50 mm) then failed to remove any more sediment from the logged catchment compared with the undisturbed control catchment. However, major storms (greater than 50 mm) still picked up sediment from channel stores, gullied tracks and minor mass movements, even 5 years after logging had ceased. The export of sediment and associated nutrients thus persists long after the end of timber harvesting.

Unfortunately, detailed analyses of the chemical composition of suspended sediments are not readily available and at this stage it is impossible to quantify the sediment-associated nutrient outputs from selectively logged catchments. Work was initiated in 1994 using a pumping centrifuge system to abstract sufficient volumes of suspended sediment for chemical analysis from experimental rainforest catchments in Malaysia and it is hoped that this gap in our knowledge can be filled in the near future.

POST-LOGGING MANAGEMENT FOR FOREST SUSTAINABILITY

While conditions vary greatly in different parts of the humid tropics, depending on terrain and logging procedures, a general survey suggests that in logged-over forest 41% of residual trees are damaged; fewer tree species are present than in primary forest; 30% of the ground is bare; the soil is compacted, with slow infiltration rates, and is subject to erosion. Burnt and unburnt plots of clear-cut primary forest of the type of size cleared by shifting cultivators are completely covered by pioneer tree species within 6 months; both seedlings and resprouts play an important role in recovery. Growth is better in the burnt plot. A similar situation was observed on a 6-month-old abandoned dryland farm. In a 2-year-old dryland farm, the undergrowth was dominated by herbaceous plants, whose species composition was affected by kinds of crops cultivated and by topography (Kartawinata et al., 1981).

The importance of post-logging management of forest is well recognised by the main agencies involved. If forests are to be harvested on areas prone to shallow slides, cutting alone, even without the disturbance associated with wood removal, can alter slope stability. Where stumps coppice, so that root systems remain alive, there is no problem. Thus in short-rotation fuelwood plantations of fast-growing coppicing species, the root systems would continue to provide strength to the soil. Where this does not occur, loss of soil shear strength depends on the rate of root decay following cutting, compared to the growth of new roots from uncut trees or forest reproduction. Thus, in many areas, post-cutting landslide hazards do not peak until 2 to 10 years after logging (Hamilton, 1986). Where roads exist, the landslide situation becomes more complex, as many unstable roadcuts are made, particularly in shale and mudstone areas of Borneo, where the inherent friability and dispersibilty of the rock leads to mudflows and debris flows. Elsewhere, rotational slumps in deeply weathered regoliths often occur as roadcuts fail.

Sometimes roads impede the natural slope drainage system by intercepting both surface and subsurface water movement. On abandoned tracks, where no maintenance occurs, this local ponding of water can lead to eventual saturation and even gleying of the substrate, so altering the pore-water pressures that failure occurs and a landslide results. Extreme rainfall events produce greater destabilisation of logged-over catchments than of natural catchments. An example of the long-lasting impacts of logging road construction arose during a 1 in 10 year magnitude storm in January 1996 in the Danum Valley area. Six landslides occurred along the abandoned logging access road to the head of the Steyshen Baru catchment. All were due to the way the road impeded downslope water movement from the 178 mm of rain in 11 hours, creating high pore-water pressures in the weathered mantle and fill beneath the road. In the nearby undisturbed forest catchment, only one small landslide occurred, affecting only $12 \, m^2$, compared to the six 30 to $200 \, m^2$ slides along the abandoned road.

MANAGEMENT OF LOGGING SYSTEMS

Controlling erosion and sediment input to stream channels should begin with planning to construct no more roads than are absolutely necessary. Building few roads saves both money and the impacts of excessive disturbance. Planning to construct fewer roads is particularly difficult in tropical steeplands because of questions of access and log removal. If there are fewer roads it is more costly to haul logs to collection points (log landings).

Suitable positioning of roads is extremely important as the siting of a road can determine whether or not it sheds sediments to streams. For example, some ridgetop roads in Peninsular Malaysia are highly eroded, but little of the sediment actually reaches the streams (O'Loughlin, 1985; Adams and Andrus, 1990), while in the upper Rajang catchment of Sarawak, roads running along valley sides have runoff feeding directly into small streams which supply sediment directly to rivers. Snig tracks running up or downslope from these roads are gullied soon after abandonment. In the Sarawak case the roads have effectively become an extension of the ephemeral storm-runoff channel network. Many abandoned snig tracks have become active gullies, and in some cases almost debris flow systems. In steep terrain, roads are often pushed along valley floors, but it is important to maintain a buffer-strip of vegetation between the road and the stream and to design drains so that they do not carry sediment directly from the road surface into the stream. Roads should be located to avoid unstable and erodible slopes, although precise identification of these areas is often challenging even for expert observers (Adams and Andrus, 1990). Reconnaissance techniques to delimit such "critical" areas (Megahan and King, 1985) should be used to reduce the impact of roads on sediment yields.

The design of roads involves a balance between opening a wide road clearing to allow the roads to dry out quickly after rain and on the other hand minimising the area of disturbance. The reduced-impact logging scheme trialled at Danum Valley, Sabah, has narrow road clearings and overcomes the problems of trafficability in rainy periods by not permitting logging at the wettest times of year.

Drainage is essential to good road maintenance but, all too often, there are too few cross-drains and large roadside gullies develop as long lines of flow occur down sloping road sections. The first step in managing for drainage is to ensure that water runs off the road surface itself by having a crowned road surface. Then if the water runs to both sides of the road, good roadside drains are needed, with cross-drains on steep slopes. Design of these cross-drains should be such that water is not diverted into areas which might be subject to mass movement if wetted. Many schemes for cross-drains exist, some of the most effective being those which involve split logs running across the road leaving a small gap between them. This gap allows any runoff coming down the road to enter the drain and join the water being transferred from the upslope to the downslope side of the road. Design of stream crossings is particularly important. Culverts carrying roads must be large enough and durable, but well-designed low-water fords and bridges can help avoid washouts and erosion problems. In many cases, culverts are not designed to allow for siltation. Where severe roadside erosion occurs, sediment chokes the culvert, forcing excess runoff to erode the road or the base of the adjacent slope.

Traffic volumes and weights have serious effects on the erosion of road surfaces, and changing traffic regimes, vehicle weights and tyre pressures can help to minimise erosion. Many types of log-extraction equipment have wider tyres and less weight per axle or wheel to reduce compaction and so lessen the erosion risk. Few systematic studies of their success in achieving these objectives have been made, however. Overall, much knowledge on good practice in road design and maintenance exists, but its adoption when logging tropical forest catchments is limited. Traditional, conventional vehicles are used, including the ubiquitous caterpillar tractor, and, in steeplands, the old military style breakdown vehicle known in Malaysia as the Sam Tai Wong. Road design might pay some attention to slope, but drainage is often inadequate.

POST-LOGGING TREATMENT OF ROADS

A major issue is the effective closure of logging tracks and snig tracks after logging has ceased. The worst erosion can often occur once logging is over and all too often in South-east Asia no remedial action is taken. Roads must be carefully stabilised before closure, with measures such as water bars and actions to promote revegetation on exposed surfaces (Adams and Andrus, 1990). Simply bulldozing earth barriers across the tracks and breaking up the compacted surface to allow infiltration and encourage vegetation regeneration can reduce runoff and erosion considerably.

Water bars were adopted as a standard practice by contractors logging the Yayasan Sabah (Sabah Foundation) concession in eastern Sabah after the plot experiments described above had demonstrated their effectiveness in reducing the length of continuous concentrated flow and associated erosion on abandoned snig tracks. As the caterpillar tractors left the area, mounds of earth about 1 m high were pushed up at approximately 20 m intervals along the snig tracks. The first rains after their creation led to the beginnings of accumulation of fine silt on the upslope side of the barriers. Small pools of water were left behind and were gradually depleted by evaporation as the fine clay deposits sitting on the compacted surface of the abandoned tracks prevented much water from infiltrating. Both runoff and sediment transport to the rivers were thus impeded. No gullies of significant length developed. On the other hand, a steep access road, descending down to the Ulu Jauh stream at Danum Valley, had no such treatment, and, despite having been graded and gravelled, became severely gullied and eroded within 18 months of logging ceasing. Water bars would have prevented this happening so quickly.

As the water bars in eastern Sabah were only mounds of loose earth, they were subject to rainsplash erosion and subsequent rilling. Some were breached after a few months, others lasted considerably longer. However, while they were intact, they prevented gullying of the tracks and allowed vegetation regeneration to proceed. Plant regrowth was most rapid towards the lower ends of tracks where they were closest to streams, the logs having been hauled upslope to landings close to the ridge top. As the lower ends had had less traffic, they were less compacted and seeds from the surrounding forest therefore germinated more quickly. After 18 months, some of the regeneration was thick enough and close enough to the ground to intercept raindrops and reduce their erosive impact on the ground. Simply by allowing this rapid regeneration, the water bars have contributed to a reduction in sediment supply. However, the effectiveness of the water bars could have been increased by throwing vegetation debris across them to protect them against rainsplash erosion. The small extra labour cost involved might well be justified in terms of the reduction in off-site, downstream sedimentation impacts.

While water bars are suitable for snig tracks, the main log haulage roads probably require, but seldom receive, more extensive treatment. Breaking up their compacted surfaces and attention to drainage systems would prevent long roadside gullies from developing. Where weathering profiles are deep, gullies on steepland logging roads can incise to great depths. One at the side of a road in the Sungei Batu Catchment in the Main Range of Peninsular Malaysia cut down to a depth of 5 m in deeply weathered granite and diverted water across a divide between sub-catchments at a col crossed by the road. Such extreme gullying need not occur if cross-drains are provided. On abandoned roads, they could be bulldozed easily and would probably have the added advantage of deterring motorised access by illegal hunters and would-be land settlers. Breaking up the surface of such roads would also give an opportunity for plant regeneration and so prevent the roads from remaining the bare scars in the landscape now so readily visible in many tropical rainforest regions. New Malaysian guidelines on soil erosion (Department of Environment, 1996) include comments on earth bunds (water bars), revegetation of abandoned roads and the importance of cross-drains.

REPLANTING AND PROTECTION FORESTRY

Sustainable yield under selective logging can be enhanced by deliberate planting. Many schemes have been attempted, but probably the most successful in the long term will be those which use native forest trees, as in the enrichment planting scheme operated in the Yayasan Sabah concession in Sabah. This project involves the raising of dipterocarp seedlings in large nurseries and, when they are of sufficient size, planting them along transects through the logged-over forest. As the seedlings are raised on site from local stock, they have a better chance of survival and resistance to possible predators than plant material introduced from outside the concession. The natural diversity of the forest is preserved, the regeneration of the most merchantable species is enhanced and the post-logging activity in the concession helps to keep access roads maintained and illegal access under control.

Protection forestry is designed to restore the ground cover in highly disturbed areas such as log landings and former camp sites. Fast-growing trees, often exotics, such as *Acacia mangium*, are planted and, if close enough to each other, within 24 to 36 months provide a complete canopy cover over the former bare area. Gradually the fallen leaves from these trees create a litter layer which protects the ground from rainsplash erosion. Problems occur during the early months of such plantings. Overland flow on the compacted soil surface continues to entrain sediment and often carries particles to the rills and gullies which developed during and immediately after logging activity. The gullies enlarge and the surface erosion continues. Heavy rains lead to further erosion of gully walls and floors, despite, in some cases, an overarching vegetation canopy, so continuing to act as a source of the sediment carried by streams in major storms for many years after selective logging has ceased.

Protection forestry therefore needs to be combined with simple earthworks in an integrated programme of soil erosion control. The bare areas should be broken up by simple water bars or contour banks. Available plant residues such as bark and branches should be spread across the area to act as a mulch, a diverter of overland flow and a breaker of raindrop impact. Plantings should be close enough together to provide canopy cover in 2 to 3 years. Any existing gullies should be blocked to prevent long lines of continuous flow. Such simple measures will stabilise the affected areas rapidly and help the regeneration of native species.

CONCLUSIONS: ACTION DURING SELECTIVE LOGGING OR REMEDIAL MEASURES AFTER LOGGING?

Despite the lessons learnt in the temperate rainforests of the Pacific North-West of the USA and British Columbia, and in the tropical rainforests of North Queensland, techniques to minimise the impact of logging have seldom been applied in South-east Asia, which has been the world's major producer of sawn logs for export from tropical forests. The legitimate and natural urge to exploit the timber resources and use the income they generate to raise national living standards has driven selective logging at a faster rate than concern about sustainability would justify. The scars of roadside landslips, gullied log landings, widened and silted stream channels and severely gullied former snig tracks cover large areas of Borneo, Sumatra and the smaller islands of South-east Asia. Now that Indonesian and Malaysian timber companies are investing in logging concessions in the Pacific Islands and South America, firm administration and widespread adoption of good sustainable selective logging practices is urgently needed.

As in many other environmental issues in tropical countries, the problem of environmental management is not one of lack of knowledge or availability of suitable technology. The problem is willingness to adopt good practice and to supervise and police contractors to ensure that it is applied and maintained properly. Experience in Sabah has shown that, when policies and needs are properly explained, chain-saw operators and caterpillar tractor drivers are fully co-operative. As in so many other spheres, it is a question of leadership and good management.

While the range of actions involved in reduced-impact logging and other good timber extraction techniques may be applied, they raise the cost of timber extraction, at least in the short term. For many secure concession holders who will harvest timber again in the future, these extra costs may be supportable and thus the greater value of the future harvest may offset the present extra costs. However, many concessions lack such a long-term perspective and immediate, single-opportunity profit drives the concession holder. Under such circumstances, reduced-impact logging techniques are highly unattractive and post-logging remedial measures may be a more attractive alternative.

Great emphasis has been placed on logging roads and snig tracks in this chapter. Roads are so much the focal point of erosion and sedimentation problems that they offer the greatest opportunity for making selective logging more sustainable. Even a small increase in the care with which access roads are designed, especially by improving drainage and avoiding excessively steep grades, would greatly reduce erosion. Government Forestry Departments could set tougher standards for road design and concession holders could ensure that contractors meet those standards. Post-logging remedial action on snig tracks and on access roads which are being closed and abandoned should be mandatory. Such measures would involve little extra cost, but would avoid both on-site degradation and off-site, downstream consequences.

In the humid tropics, the more we keep areas like the natural forest, the more sustainable our environment remains. We know what to do, but changing people's behaviour will remain a challenge, at least until long-term sustainability becomes as desirable as quick profits.

ACKNOWLEDGEMENTS

The work in Sabah used here has been the product of a team effort and a great debt is owed particularly to Kawi Bidin, Nick Chappell, Tony Greer, Waidi Sinun,

Mike Spilsbury, Jadda Suhaimi, Azman Sulaiman and Rory Walsh. The work has been supported by NERC, ODA and the Royal Society, and has had the benevolent advice and approval of the Danum Valley Management Committee, to whose former Secretary, Clive Marsh, a particular debt is owed. This paper is Number A/287 of the Royal Society's Southeast Asian Rainforest Research Programme.

REFERENCES

Adams, P.W. and Andrus, C.W. 1990. 'Planning secondary roads to reduce erosion and sedimentation in humid tropic steeplands'. *International Association of Hydrological Sciences Publication,* **192**, 318–327.

Anderson, J.M. and Spencer, T. 1991. 'Carbon, nutrient and water balances of tropical rain forest ecosystems subject to disturbance: management implications and research proposals'. *MAB Digest,* **7**, UNESCO, Paris.

Baharuddin, K. 1988. Effect of logging on sediment yield in a hill dipterocarp forest in Peninsular Malaysia. *Journal of Tropical Forest Science,* **1**, 56–66.

Bruijnzeel, L.A. 1993. 'Land-use and hydrology in warm humid regions: where do we stand?'. *International Association of Hydrological Sciences Publication,* **216**, 3–34.

Burgess, P.F., 1973. 'The impact of commercial forestry on the hill forests of the Malay Peninsula'. *Proceedings of the Symposium on Biological Resources and National Development,* 237–258.

Cassels, D.S., Gilmour, D.A. and Gordon, P. 1982. 'The impact of plantation forestry on stream sedimentation in tropical and subtropical Queensland – an initial assessment'. *Agricultural Engineering Conference 1982. Resources – Efficient Use and Conservation,* Institution of Engineers Australia, 138–142.

Department of Environment, Malaysia, 1996. *'Guidelines for Prevention and Control of Soil Erosion and Siltation in Malaysia'.* Jabatan Alam Sekitar, Department of Environment, Ministry of Science, Technology and Environment, Kuala Lumpur.

Dixon, J. A. 1990. 'Monetising erosion and sedimentation costs where steeplands meet the sea'. *International Association of Hydrological Sciences Publication,* **192**, 195–207.

Douglas, I., Greer, T., Wong, W.M., Spencer, T. and Sinun, W. 1990. 'The impact of commercial logging on a small rainforest catchment in Ulu Segama, Sabah, Malaysia'. *International Association of Hydrological Sciences Publication,* **192**, 165–173.

Douglas, I., Spencer, T., Greer, T., Bidin, K., Sinun, W. and Wong, W.M. 1992a. 'The impact of selective commercial logging on stream hydrology, chemistry and sediment loads in the Ulu Segama Rain Forest, Sabah'. *Philosophical Transactions of the Royal Society, London B,* **335**, 397–406.

Douglas, I., Greer, T., Wong, W.M., Bidin, K., Sinun, W. and Spencer, T. 1992b. 'Controls of sediment discharge in undisturbed and logged tropical rain forest streams'. *Proceedings of the 5th International Symposium on River Sedimentation,* Karlsruhe, April 1992, 1019–1024.

Douglas, I., Greer, T., Bidin, K. and Sinun, W. 1993. 'Impact of roads and compacted ground on postlogging sediment yield in a small drainage basin, Sabah, Malaysia'. *International Association of Hydrological Sciences Publication,* **216**, 213–218.

Douglas, I., Greer, T., Sinun, W., Anderton, S., Bidin, K., Spilsbury, M., Jadda Suhaimi and Azman bin Sulaiman, 1995. 'Geomorphology and rainforest logging practices'. In *Geomorphology and Land Management in a Changing Environment* (eds D.F.M. McGregor and D.A. Thompson), pp. 309–320. Wiley, Chichester.

Ewing, A.J. and Chalk, R. 1988. 'The forest industries sector: an operational strategy for developing countries'. *World Bank Technical Paper Industry and Energy Series,* **83**.

Fox, J.E.D. 1970. 'The natural vegetation of Sabah and natural regeneration of the Dipterocarp forests'. PhD Thesis, University of Wales.

Fritsch, J.M. 1983. 'Évolution des écoulements, des transports solides a l'exutoire et de l'erosion sure les versants d'un petit bassin après défrichement mécanisée la forêt tropicale humide'. *International Association of Hydrological Sciences Publication,* **140**, 197–214.

Fritsch, J.M. 1993. 'The hydrological effects of clearing tropical rainforest and of the implementation of alternative land uses'. *International Association of Hydrological Sciences Publication,* **216**, 53–66.

Gilmour, D.A., Cassells, D.S. and Bonell, M. 1982. 'Hydrological research in the tropical rainforests of north Queensland: some implications for land use management'. *Proceedings of the First National Symposium on Forest Hydrology,* Melbourne, 145–152.

Greer, T., Douglas, I., Bidin, K., Sinun, W. and Jadda Suhaimi, 1995. 'Monitoring geomorphological disturbance and recovery in commercially logged tropical forest, Sabah, East Malaysia, and implications for management'. *Singapore Journal of Tropical Geography,* **16**, 1–21.

Hamilton, L.S. 1986. 'Towards clarifying the appropriate mandate in forestry for watershed rehabilitation and management'. In *Strategies, Approaches and Systems for Integrated Watershed Management.* FAO, New York, 136–153.

Helvey, J.D. and Kochendorfer, J.N. 1990. *'Soil Density and Moisture Content on Two Unused Forest Roads during the First 30 Months after Construction'.* Research Paper NE 629. US Department of Agriculture, Forest Service, Northeastern Forest Experiment Station, Radnor, PA.

Henderson, G. S. and Witthawatehutikul, P. 1985. 'The effect of road construction on sedimentation in a forested catchment at Rayong, Thailand'. In *Symposium on Effects of Forest Land Use on Erosion and Slope Stability* (eds C.L. O'Loughlin, and A.J. Pearee), pp. 247–253. East–West Center, Environment & Policy Institute, Honolulu, Hawaii.

Johns, A.D. 1992. 'Vertebrate responses to selective logging: implications for the design of logging systems'. *Philosophical Transactions of the Royal Society of London B,* **335**, 437–442.

Kartawinata, K., Adisoemarto, S., Riswan, S. and Vayda, A.P. 1981. 'The impact of man on a tropical forest in Indonesia'. *Ambio,* **10**, 115–119.

Lai, F.S. 1993. 'Sediment yield from logged, steep upland catchments in Peninsular Malaysia'. *International Association of Hydrological Sciences Publication,* **216**, 219–229.

Lai, F.S., Lee, M.J. and Mohd. Rizal, S. 1995. 'Changes in sediment discharge resulting from commercial logging in the Sungai Lawing basin, Selangor, Malaysia'. *International Association of Hydrological Sciences Publication,* **226**, 55–62.

Lambert, F.R. 1992. 'The consequence of selective logging for Bornean rainforest birds'. *Philosophical Transactions of the Royal Society of London B,* **335**, 443–457.

Luce, E. 1995. 'Policing Asia's last frontier'. *Financial Times Weekend FT,* 18/19 November 1995, ix.

Malmer, A. 1990. Stream suspended sediment load after clear-felling and different forestry treatments in tropical rainforest, Sabah, Malaysia. *International Association of Hydrological Sciences Publication,* **192**, 62–71.

Malmer, A. 1992. 'Water-yield changes after clear-felling tropical rainforest and establishment of forest plantation in Sabah, Malaysia'. *Journal of Hydrology,* **134**, 77–94.

Malmer, A. 1993. 'Dynamics of hydrology and nutrient losses as response to the establishment of forest plantation. A case study on tropical rainforest land Sabah, Malaysia'. PhD Thesis, Swedish University of Agricultural Sciences, Umea.

Malmer, A. and Grip, H. 1990. 'Soil disturbance and loss of infiltrability caused by mechanized and manual extraction of tropical rainforest in Sabah, Malaysia'. *Forest Ecology and Management,* **38**, 1–12.

Marsh, C. W. and Greer, A. G. 1992. 'Forest land use in Sabah Malaysia: an introduction to the Danum Valley'. *Philosophical Transactions of the Royal Society of London B,* **335**, 331–339.

Megahan, W.F. and King, P.N. 1985. 'Identification of critical areas on forest lands for control of nonpoint sources of pollution'. *Environmental Management,* **9**, 7–18.

Nicholson, D.I. 1979. *'The Effects of Logging and Treatment on the Mixed Dipterocarp Forest of South East Asia'.* FAO:MISC/79, Rome.

Nik, Abdul Rahim and Harding, D. 1990. 'Management of forested watersheds: a case study in Peninsular Malaysia'. *Paper presented at the UNESCO/MRP Workshop on Watershed Development and Management*, 19–23 February 1990, Kuala Lumpur, Malaysia.

O'Loughlin, C. 1985. '*The Influence of Forest Roads on Erosion and Stream Sedimentation – Comparisons Between Temperate and Tropical Forests*'. East–West Center, Environment & Policy Institute, Honolulu, Hawaii.

Roche, M.A. 1981. 'Watershed investigations for development of forest resources of the Amazon region in French Guyana'. In *Tropical Agricultural Hydrology* (eds R. Lal and R.W. Russell), pp. 75–82. Wiley, Chichester.

Ruangpanit, N. 1985. 'Percent crown cover related to water and soil losses in mountainous forest in Thailand'. In *Soil Erosion and Conservation* (eds S.A. El-Swaify, W.C. Moldenhauer and A. Lo), pp. 462–471. Soil Conservation Society of America, Ankeny, Iowa.

Sundberg, U. 1983. 'Logging in broadleaved tropical forests'. *FAO: RAS/78/010 Working Paper No. 27*, FAO, Rome.

Unstead, J.F. 1950. '*A Systematic Regional Geography. Volume III, A World Survey from the Human Aspect*'. University of London Press, London.

Van der Plas, M.C. and Bruijnzeel, L.A. 1993. 'Impact of mechanised selective logging of forest on topsoil infiltrability in the upper Segama area, Sabah, Malaysia'. *International Association of Hydrological Sciences Publication, 216*, 203–211.

Whitmore, T.C. 1984. '*Tropical Forests of the Far East*' (2nd edition). Oxford University Press, Oxford.

Zulkilfi Yusop, Anhar Suki and Baharuddin Kassan, 1990. 'Postlogging effects on suspended solids and turbidity – five years' observation'. *UNESCO/MRP Workshop on Watershed Development and Management*, 19–23 February, 1990, Kuala Lumpur, Malaysia.

CHAPTER 10

Environmental Limits to Sustainable Coffee Cultivation on Tropical Soils: The Sungai Pemadang Catchment, Brunei

DAVID J. MITCHELL
School of Applied Sciences, University of Wolverhampton, UK

INTRODUCTION

Tropical evergreen rainforest covers approximately 70% of Brunei Darussalam, therefore the agricultural area is relatively small. Agricultural production accounts for only about 1% of the gross domestic product of Brunei and 80% of food is imported. To increase food self-sufficiency, the Brunei Government's National Plan has given priority to the development of agriculture. Besides traditional crops of rice, sago, rubber, pepper and citrus fruits, the Department of Agriculture is experimenting with new crops, such as fruits and other tree crops, at four research stations.

Since the fourteenth century coffee has been an important beverage throughout the world. The most common variety is *Coffea arabica*, which is cultivated in elevated tropical and sub-tropical zones. Owing to susceptibility to disease, the cultivation of *arabica* coffee in humid equatorial lowlands is unsuitable (Williams, 1975). In the humid tropics the more resistant species, *C. robusta* and *C. liberica*, are cultivated. In the Malayan Peninsula, *C. liberica* is grown on acid peat soils (Williams, 1975). A 1968 survey of land resources and land capability for the Government of Brunei considered the introduction of new crops in undeveloped areas (Hunting Technical Services, 1969). The survey identified the economic value of *Coffea liberica* and stated: "The climate and soils would appear to be suitable for this crop, with its requirements for a fairly high rainfall and acid soils of pH 4.2–5.1" (Hunting Technical Services Vol. II, 1969, p. 139). Research into the environmental effects of the growth of coffee has been mainly restricted to studies on *C. arabica* coffee as reviewed by Maestri and Barros (1977). The interest in *C. liberica* and *C. robusta* has been mainly restricted to Asia and West and Central Africa, with production aimed at a more localised market.

The Sustainable Management of Tropical Catchments. Edited by David Harper and Tony Brown.
© 1998 John Wiley & Sons Ltd.

Table 10.1   Average monthly rainfall and rain days at Sungai Liang Meteorological Station, 1973–91

Month	Average rainfall (mm)	Average no. rain days
January	311	12.3
February	181	8.2
March	122	7.8
April	231	10.4
May	208	11.6
June	195	11.4
July	203	11.6
August	172	10.9
September	230	13.7
October	298	14.7
November	367	17.1
December	308	14.9
Annual	2827	144.5

This chapter examines the sustainability of coffee plantations in the humid tropics of Brunei, with special reference to changes in soil nutrients.

SUSTAINABLE AGRICULTURE IN HUMID TROPICAL FORESTS

Humid tropical forests are characterised by year-round growth, high species diversity, tightly coupled cycling of nutrients between plant and soil, rapid weathering and high rates of soil biological activity (Cooper, 1981). The intricate processes operating in tropical rain forests have been analysed and discussed by Sutton et al. (1983). Research in South-east Asia has examined these processes, for example, litter production and decomposition (Ogawa 1978; Anderson et al. 1983; Gong and Ong, 1983), pedological processes and nutrient supply (Baillie and Ashton, 1983; Baillie, 1989; Burnham, 1989), litter removal by termites (Collins, 1983), nutrient dynamics in forest fallows (Nakano and Syahbuddin, 1989), and forest ecosystems (Whitmore, 1978). Reviewing definitions on sustainability, Greenland (1994) stated "A sustainable land management system is one that does not degrade the soil or significantly contaminate the environment, while providing necessary support to human life". Clearance and cultivation can have a dramatic influence on the soils and microclimate of an area, and therefore management practices need to harmonise with environmental conditions, if sustainable yields are to be achieved.

SUSTAINABLE MANAGEMENT OF COFFEE IN BRUNEI

In 1979, as a consequence of the decline of profitability of natural rubber, a Brunei farmer, Ling Joon Yee, started planting C. liberica coffee seedlings in the Pemadang catchment. Encouraged by early success of the crop, he applied for and was granted undeveloped land adjacent to his farm by the Tutong District Office (Department of Agriculture, 1989). Further assistance in the form of planting materials, subsidised fertilisers and crop protection services were provided by the Department of Agriculture. With this Government encouragement, Bukit Guayan Coffee Plantation, the only coffee plantation in Brunei, expanded the planted area from 5 ha in 1979 to 72 ha in 1992.

CLIMATE

Brunei has a humid equatorial climate, with constant warm temperatures, high humidity and rainfall. Two main seasons occur, associated with the north-east and south-west monsoons, but they are poorly defined. There is a bimodal rainfall regime, with the main rainy season caused by the north-east monsoon between November and January and a minor wet season between May and October, during the south-west monsoon. At Sungai Liang Meteorological Station, 17 km west of the catchment, the average annual rainfall (1973–91) was 2827 mm, with 45% occurring between October and January (Table 10.1). February and March are usually the driest months, although rainfall amounts can vary considerably. Slow-moving tropical air crossing the warm South China Sea produces heavy convectional rainfall. Rainfall occurs, on average, only 145 days per year, giving an average over 19 years of 20 mm per rainfall day. Air temperatures show little seasonal or annual variation.

Temperatures tend to be higher during the south-west monsoon and lower during the north-east monsoon, due to the effects of cooler onshore rain-bearing winds. Monthly climatic data from Sungai Liang, prior to the field research, indicated the relatively constant maximum and minimum temperatures throughout the year (Table 10.2). Percentage relative humidity in excess of 80% and an average daily 7.1 hours of sunshine related closely to these constant air temperatures.

PEMADANG CATCHMENT

Sungai (River) Pemadang (Lat. 4°12′ N, Long. 114° 9′ E) is a tributary of the Sungai Tutong which drains from the Bedawam Mountains (highest peak 529 m) northwards to the South China Sea (Figure 10.1). The soils, relief and geology of Pemadang catchment (31.5 km^2) are closely related to the vegetation and consequent development. Higher areas, underlain by Tertiary sands of the Liang Formation, consist of poorly consolidated sands, sandy clays and clays. Some of the sands are coarse with pebbles. The soils of the higher areas are dominated by the Nyalau family, red–yellow podsolic soils (Figure 10.2) which are deeply weathered, leached acid soils, described as Orthic Acrisols (FAO/UNESCO, 1974). A representative profile has been described and analysed by Hunting Technical Services (1969). The profiles of the Nyalau soils of the catchment were found to be much shallower due to the steeper slopes. The yellowish-red B/C horizon occurred between 18 and 40 cm, averaging 24.5 cm for the six study profiles. Lowland areas, composed of more recent alluvium and mangrove deposits, are dominated by Anderson organic peat soils, which extend along the valley bottoms and Bijat gley soils. A representative profile from the peat swamps in Sungai Belait, west of the catchment, was composed almost entirely of peat to a depth of 102 cm (Hunting Technical Services, 1969). Loss-on-ignition was between 95.99 and 98.65%. Peat deposits in the central part of the swamp were as

Table 10.2   Monthly climate data January 1991 to June 1992 from Sungai Liang Meteorological Station

Year and month	Temperature (°C) Mean	Max.	Min.	Relative humidity (%)	Average daily sunshine (hours)	Rainfall total (mm)	Rainfall days No.	Daily mean (mm)	Daily max. (mm)
1991									
January	27.8	32.5	21.5	87	7.4	121.6	5	24.3	88.9
February	27.9	32.3	22.5	87	6.2	101.6	9	11.3	24.0
March	28.2	33.5	20.5	81	8.3	25.0	2	12.5	21.9
April	28.0	33.8	23.0	83	7.3	363.0	10	36.3	107.6
May	28.6	35.0	23.0	81	6.8	198.2	10	19.8	64.5
June	28.6	34.5	22.0	80	8.1	246.5	10	24.7	74.9
July	28.4	35.0	22.5	81	6.7	36.5	4	9.1	11.8
August	28.4	35.8	22.0	82	6.1	131.8	8	16.5	36.6
September	28.2	34.0	22.5	85	5.2	202.5	11	18.4	45.0
October	28.2	34.9	21.8	82	6.6	278.8	10	27.9	123.9
November	–	–	–	–	–	352.7	19	18.6	82.9
December	28.0	33.8	22.5	84	6.4	292.7	11	26.6	115.9
1992									
January	27.7	33.0	21.0	78	7.0	93.2	7	13.3	42.5
February	27.6	33.4	20.8	85	8.0	67.9	3	22.6	61.0
March	28.5	34.0	22.0	81	8.7	0.0	0	0.0	0.0
April	28.8	34.5	22.5	79	7.8	150.3	4	37.6	75.2
May	28.8	35.0	22.5	81	–	350.2	13	26.9	87.2
June	28.3	34.9	22.0	82	–	275.7	16	17.2	66.2

much as 15 m deep (Anderson, 1964). The Anderson soils in the Pemadang catchment are regarded as a shallow phase. The four representative profiles had a surface peat horizon 15–27 cm thick, followed by a gleyed clay horizon. In three of the profiles, a further layer of fibrous peat was followed by a lower gleyed clay. Percentage loss-on-ignition on samples from three representative profiles were 80.4, 67.5 and 46.1% (mean 64.7%), distinctly lower and much more variable than the true Anderson soil. Peat accumulation on the Tutong flood plain has been interrupted by periodic flooding of fine fluvial sediments and deposition of valley-side colluvium.

The upper catchment is located within Andulau National Forest Reserve, therefore 75% is densely covered by mixed Dipterocarp and secondary forest (Figure 10.3). The lowland area is covered by secondary forest, generally over 25 years old, or peat swamp forest dominated by *Shorea albida* (Ashton, 1964). Generally, the lowland areas have been cleared for agriculture.

Bukit Guayan consists of undulating hills, which slope south and south-east to the peat swamps of Sungai Damit Pemadang valley. The Nyalau soils, found on these hilly areas, have a light yellowish-brown, sandy loam topsoil, medium to low fertility and are slightly acid (pH 4.3–5.0). On the floodplain of the Sungai Damit, peat soils of the Anderson series occur, which are relatively fertile, but very acid (pH 3.2–3.8).

Seedlings from *C. liberica* coffee are known to be genetically variable; therefore, the

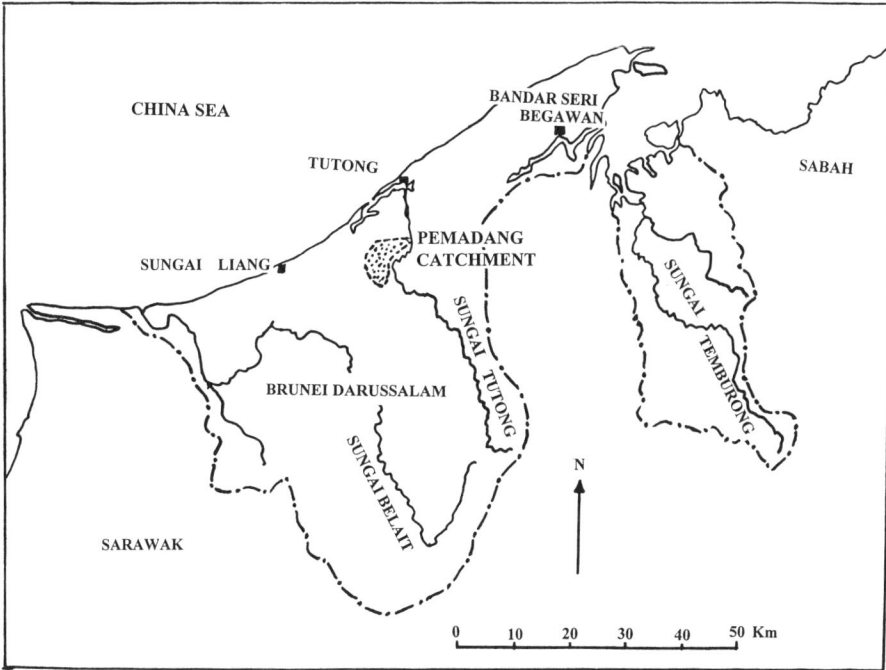

Figure 10.1   Location of Pemadang catchment in Brunei Darussalam

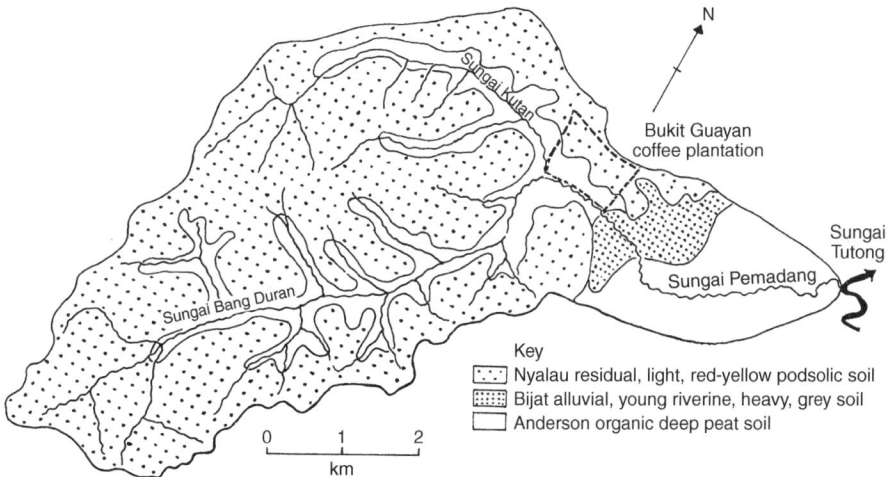

Figure 10.2   Soils of the Pemadang catchment (from Hunting Technical Services, 1969)

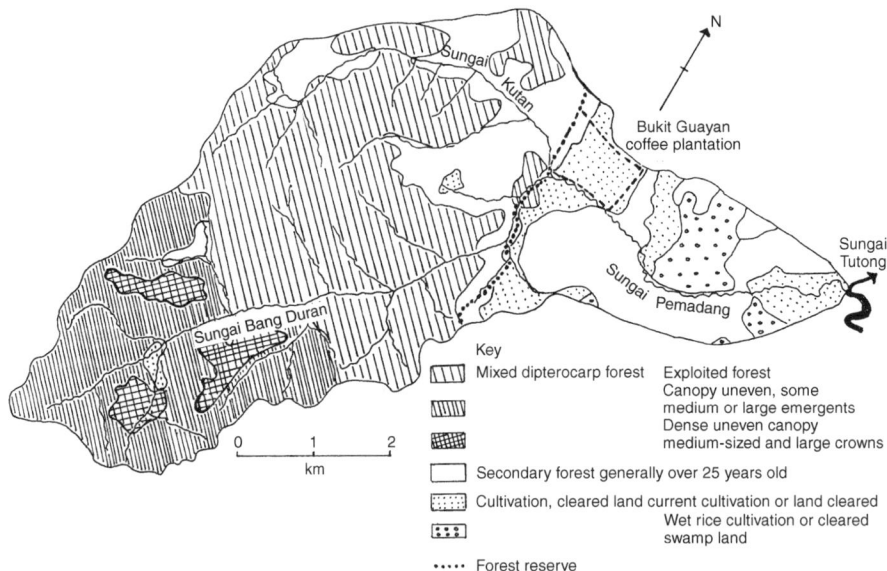

Figure 10.3   Vegetation of the Pemadang catchment (from Hunting Technical Services, 1969)

performance of coffee trees on the plantation is variable. Yields are increased by selection of cuttings from high performance trees noted for large beans and thin skins. As trees mature, usually 5 years after planting, yields of coffee increase with selective planting and replacement of non-productive trees (Figure 10.4).

ASSESSMENT OF THE SUSTAINABILITY OF THE COFFEE PRODUCTION

Cultivation in tropical catchments can present many difficulties and often creates constraints on further development. Sustainability of coffee growing in the Pemadang catchment depends on how these difficulties and constraints are overcome. Sustainability can be considered under two categories: problems with establishment and management of the crop (for which solutions can often be found) and physical constraints (which are difficult to measure and even more difficult to solve).

The majority of the crop management issues are readily solved on Bukit Guayan Plantation. The plantation has the advantage of a farmer, with indigenous forest skills, advised and funded by a supportive Department of Agriculture. It is useful to examine these problems, because many of them are fundamental issues in the development of tropical catchments. The co-operation between Government Departments and Ling Joon Yee is vital in the success of the project, which is equally important for both parties.

With the continual success of the enterprise, the farmer was periodically granted extra land for expansion. Consequently, coffee has been planted in a sequential development (Figure 10.5). The same methods of development and cultivation have been used; therefore a useful chronosequence of cultivation is available for study.

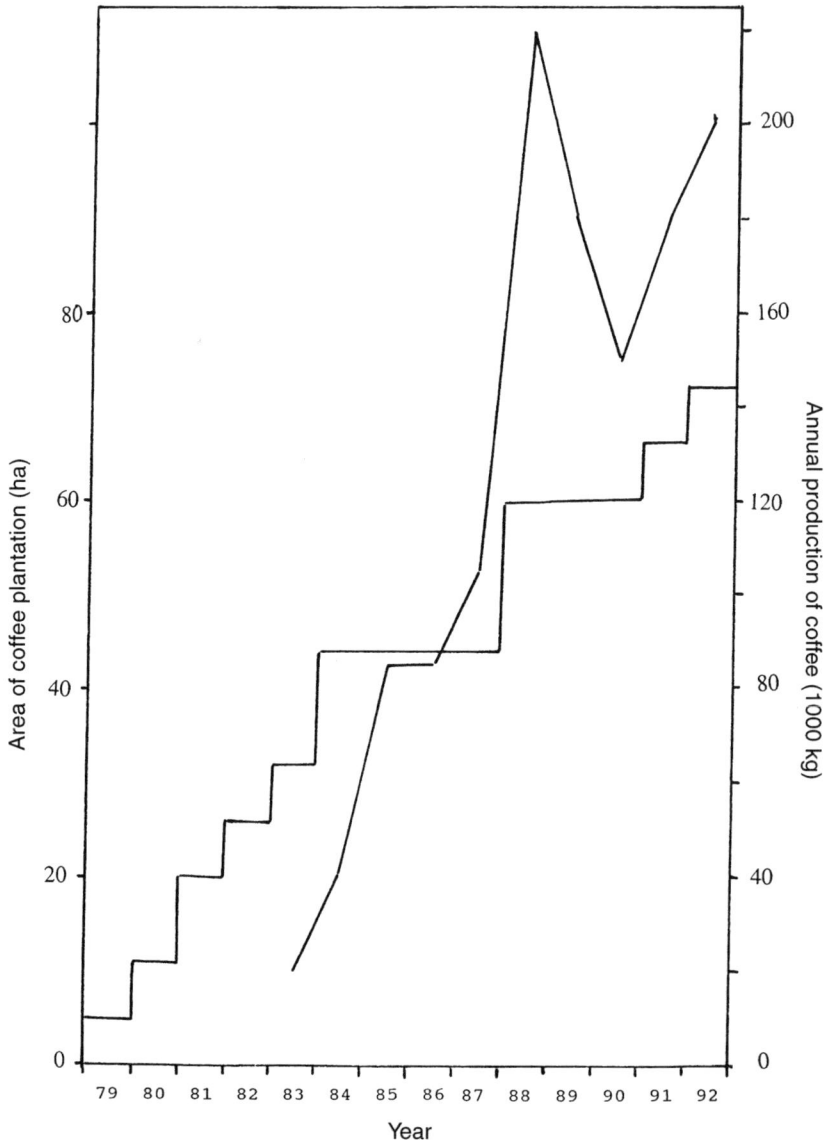

Figure 10.4   Coffee planted and production at Bukit Guayan Plantation, 1979–92 (Department of Agriculture 1989)

Soil profiles were described and sampled in each dated plot and 10 replicated topsoil (0–15 cm) samples were taken in September 1992. Samples were analysed for organic content and geochemical composition. The soil organic matter content of the fine-earth fraction was determined by loss-on-ignition (375°C for 16 h; Ball, 1964). Total element concentrations were analysed using X-ray fluorescence spectrometry on a Fisons Applied Research Laboratory 8410 Spectrometer.

BUKIT GUAYAN COFFEE PLANTATION

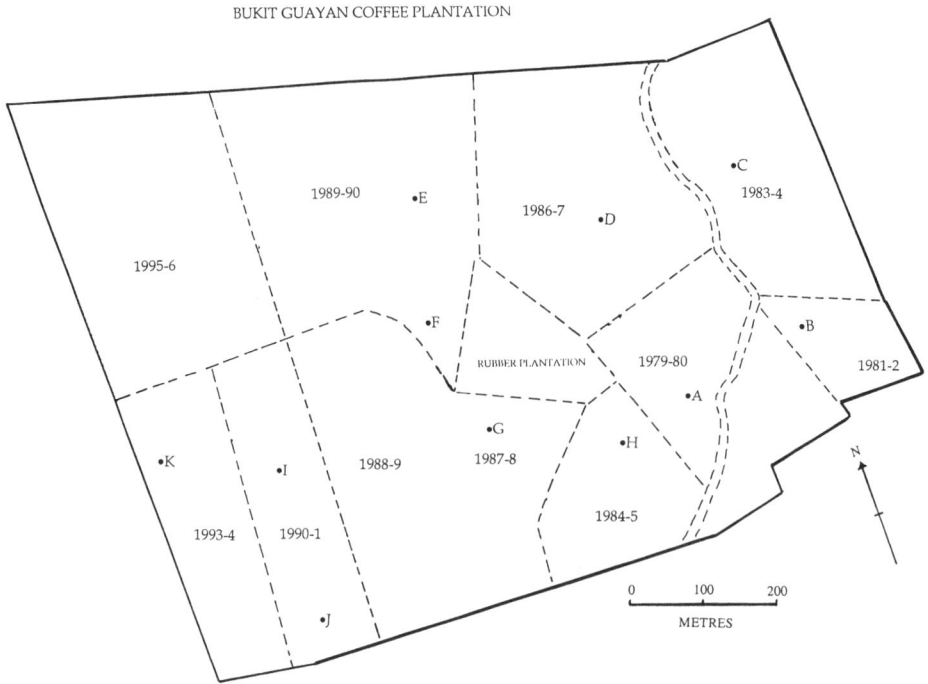

Figure 10.5   Coffee planting sequence at Bukit Guayan. A–K are soil profile sites

FOREST CLEARANCE

Nutrient recycling related to leaf-fall and litter layer is an important part of the forest ecosystem (Anderson and Swift, 1983). Loss of litter influx and hence nutrients from *in situ* cycling have major effects on soil organic and nutrient budgets (Ross, 1993). Initially, following clearance, organic matter input is as much as $100 \, t \, ha^{-1}$ (Anderson and Spencer, 1991), due to the increase of leaf-fall and decay of *in situ* roots. After this initial large input, soil organic matter declines. Therefore, methods of forest clearance are crucial in successful cultivation. The most widely used techniques are clear-felling and burning (including slash and burn), bulldozing and selective logging. Large-scale forest clearance by burning results in loss of carbon, nitrogen and sulphur in smoke (Ewel et al., 1981). Furthermore, the resulting bare slopes are vulnerable to severe soil erosion and loss of soil fertility in tropical areas. Smaller scale "slash and burn", which is still widely practised by indigenous tribes in neighbouring Sarawak, but is of minor importance in Brunei, is a feature of shifting cultivation rather than sedentary plantation farming. Slash and burn causes a sudden mineralisation of the nutrient store, which is then vulnerable to leaching. Plantations established in Africa, using this method of clearance, have to be abandoned after 3 years (Walter, 1971), unless there are large inputs of artificial fertilisers. In Sarawak, equatorial forest cleared in this way can only support crops for 2 years. If cultivation does not take place after forest clearance, a well-structured soil is left with macropores from cracks, faunal

Figure 10.6   Shade trees on Bukit Guayan Plantation

burrows and holes from decaying roots (Sollins, 1989). Although these macropores provide pathways for rapid infiltration, leaching in the soil matrix is relatively slow. On Bukit Guayan Plantation, a carefully planned approach has been adopted. Initially the undergrowth is cut and burnt in controlled fires. Larger trees are retained to provide shade for the young coffee seedlings (Figure 10.6). As the coffee bushes develop, the shade trees are gradually removed by "ring barking". If slash and burn was practised on these relatively steep slopes, soil erosion and nutrient loss would be severe. The labour-intensive methods employed on Bukit Guayan Plantation limit the potential risk of soil erosion on the relatively steep slopes (Figure 10.7).

WEED INFESTATION

Competition for light in undisturbed equatorial forests restricts the growth of ground flora. Removal of the tree canopy and planting of young seedlings creates open sites for weed infestation. In order to allow the growth of the young coffee trees, weeding is essential, but once the trees are established, sufficient cover from the dense foliage of the coffee trees restricts weed growth. Manual weed control is continued in order to remove perennial plants and unwanted trees and creepers (Figure 10.8). At Bukit Guayan Plantation, different planting densities have been explored. It was found that 6×6 m spacing (300 trees ha^{-1}) of coffee trees was too open, while 4×4 m spacing (550 trees ha^{-1}) was too dense. Since 1988, 5×5 m spacing (430 trees ha^{-1}) has been found to be the best compromise to avoid trees overlapping and to restrict weed growth.

Figure 10.7   Overview of Bukit Guayan Plantation

Figure 10.8   Manual weeding at Bukit Guayan Plantation

VARIABILITY OF COFFEE TREES

Coffea liberica is genetically variable; consequently 10–20% of planted seedlings are non-productive. Besides using seedlings from selected high-yielding, superior quality trees, the farmer has used grafting skills. Hybrids produced by grafting *C. robusta* onto *C. liberica* roots gave 10-fold increase in yields. These new hybrids are being used in new areas, as replacement trees and gap-filling throughout the plantation.

PRUNING

In warm humid tropical climates *Coffea liberica* trees can grow to 20 m, becoming too large for harvesting, therefore, the trees are topped at 1.7 m. Topping stimulates the growth of orthotropic shoots, which are in turn topped (Halle et al., 1978). While trees are young, double stems are encouraged, increasing the number of lateral branches, while in turn enhancing yields. A programme of pruning was adopted in the plantation in 1989, especially with some of the older trees. This action temporarily reduced yields in 1989.

PESTS

The close proximity of the plantation to undisturbed forest reserve results in a problem with monkeys damaging berries and trees. Ants also present problems, especially during harvesting. Ants are controlled using Dieldrex or BHC Gumaxane. In comparison with *C. arabica*, *C. liberica* is relatively disease free. Peregrine and Ahmad (1982) have listed 10 diseases for *C. liberica* and 18 for *C. robusta* in Brunei. As protection against diseases, trees are sprayed at 20-day intervals after harvesting. Diseases have not been a problem at the plantation, due to its isolation.

SOIL ORGANIC MATTER

The influence of soil organic matter on soil fertility and the sustainability of plant nutrients in tropical soils has been reviewed by Ross (1993). In an undisturbed equatorial rainforest, the leaf litter fall is large and the rate of decay is rapid; therefore in these natural conditions the organic composition of the topsoils is high. Leaf litter, litter standing crop and turnover coefficents (K_L) for alluvial and Dipterocarp forests have been studied in several locations in South-east Asia. In the Mulu National Park, 70 km south of the Pemadang catchment, leaf and total (small) litter fall from the alluvial forests were 6.6 and 9.4 t ha^{-1} yr^{-1} with turnover coefficients of 1.8 and 1.7 respectively (Anderson et al., 1983). In the same study, Dipterocarp forest leaf and total litter amounted to 5.4 and 7.7 t ha^{-1} yr^{-1}, giving turnover coefficients of 1.7 and 1.3 respectively. It is generally considered that soils degrade quickly after deforestation, partly due to the loss of organic matter. Anderson et al. (1983) found that decomposition rates in the Mulu National Park were similar to temperate deciduous forest. The active mass of decomposing litter forms the majority of the organic matter in tropical forest soils (Whitmore, 1975). Materials more resistant to decomposition tend to form soil humus, which only amounts to 10–20% of the input (Nye and Greenland, 1960). Once the supply of organic material is reduced, the microbial habitat for decay is destroyed. The role of termites in litter removal in Mulu National Park has been assessed by Collins (1983). He found that in forest areas cleared for cultivation, the number of species of termites were reduced to 6 compared with 25 species in virgin forest. Large changes can occur due to decreases in soil feeders, resulting in nutrient changes, for example decreasing faecal phosphorus. In Bukit Guayan the organic content of the Ah horizon decreased from 11.3% (forest cleared in 1989–90) to 3.1% (soil cultivated for 12 years since 1979–80) (Figure 10.9). There was distinct contrast in the habitat of organic decay from humid forest floor,

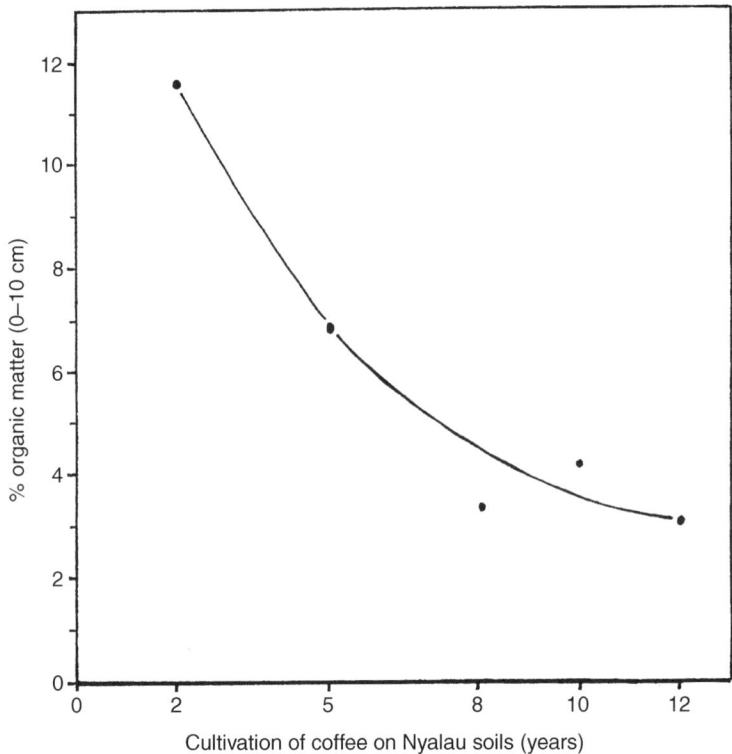

Figure 10.9   Change in soil organic matter at Bukit Guayan Plantation

compared with the "dry" open cultivated slopes. Greenland (1994) considered that norms should be set for different soils as an "index of sustainability". In a comparison between cultivated fields and forest plots in Nigeria, Swift et al. (1981) showed the temperature range of cultivated soil was 18°C, six times greater than the forest. Besides the loss of nutrients from decaying litter, removal of the forest reduces the loss of canopy leaching, which is recognised as an important pathway (Kenworthy, 1971). In a study of Malayan forest, Kenworthy (1971) estimated that of the total nutrients leached, 15% of calcium, 12% of magnesium and 220% of the potassium was leached from the canopy. Hence the turnover of potassium in the forest is more than twice the stored content.

CHANGES IN SOIL NUTRIENTS

The nutrient cycle of the equatorial forest is intricately linked to a mature ecosystem. Burnham (1989) identified two types of nutrient cycling in tropical forests. Firstly "closed" nutrient cycling, which tends to occur in undisturbed forest. Nutrients from organic material are recycled by shallow-rooted plants with little or no influence from parent material. Secondly "open" nutrient cycling, where rock weathering contributes to nutrient supply. This latter form was found in mixed Dipterocarp forests in

Borneo (Baillie, 1989). Using a series of soil profiles, Baillie and Ashton (1983) analysed the nutrient cycles of mixed Dipterocarp forest established on red–yellow podsolic soils of Sarawak. They confirmed that phosphorus is a critical nutrient in the forest with magnesium as the most important mineral. They also found that exchangeable cation amounts were more variable than reserve levels. In conclusion, they considered that the size of the reserves was more important to the long-term nutrient supply than the ephemeral cycle of available nutrients. The role of natural disturbances due to tree fall (Hartshorn, 1978) substantiates these conclusions. Erosion and fresh weathering occurs within the rooting zone, influencing nutrient supply as a partially open nutrient cycle (Baillie and Ashton, 1983). Removal of forest results in an end to the closed cycling of nutrients. Therefore, with successful cultivation and sustainable tree cropping, a more open nutrient cycling occurs, usually requiring the use of artificial fertilisers. Continuously moist soils tend to have closed nutrient cycles with shallow roots (Burnham, 1989). In Sarawak with weathered rock at 1.1 m, Baillie and Mamit (1983) found that mean rooting depth was 1.8 m with few roots below 3 m. With forest clearance and drying of the soil, deeper roots are likely to reach nutrients from newly weathered rock. Using a chronosequence of shifting cultivation, Nakano and Syahbuddin (1989) found that while exchangeable cations and pH decreased, available phosphorus increased again after 4 years of decline. They considered that, in secondary forests, more nutrients were held in the top 30 cm of soil than in the litter layer, as shown by Nye and Greenland (1960). Organic matter decreased due to less leaf-fall, removal of nutrients by soil water leaching and rapid runoff, reduction in faunal microbial activity due to surface drying and decreases in litter.

In Bukit Guayan, the chronosequence of soils demonstrated that a decrease in litter inputs and increased leaching, decreased organic matter and also decreased the $K_2O : SiO_2$ ratio (Figure 10.10). The geochemistry of the chronosequence of soils, cultivated at different times, indicated some definite trends with depth and time (Table 10.3). Aluminium and iron showed marked increases with depth and a decrease with cultivation. Calcium was, as expected, low with minimal changes. Only small quantities of potassium and magnesium were measured, but distinct patterns of these important cyclic nutrients were evident. Both elements decreased with time and increased with depth, indicating leaching processes in operation. In contrast phosphorus declined with depth, possibly associated with the input of fertilisers and surface faunal activity. The horizons in the Anderson peaty soils were more complex with zones of gleyed clays between zones of fibrous peat, influencing the geochemistry (Table 10.4).

Compared with the Nyalau soils, aluminium and iron concentrations in the Anderson soils were higher. Iron decreased from the surface in these soils, unlike the acrisols. Increased organic matter resulted in increases in calcium, potassium and magnesium, with less evidence of leaching. One of the most noticeable features of these soils was the high sulphur content. In the four profiles, the greatest sulphur concentrations were found at approximately the same depth at 31–62, 45–83, 47–49 and 60+ cm respectively. In order to maintain soil fertility, fertilisers are applied to each tree at a specified rate (Table 10.5). Besides the precautionary use of general fertilisers, more selective replacement of macronutrients and trace elements are needed, meeting the special requirements of coffee.

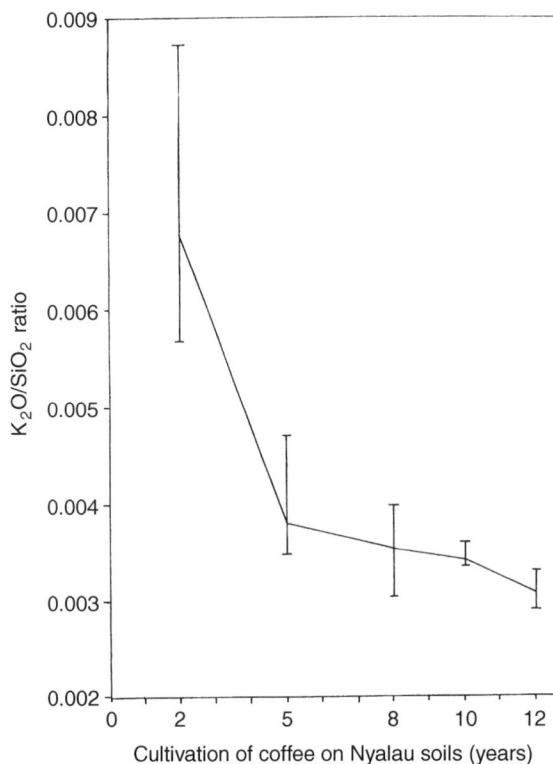

Figure 10.10   Decrease in K_2O/SiO_2 ratio in soils of Bukit Guayan Plantation. $N = 10$, vertical bars denote range

RAINFALL VARIABILITY

Although annual rainfall is high, between 2200 and 3600 mm per year at Sungai Liang Meteorological Station, rainfall variability creates cultivation problems. These problems become more critical with changes in the ecosystem. In the dense forest ecosystem, soil moisture from intense rainstorms is retained, but when the forest is cleared, runoff and evaporation increases. Consequently, humid tropical conditions can be replaced by a more variable pattern, which can include periods of severe drought. Monthly rainfall data, 1991–92, prior to the sampling period, emphasised these variable conditions (Table 10.2). For each month, daily maximum rainfall was often high, with rain falling on only a relatively few days in each month. The highest daily rainfall occurred in October 1991 with 123.9 mm (46% of the monthly total). These intense storms result in high runoff and severe leaching on the Nyalau soils and waterlogging on the Anderson soils. Probably more injurious to the coffee tree, was the occurrence of dry conditions. Between January and April 1992, only 14 days with rain occurred, with no rain at all in March. These dry periods recur quite frequently associated with El Niño, as shown by the 3-month moving averages for 1979 to 1992 (Figure 10.11). Even in the forest lowlands of Sarawak, Brunig (1969) found that low rainfall over 30-day periods result in drought stress. In a study in East Africa, the

Table 10.3   Geochemical analysis of soil profiles on the Nyalau soils

Profile (year coffee tree species planted)	Depth (cm)	Al_2O_3	CaO	Fe_2O_3	K_2O	MgO	MnO	P_2O_5	SiO_2	Organic
A (1979–80)	3–10	3.94	0.14	1.09	0.006	0.26	0.008	0.10	49.81	3.05
	10–40	4.58	0.13	1.40	0.006	0.24	0.007	0.08	48.97	2.42
	40+	7.29	0.13	2.30	0.093	0.31	0.008	0.08	47.97	1.62
B (1981–82)	3–14	7.02	0.15	1.67	0.087	0.38	0.010	0.09	47.37	4.18
	14–26	8.18	0.14	1.99	0.112	0.38	0.010	0.08	47.39	3.04
	26+	8.79	0.13	2.14	0.168	0.40	0.010	0.07	45.75	1.93
C (1983–84)	0–2	4.04	0.13	1.10	0.050	0.27	0.015	0.07	48.64	3.97
	2–9	5.84	0.13	1.68	0.191	0.32	0.011	0.08	46.72	2.73
	9–20	5.53	0.13	1.66	0.162	0.31	0.041	0.06	45.82	1.72
	20+	7.99	0.12	2.15	0.239	0.39	0.008	0.05	44.87	1.39
D (1986–87)	0–2	5.51	0.19	1.53	0.168	0.39	0.022	0.09	44.29	10.49
	2–10	6.21	0.13	1.72	0.206	0.35	0.019	0.08	44.41	3.19
	10–22	7.58	0.13	1.91	0.266	0.38	0.022	0.08	45.52	2.04
	22+	7.09	0.13	1.97	0.201	0.36	0.014	0.06	45.45	1.19
E (1989–90)	0–2	9.14	0.17	1.99	0.682	0.59	0.016	0.11	40.19	10.92
	2–10	9.46	0.13	2.05	0.604	0.56	0.019	0.08	43.81	11.31
	10–18	10.80	0.13	2.12	0.713	0.64	0.010	0.08	43.44	3.21
	18+	11.29	0.13	2.22	0.765	0.63	0.010	0.07	44.85	1.91

spacing of rows influenced evaporation, with 15–20% higher evaporation between rows and 30–50% less within coffee bushes (Walter, 1971). Although drought stress can be problematic, cell sap concentration controlled by humidity can suppress flowering of coffee. In an experiment in a coffee plantation in Peru, Alvim (1973) showed that bushes in soils dried almost to wilting point flowered after watering, while bushes watered regularly did not produce flowers. It is possible, these drought conditions also result in soil mineral deficiencies (Whitmore, 1975). Between 1989 and 1992, these dry periods became more frequent, resulting in decreased annual yields from the Bukit Guayan Plantation (Table 10.6).

As a consequence, severe drought conditions were visible in the coffee trees during 1992. Tree damage, resulting in lower yields, was found on the ridges and unshaded areas due to soil drying. In contrast, the Anderson peat soils retained more moisture, resulting in higher yields. Different problems occurred in the peat soils, however. Drying conditions resulted in peat shrinkage and the establishment of a network of aridity cracks. These cracks caused atmospheric oxidation of the sulphur and increased acidity. In profile G, the surface sulphur content was 1.9% with the lower horizon 6.9%, resulting in increase of soil acidity and damage to the tap root.

HIGH TEMPERATURES AND SUNSHINE

Often leaf scorching and tree damage is associated with increased temperatures and high sunshine during drought periods, but more critical is the indirect response of

Table 10.4   Geochemical analysis of soil profiles on the Anderson soils

Profile (year coffee trees species planted)	Depth (cm)	Al_2O_3	CaO	Fe_2O_3	K_2O	MgO	MnO	P_2O_5	SiO_2	SO_3	Organic
A (1984–85)	0–3	7.14	0.24	9.58	0.42	0.33	0.02	0.40	17.99	2.57	67.14
	3–11	7.14	0.21	8.55	0.42	0.37	0.02	0.32	17.83	2.99	67.77
	11–27	13.00	0.15	1.69	0.84	0.48	0.01	0.26	18.05	4.13	62.32
	27–31	1.05	0.11	0.11	0.01	0.25	0.01	0.08	2.76	0.01	28.39
	31–62	15.93	0.15	1.05	1.24	0.53	0.01	0.13	19.21	4.74	57.88
	62–87+	26.54	0.13	1.18	2.25	0.87	0.01	0.10	25.88	1.63	23.52
B (1987–88)	0–3	6.77	0.00	1.73	0.36	0.00	0.02	0.37	17.85	1.93	85.22
	3–15	10.66	0.00	1.54	0.59	0.00	0.01	0.27	24.22	3.36	75.56
	15–33	26.27	0.13	1.10	1.53	0.81	0.01	0.09	31.98	0.18	16.02
	33–45	13.90	0.16	0.73	0.60	0.47	0.01	0.23	10.07	4.86	78.32
	45–83	14.94	0.14	0.51	0.32	0.26	0.01	0.21	7.08	6.91	79.93
C (1990–91)	0–1	18.39	0.30	7.82	1.21	0.65	0.01	0.34	21.05	1.29	45.16
	7–9	17.17	0.40	5.10	1.11	0.69	0.01	0.24	20.63	1.64	46.99
	17–19	17.21	0.54	5.08	1.15	0.72	0.01	0.24	20.92	1.80	49.32
	27–29	25.61	0.32	2.21	1.55	0.89	0.01	0.12	28.05	0.53	27.15
	34–36	13.75	0.73	4.31	1.08	0.68	0.01	0.17	19.09	2.81	59.73
	47–49	11.20	1.09	3.43	0.84	0.83	0.02	0.20	14.60	4.45	63.81
	55–57	15.55	0.77	3.09	1.14	0.81	0.02	0.23	19.35	2.50	55.38
	61–62	21.42	0.48	2.79	1.46	1.00	0.01	0.11	27.52	0.65	16.70
	71–78	27.31	0.25	3.28	1.73	1.08	0.01	0.07	31.69	0.08	7.48
D (Forest)	0–9	22.58	0.14	14.14	1.24	0.60	0.01	0.16	21.63	0.38	40.38
	9–21	24.04	0.14	2.59	1.43	0.79	0.01	0.14	26.08	0.92	27.45
	21–60	1.04	0.11	0.11	0.01	0.25	0.00	0.07	2.77	0.01	6.67
	60+	17.56	0.15	1.22	1.21	0.67	0.01	0.11	25.91	2.89	40.67

Table 10.5   Fertiliser application regime on Bukit Guayan Coffee Plantation

	Rate of application per tree	
Fertiliser type	Anderson peat soil	Nyalau Acrisol
NPK blue	2 kg, 4 × per year	2 kg, 4 × per year
Urea	–	0.1 kg, 4 × per year
CIRP*	0.5 g, 2 × per year	–

* Proprietary treatment for acidic peat soils

Coffea to sunlight. Coffee is one of the least efficient plants, with the maximum rate of net photosynthesis of *C. arabica* of the order of 4.5 mg CO_2 dm^{-2} h^{-1} (Nutman, 1937). Direct sunlight causes rapid stomatal closure, therefore partial shade is important. Although isolated trees were retained for shade, these were generally removed as the coffee trees matured (Figure 10.6). If shade is too great, yields are reduced because trees compete for moisture. In places the petai tree (*Parkia speciosa*)

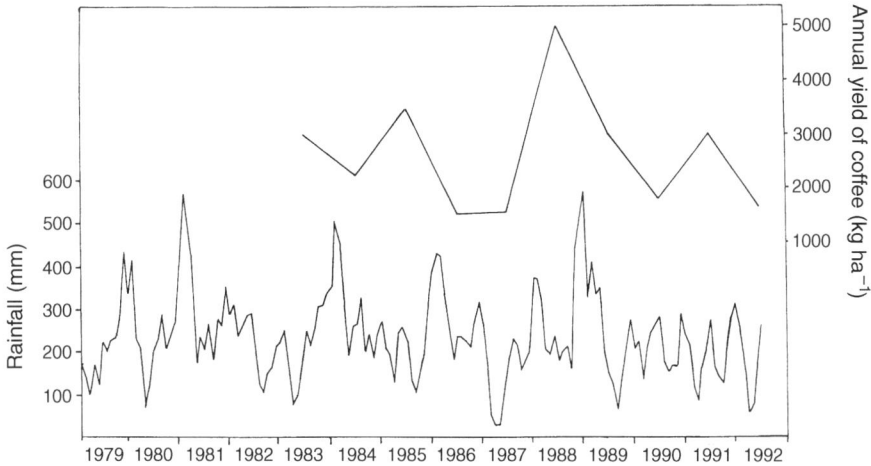

Figure 10.11   Three-month moving averages of rainfall at Sungai Liang compared with coffee yield at Bukit Guayan

Table 10.6   Annual comparison of rainfall at Sungai Liang with coffee yield at Bukit Guayan.

Year	Area (ha) of productive coffee trees	Total yield (kg)	Average yield (kg ha^{-1})	Annual rainfall (mm)	Days with rain
1983	5	20 000	4000	2951.4	141
1984	11	40 000	3636	3270.8	131
1985	20	85 000	4250	2721.7	140
1986	26	85 000	3269	3125.4	150
1987	32	105 000	3281	2217.0	126
1988	44	220 000	5000	3638.2	183
1989	44	180 000	4091	2492.7	136
1990	44	150 000	3409	2674.2	129
1991	44	180 000	4091	2350.9	110
1992	60	200 000	3333	–	–

has been planted. This species is ideal for shade provision, because of the open foliage, and it is also a nitrogen-fixing tree. In order to provide shade and to improve crop diversity, the farmer has additionally planted 8000 fruit trees in recent years, including 5000 durian (*Durio zibethinus*), 1000 jackfruit (*Artocarpus heterophyllus*) and 2000 rambutan (*Nephelium lappaceum*).

SUSTAINABLE AGRICULTURE IN TROPICAL RAINFOREST

The survey of the whole of Brunei by Hunting Technical Services (1969) subdivided Pemadang catchment into three land classifications (Classes IV, V and VI) (Figure 10.12). The upland Nyalau soils possessed a moderate potential for forest development, with a serious risk of soil erosion due to the steep slopes. The lowland area of

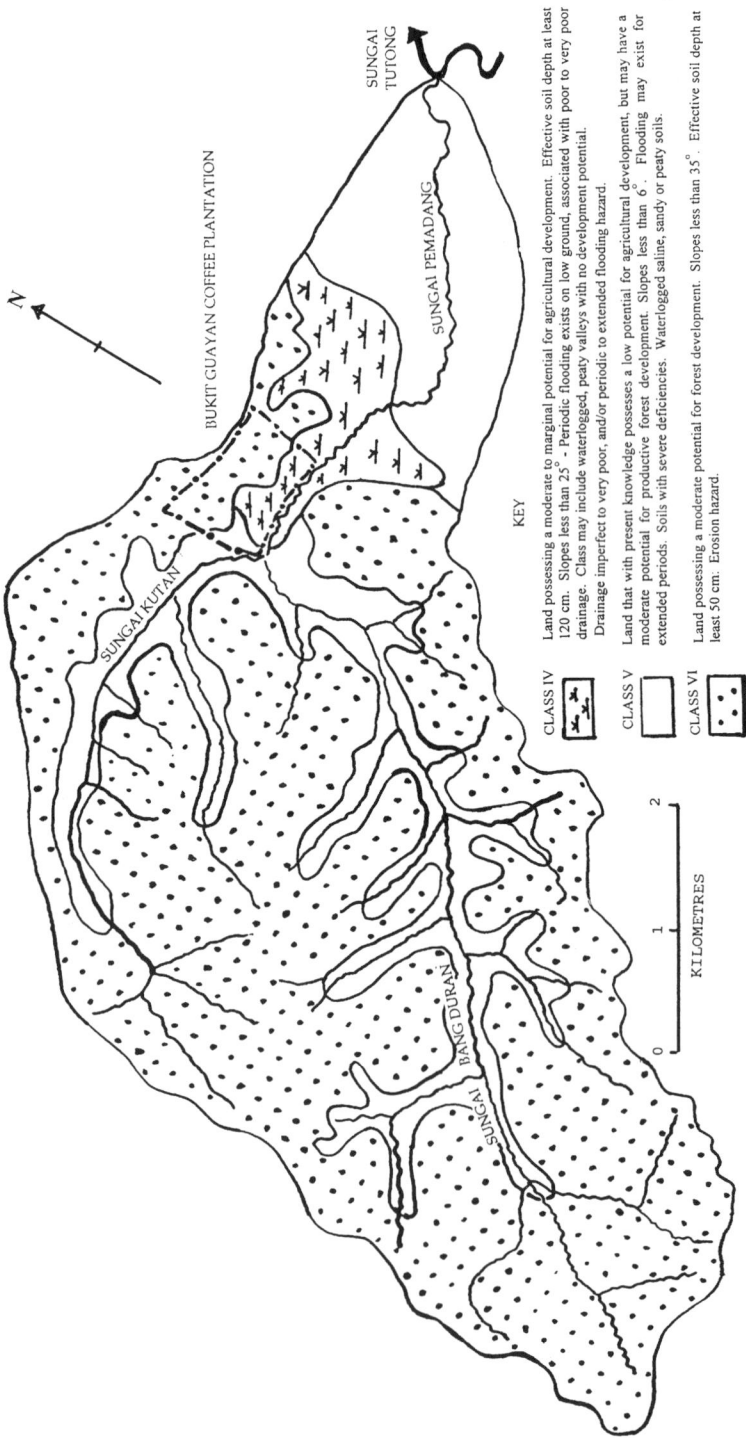

KEY

CLASS IV — Land possessing a moderate to marginal potential for agricultural development. Effective soil depth at least 120 cm. Slopes less than 25° . Periodic flooding exists on low ground, associated with poor to very poor drainage. Class may include waterlogged, peaty valleys with no development potential. Drainage imperfect to very poor, and/or periodic to extended flooding hazard.

CLASS V — Land that with present knowledge possesses a low potential for agricultural development, but may have a moderate potential for productive forest development. Slopes less than 6° . Flooding may exist for extended periods. Soils with severe deficiencies. Waterlogged saline, sandy or peaty soils.

CLASS VI — Land possessing a moderate potential for forest development. Slopes less than 35° . Effective soil depth at least 50 cm. Erosion hazard.

Figure 10.12 Land classification of the Pemadang catchment

Bukit Guayan Plantation was placed in a category with moderate to marginal potential for agricultural development, with waterlogging and flood hazards. Considering the marginal nature of the land for agriculture, the development of the coffee plantation has been highly successful.

Within limits set by economic and social constraints, sustainable land management systems need to preserve the soil's productivity (Greenland, 1994). This study of Bukit Guayan Coffee Plantation in the Pemadang catchment indicates a number of requirements which are necessary for the establishment of sustainable agriculture in a tropical rainforest. These are:

1. Good practice in forest clearance is essential to protect the soil from rapid nutrient depletion and soil erosion. Besides ensuring constant level of organic matter, the soil's physical structure should be maintained.
2. The provision of shade is important for coffee; this can be provided by retaining selected trees from the original forest, planting special trees which have nitrogen fixing qualities and planting fruit trees which can also increase the economic diversity of the cropping system.
3. Utilisation of agroforestry skills, especially those of indigenous people, is necessary because labour intensive methods are required in grafting hybrid trees, selection of productive trees with high-quality yields and continual replacement of damaged or unproductive trees.
4. Improvement in the genetic quality of tree crops is necessary and can be achieved by obtaining higher yielding varieties from other parts of South-east Asia.
5. Tree cropping (agroforestry) is more suited to humid tropical regions than mechanical cultivation of row crops, especially in steeply sloping terrain.

A number of inherent problems remain, resulting from the change from forest to tree cropping. These are:

1. The long-term effects of changing from a closed nutrient system of the rainforest to an open nutrient system of the plantation.
2. Changes in the soil moisture conditions can become more critical. The occurrence of drought conditions becomes a greater problem in the more open plantation environment.
3. The maintenance of soil fertility associated with the geochemical features of each soil type.
4. Coffee plantations are very labour intensive, because of the terrain and need for individual crop maintenance, but agricultural labour is scarce in Brunei.

ACKNOWLEDGEMENTS

The author is grateful to Teck Wong Wan and his family for hospitality, PG Ibrahim Bin PG Mohd Salleh (Chief Soil Scientist, Department of Agriculture) for background information on soils and the plantation, Hj Zaini Hj Punant (Meteorologist,

Civil Aviation Department) for climatic data, Ling Joon Yee for permission to study Bukit Guayan Plantation, Bunthai Meesab for help with preparation of soil pits, Dr Michael Fullen for commenting on the text, Brian Bucknall and Dr Craig Williams for the XRF geochemical program, Dr David Halsey and Marie Roberts for help with the tables and Malcolm Inman for reproducing the photographs.

REFERENCES

Alvim, P.T. 1973. 'Factors affecting flowering of coffee'. In *Genes, Enzymes and Population,* vol. 2 (ed. A.M. Srb), pp. 193–202. Plenum, New York.

Anderson, J.A.R. 1964, 'The structure and development of the peat swamps of Sarawak and Brunei'. *Journal of Tropical Geography,* **18,** 7–16.

Anderson, J.M. and Spencer, T. 1991. *'Carbon, Nutrient and Water Balances of Tropical Forest Ecosystems Subject to Disturbance: Management Implications and Research Proposals'.* MAB Digest No. 7, UNESCO, Paris.

Anderson J.M. and Swift M.J. 1983. 'Decomposition in tropical forests'. In *Tropical Rain Forest: Ecology and Management* (eds S.L. Sutton, T.C. Whitmore and A.C. Chadwick), pp. 287–309. Blackwell Scientific Publications, Oxford.

Anderson, J.M., Proctor, J. and Vallack, H.W. 1983. 'Ecological studies in four contrasting lowland rain forests in Gunung Mulu National Park, Sarawak. II Decomposition processes and nutrient losses from leaf litter'. *Journal of Ecology,* **71,** 503–527.

Ashton, P.S. 1964. 'Ecological studies in the mixed Dipterocarp forests of Brunei State'. *Oxford Forest Memoir,* **25.**

Baillie, I.C. 1989. 'Soil characteristics and classification in relation to the mineral nutrition of tropical wooded ecosystems'. In *Mineral Nutrients in Tropical Forest and Savanna Ecosystems* (ed. J. Proctor), pp. 15–26. British Ecological Society Special Publication No. 9, Blackwell Scientific Publications, Oxford.

Baillie, I.C. and Ashton, P.S. 1983. 'Some soil aspects of the nutrient cycle of mixed Dipterocarp forests in Sarawak'. In *Tropical Rain Forest: Ecology and Management* (eds S.L. Sutton, T.C. Whitmore and A. C. Chadwick), pp. 347–356. Blackwell Scientific Publications, Oxford.

Baillie, I.C. and Mamit, J.D. 1983. 'Observations on rooting in mixed Dipterocarp forest, Central Sarawak'. *Malayan Forester,* **46,** 369–374.

Ball, D.F. 1964. 'Loss on ignition as an estimate of organic matter and organic carbon in non calcareous soils'. *Journal of Soil Science,* **15,** 84–92.

Brunig, E.F. 1969. 'On seasonality of droughts in the lowlands of Sarawak (Borneo)'. *Erdkunde,* **23,** 127–133.

Burnham, C.P. 1989. 'Pedological processes and nutrient supply from parent material in tropical soils'. In *Mineral Nutrients in Tropical Forest and Savanna Ecosystems* (ed. J. Proctor), pp. 27–44. Special Publication of the British Ecological Society, No. 9, Blackwell Scientific Publications, Oxford.

Collins, N.M. 1983. 'Termite populations and their role in litter removal in Malaysian rain forests'. In *Tropical Rain Forest: Ecology and Management* (eds S.L. Sutton, T.C. Whitmore and A.C. Chadwick), pp. 311–326. Blackwell Scientic Publications, Oxford.

Cooper C.F. 1981. 'Climatic variability and sustainability of crop yield in the moist tropics'. In *Food and Climate Interaction* (eds W. Bach, J. Pankrath and S.H. Schneider), pp. 167–186. D. Reidel Press, London.

Department of Agriculture, 1989. 'Status of coffee in Brunei Darussalam'. Unpublished report of the Ministry of Industry and Primary Resources, Brunei.

Ewel, J.J., Berish, C., Brown, B., Price, N. and Raich, J. 1981. 'Slash and burn impacts on a Costa Rican wet forest site'. *Ecology,* **63,** 816–829.

FAO/UNESCO, 1974. *'Soil Map of the World (1:5,000,000) Vol. 1 Legend, Map Sheet IX Southeast Asia.'* UNESCO, Paris.

Gong, W.K. and Ong, J.E. 1983. 'Litter production and decomposition in a coastal hill Dipterocarp forest'. In *Tropical Rain Forest: Ecology and Management* (eds S.L. Sutton, T.C. Whitmore and A.C. Chadwick), pp. 275–287. Blackwell Scientic Publications, Oxford.

Greenland, D.J. 1994. 'Soil science and sustainable land management'. In *Soil Science and Sustainable Land Management in the Tropics* (eds J. K. Syers and D. L. Rimmer), pp. 1–15. CAB International, Wallingford.

Halle, F., Oldeman, R.A.A. and Tomlinson, P.B. 1978. *'Tropical Trees and Forests: An Architectural Analysis'*. Springer-Verlag, Berlin.

Hartshorn, G.S. 1978. 'Treefalls and tropical forest dynamics'. In *Tropical Trees as Living Systems* (eds P.H. Tomlinson and M.H. Zimmermann), pp. 617–838. Cambridge University Press, New York.

Hunting Technical Services, 1969. *'Land Capability Study.'* Government of Brunei, Hunting Technical Services Ltd, Boreham Wood, Herts.

Kenworthy, J.B. 1971. 'Water and nutrient cycling in a tropical rain forest'. In *The Water Relations of Malesian Forests* (ed. J.R. Flenley), pp. 49–65. Transactions of the First Aberdeen–Hull Symposium on Malesian Ecology, Department of Geography, University of Hull.

Maestri, M. and Barros, R.S. 1977. 'Coffee'. In *Ecophysiology of Tropical Crops* (eds P.T. Alvim and T.T. Kolzlowski), pp. 249–278. Academic Press, New York.

Nakano, K. and Syahbuddin, 1989. 'Nutrient dynamics in forest fallows in South-east Asia'. In *Mineral Nutrients in Tropical Forest and Savanna Ecosystems* (ed. J. Proctor), pp. 325–336. British Ecological Society Special Publication No. 9, Blackwell Scientific Publications, Oxford.

Nutman, F.J. 1937. 'Studies on the physiology of *Coffea arabica*. I Photosynthesis of coffee leaves under natural conditions'. *Annals of Botany (London)*, **1**, 353–367.

Nye, P.H. and Greenland, D.J. 1960. *'Soils Under Shifting Cultivation'*. Technical Communication No. 51, Commonwealth Bureau of Soils, Harpenden.

Ogawa, H. 1978. 'Litter production and carbon cycling in Pasoh Forest'. *Malayan Nature Journal*, **30**, 367–373.

Peregrine, W.T.H. and Kassim Bin Ahmad, 1982. *'Brunei: a First Annotated List of Plant Diseases and Associated Organisms'*. Phytopathological Paper No. 27, Commonwealth Mycological Institute, Kew, England.

Ross, S.M. 1993. 'Organic matter in tropical soils: current conditions, concerns and prospects for conservation'. *Progress in Physical Geography*, **17**, 265–305.

Sollins, P. 1989. 'Factors affecting nutrient cycling in tropical soils'. In *Mineral Nutrients in Tropical Forest and Savanna Ecosystems* (ed. J. Proctor), pp. 85–95. British Ecological Society Special Publication No. 9, Blackwell Scientific Publications, Oxford.

Sutton, S.L., Whitmore, T.C. and Chadwick, A.C. 1983. *'Tropical Rain Forest: Ecology and Management'*, British Ecological Society Special Publication No. 2, Blackwell Scientific Publications, Oxford.

Swift, M.J., Russel-Smith, A. and Perfect, T.J. 1981. 'Decomposition and mineral nutrient dynamics of plant litter in a regenerating bush-fallow in the sub-humid tropics'. *Journal of Ecology*, **69**, 981-995.

Walter, H. 1971. *'Ecology of Tropical and Subtropical Vegetation'*. Oliver & Boyd, Edinburgh.

Whitmore, T.C. 1975. *'Tropical Rain Forest of the Far East'*. Clarendon Press, Oxford.

Whitmore, T.C. 1978. 'The forest ecosystems of Malaysia, Singapore and Brunei: description, functioning and research needs'. In *Tropical Forest Ecosystems*, pp. 641–653. UNESCO–UNEP, United Nations.

Williams, C.N. 1975. *'The Agronomy of the Major Tropical Crops'*. Oxford University Press, Oxford.

Catchment Conservation as an Integral Part of Sustainable Management

Introduction

DAVID HARPER

This section contains six chapters, most of them characterised by a catchment-scale approach to land use issues, all of them seeking in different ways to couple land and water processes in order to guide and achieve sustainable management. Many of them deal with examples which have been poorly studied or are little published.

Chapter 11 describes a rare tropical forest type, peat forest, from an example in Kalimantan, and considers its importance and its prospects for sustainable management. Needs include changes in national approaches such as towards multi-sectoral land use planning as well as simple ones such as reconsidering the boundaries of land uses to match biological and geographical realities.

Chapter 12 shows, from the sub-tropical example of Nepal, how the biodiversity of river systems reflects land uses in the catchment and may be used as an indicator of overall catchment health, providing targets for return to sustainable management. There are relatively few examples of this approach anywhere in the world, as river health is often considered from only a single aspect such as point-source pollution or fishing quality, and so the chapter provides a guide to what may be achieved in temperate as well as the tropical zones.

Chapter 13 presents a catchment type which is uniquely tropical – an inland delta – and illustrates the range of environmental and social problems which have to be resolved if water is to be sustainably allocated between conflicting human agricultural uses. Of particular importance is the clear need for substantial social change among the people who use the delta's flood waters, regardless of the nature of "development" assistance from outside.

Chapter 14 presents a contrasting example showing the importance of social factors in development and catchment planning, by examining the use made of forest products in subsistence agriculture in the Gambia. Whilst not catchment-scale in its approach, the study shows clearly how important it is that development planning incorporates those indigenous practices which *are* sustainable. The past three decades are littered with examples of failed development programmes which did not consider either the needs or the practices of indigenous people.

Chapter 15 illustrates this latter problem from the Tana, the largest catchment in Kenya. Most of its lower half flows through arid bush, with natural irrigation of its floodplain provided by the flood peaks from the high Aberdare and Mt Kenya massifs. Highly unsustainable irrigation development schemes along the river have been attempted, threatening the unique biodiversity of floodplain forest. Changing the ecology of the entire river system, and with effects throughout the catchment, is the continuing construction of a chain of hydro-power reservoirs (five built, two more planned), of immense economic importance, but with downstream ecological damage as yet unmitigated. The sustainable, multipurpose use of hydro-power reservoirs (fisheries, irrigation, maintenance of functioning floodplains) is an enormous challenge which no tropical country or its donor agencies has yet successfully met.

This section concludes in Chapter 16 with a smaller scale example than the Tana, but one no less threatened. Lake Naivasha, in the Rift Valley of Kenya, is an unusual lake in that it is endoreic yet fresh. Like the Tana, it is supplied from high-altitude rainfall, in this case the west-facing slopes of the Aberdare Mountains. The waters of the lake and its catchment are thus highly valued – for irrigation, power-station cooling and public supply. An economically important horticultural industry has flourished in the last decade. Enormous ecological changes have occurred both in the lake and its immediate floodplain, and merely monitoring these changes poses an enormous logistical problem. The chapter examines the use of simple satellite image interpretation as one way of providing the basic information needed at minimal cost.

Between them the chapters illustrate that there are no simple solutions to the problems of sustainable catchment management, and no clear success stories. Each points to ways forward, but a successful management plan will have to learn from the many mistakes and integrate the many disciplines. Arguably, sustainable management of catchments is still a long-off goal in the developed world, but there is more to lose in terms of both biodiversity and human livelihood, if it fails to be achieved in the tropics.

CHAPTER 11

Conservation of Tropical Peat Swamp Forests: A Peatland Catchment in Central Kalimantan

SUSAN E. PAGE
Department of Biology, University of Leicester, UK

JACK RIELEY
Department of Geography, University of Nottingham, UK

INTRODUCTION

Peatlands are widely distributed throughout the world, from the arctic to the tropics in both northern and southern hemispheres, covering in total about 3% of the earth's land surface. The physical and chemical properties of peat and the vegetation it supports vary with geographical location, climate, topography, hydrology and hydrochemistry. In temperate and cold climatic regions the peat is derived mainly from bryophytes and low-growing vascular plants, whilst the natural vegetation of lowland peatlands in the humid tropics is forest and the peat is composed almost entirely of wood.

Tropical peatlands occur in mainland East Asia, South-east Asia, the Caribbean, Central America, South America and central and southern Africa. A current estimate of the total area of undeveloped tropical peatland is in the range 30–45 million hectares, which is approximately 10% of the global peatland resource (Immirzi and Maltby, 1992). The largest area of tropical peatland is in South-east Asia (i.e. in the countries of Thailand, Malaysia, Indonesia, Papua New Guinea and the Philippines; Table 11.1) and this accounts for more than 60% of the world total although, in some estimates of the extent of peatland, there is no separation between the areas occupied by freshwater swamp forest and peat swamp forest. Most of the tropical peatland resource occurs at low altitudes in coastal or sub-coastal locations; a lesser unquantified amount occurs at high altitudes in the mountains of Irian Jaya and Papua New Guinea (Wayi and Freyne, 1992).

The vegetation of the lowland tropical peat swamps of South-east Asia is forest within which most of the tree families of evergreen dipterocarp rainforest are

The Sustainable Management of Tropical Catchments. Edited by David Harper and Tony Brown.
© 1998 John Wiley & Sons Ltd.

Table 11.1 Estimate of the tropical peat swamp forest resource in South-east Asia (information derived from Andriesse (1974), Driessen and Soepraptohardjo (1974), FAO (1974), Shier (1985), RePPProt (1990), Immirzi and Maltby (1992), Oraveinen et al. (1992) and Wayi and Freyne (1992))

Country	Area (ha $\times 10^3$)
Brunei	10
Thailand	68
Philippines	104–240
Papua New Guinea	500–2890
Malaysia	2250–2730
Indonesia	17 000–27 000

Total in the region could be as much as 30×10^6 ha

represented (Anderson, 1963, 1964, 1983). This varies from a mixed forest community with up to 240 species per hectare to forest of much lower tree diversity (30–55 species per hectare) (Anderson, 1963; Silvius, 1984). In some of the large peat swamp forests of South-east Asia a catenary sequence of forest types is evident, the principal feature of which is a replacement of high forest (up to 45 m) by a denser "pole forest" in which the diameter of the tree trunks is rarely greater than 30 cm and the canopy no more than 20 m high. These vegetation changes reflect both the decreasing fertility *sensu lato* of the peat substrate and adaptation of the trees themselves to limiting environmental conditions.

Whilst peat swamp forest vegetation is less diverse than dryland rainforest, it has been recognised as an important regional reservoir of plant diversity in South-east Asia (Anderson, 1963; Silvius, 1984; Whitmore, 1984). Many of the plant species are restricted to this habitat and a few endemic trees have been described from Malaysia and Thailand (Ng and Low, 1982). In contrast, several published accounts of the fauna of peat swamp forest have suggested that animal diversity is low (e.g. Janzen, 1974; Whitten et al., 1987); primate densities are said to be less than in lowland dipterocarp forests and there are no known endemic animals (Whitmore, 1984). A few studies, however, have stressed the wildlife importance of this habitat: Malaysian peat swamp forests are afforded a high conservation value since they are the habitat of endangered mammals, including *Dicerorhinus sumatrensis* (Sumatran rhinoceros) and *Nasalis larvatus* (proboscis monkey) (Anon., 1987).

In addition to biodiversity and wildlife habitat maintenance, tropical peatlands are believed to perform a range of environmental and landscape functions, although there have been few studies of pristine systems (Rieley and Page, 1997). Functions which have been attributed to them include:

- Carbon storage (Immirzi and Maltby, 1992; Sorensen, 1993); in their original state peatlands sequester atmospheric carbon but development activities result in oxidation of the carbon store and alter these systems from net carbon sinks to carbon sources.

- Climate regulation on a micro and meso scale (Silvius and Giesen, 1992).
- Water storage and supply, flow regulation and flood mitigation (Boelter, 1964; Andriesse, 1988; Prentice and Parish, 1992).
- Prevention of saline water intrusion (Andriesse, 1988); coastal peat swamps act as a buffer between salt and freshwater hydrological systems, preventing excessive saline intrusion into coastal lands.
- Resource provision to local communities; peat swamp forests provide food, fuel, timber and a range of non-timber products (Sjarkowi, 1997).

Until recently, there has been no significant permanent human settlement in tropical peat swamps. The pace of economic development in South-east Asia is, however, increasing the land use pressures on peatlands. Extensive areas have been cleared of forest, drained and converted to agriculture for crops or plantation trees, a process that is most advanced in Peninsular Malaysia. There are many problems associated with their exploitation (see Chapter 9), in particular, the adverse physical and chemical properties of tropical peats are detrimental to plant growth; their high water-holding capacity, high acidity, low nutrient content and potential toxicity, compared to mineral soils, make them an unsuitable substrate for agriculture unless lime and fertilisers are applied.

As a result of various types of exploitation, the area of pristine peat swamp forest throughout the South-east Asian region as a whole is declining rapidly. In Indonesia it has been estimated that by the year 2000 there will be no undisturbed peat swamp forest left in South Sumatra (PHPA and AWB, 1990), whilst in Kalimantan only the forests on the deep peats of the interior are relatively safe from intensive development at the present time; they are all, however, designated for timber extraction.

THE PEAT SWAMP FORESTS OF KALIMANTAN

Indonesia possesses the largest area of peat in the tropics, with about one-third (6800×10^3 ha) located in Kalimantan (Indonesian Borneo) mainly near the coast (Driessen, 1977; RePPProt, 1990) (Table 11.2).

Three major peatland types have been described from Kalimantan, distinguished by their location, mode of formation and age (Sieffermann et al., 1988; Rieley et al.,

Table 11.2 Approximate area of peatland in Indonesia (based on RePPProt, 1990)

Country	Area (ha $\times 10^3$)
Sumatra	8,300
Kalimantan	6,800
Irian Jaya	4,600
Other	300
Total	20,000

1992). Basin and coastal peats have formed in depressions in low topographic situations near to the coast and inland along river valleys; high peat occurs on interior low-altitude catchments (maximum of 50 m a.s.l.) and watersheds up to 200 km inland. High peats are most extensive in Central Kalimantan. Many of the peatlands described from Borneo are domed with a flat central bog plain and a maximum peat depth which exceeds 10 m (Anderson, 1964; Andriesse, 1972; Driessen and Rochimah, 1976). Owing to their convex shape, river flood waters do not penetrate these peatlands and their hydrology and hydrochemistry are influenced solely by aerial deposition (dryfall and rainfall); as a consequence, they are acidic and nutrient-poor.

The most important economic activities on the peatlands of Kalimantan are agriculture and timber extraction. The shallower coastal peatlands (<3 m depth) present the best opportunity for conversion to agriculture since they are potentially more fertile than deeper peats and the roots of crop plants may penetrate the mineral soils underneath. Extensive areas of coastal peatland in West Kalimantan have been converted to agricultural production. The deeper, interior peatlands of Central Kalimantan have proved to be less tractable for this purpose; whilst some clearance for cultivation has taken place around settlements, there are large areas of derelict land where attempts at agriculture have been abandoned. Virtually all of the remaining forests on peat are, or will be, exploited for timber. They are a source of several commercially valuable timbers and, as production from other forest types within the region begins to decline, they are increasingly regarded as an important timber resource. The extraction techniques in the peat swamp forests employ low technology. The use of greater mechanisation is inhibited by the soft, unstable nature of the peat surface, linked with the availability of cheap labour.

THE SUNGAI SEBANGAU CATCHMENT: A CASE STUDY

Peat-covered interfluves are important models for whole-catchment studies since they are relatively homogeneous and have well-defined natural boundaries. They provide opportunities to carry out integrated environmental assessments for incorporation into land use planning policies and implementation strategies. A detailed study of peat swamp forest is being carried out in the catchment of the Sungai (River) Sebangau, Central Kalimantan, to provide baseline environmental data from which to prepare strategies for the sustainable management and development of tropical peatlands. This part of Kalimantan contains some of the largest remaining areas of relatively undisturbed peatland in South-east Asia; the upper Sebangau catchment includes extensive areas of high peat, which are replaced by coastal peat nearer to the Java Sea. Field research is being concentrated in the upper catchment 20 km south-west of Palangka Raya, the provincial capital, with a less intensive, exploratory investigation in the middle catchment.

The Sg. Sebangau catchment (approximately 5000 km^2) is an area of low relief, covered with vast expanses of peat swamp forest in which only occasional, small, granite hills rise above the peat-covered landscape. Sg. Sebangau is a slow-flowing "blackwater" river which rises 200 km inland from the Java Sea at an approximate altitude of 20 m a.s.l.; its source is a peat-covered watershed to the west of Palangka Raya and, for most of its length, the river flows over peat. Apart from the villages of Kereng Bangkirai in the north, Bantanan in the south and a transmigration settlement on the southern side of the tributary Sg. Paduran in the lower catchment, there are no permanent settlements (Figure 11.1). The principal land uses are timber extraction, river fishing and small-scale agriculture associated with the settlements.

This region experiences fairly heavy annual rainfall and a moderately seasonal climate, with a wet season lasting from October to June and a dry season which

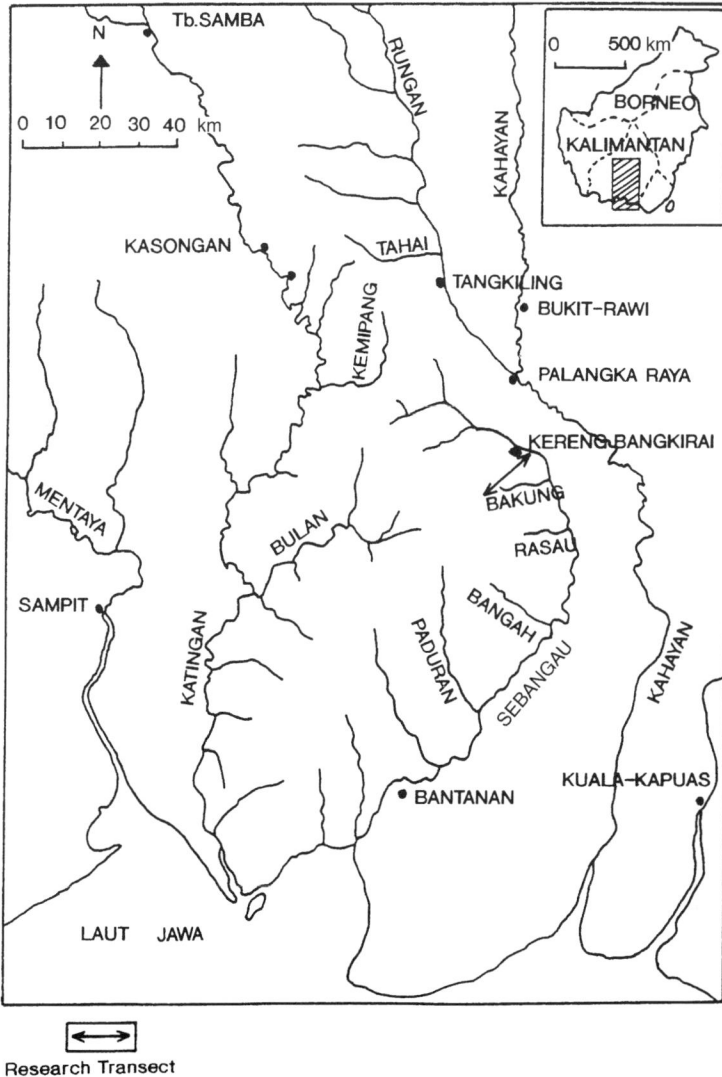

Figure 11.1 The peatland catchments of central Kalimantan, showing main rivers and drainage systems

usually extends from July to September. The annual precipitation is approximately 2500 mm; average air temperatures show little daily or annual variation with a mean daytime maximum of 35°C. The most important climatic variable is the availability of moisture, which depends largely on the seasonal distribution of rainfall rather than its total amount. In this part of Kalimantan, evapotranspiration exceeds precipitation for 2–3 months during the dry season, although in some years there may be a net water deficit for up to 5 months (Sieffermann, pers. comm.).

Access to the principal study site, which is located 2° latitude south of the equator, was by the extraction railway of the Setia Alam Jaya logging concession. In 1993, an 11 km reference transect was established due south-west from the river towards a small granite hill in the interior; in 1994 a second transect 16 km long was set out along a new extraction route parallel to and north of the first transect (Figure 11.2). These transects formed the templates for studies of peat depth and stratigraphy, peat and peat water chemistry, vegetation structure, tree species composition, faunal diversity, hydrology and microclimate. Permanent plots for the study of the vegetation were established in unlogged forest at distances of 1.3, 2.5, 6.3, 7.9, 9.4 and 15 km from the river.

PEAT DEPTH, SURFACE ELEVATION AND STRATIGRAPHY

Initial surface levelling and peat depth determinations showed that peat depth increased progressively away from the river. The peat mass appeared to be biconvex

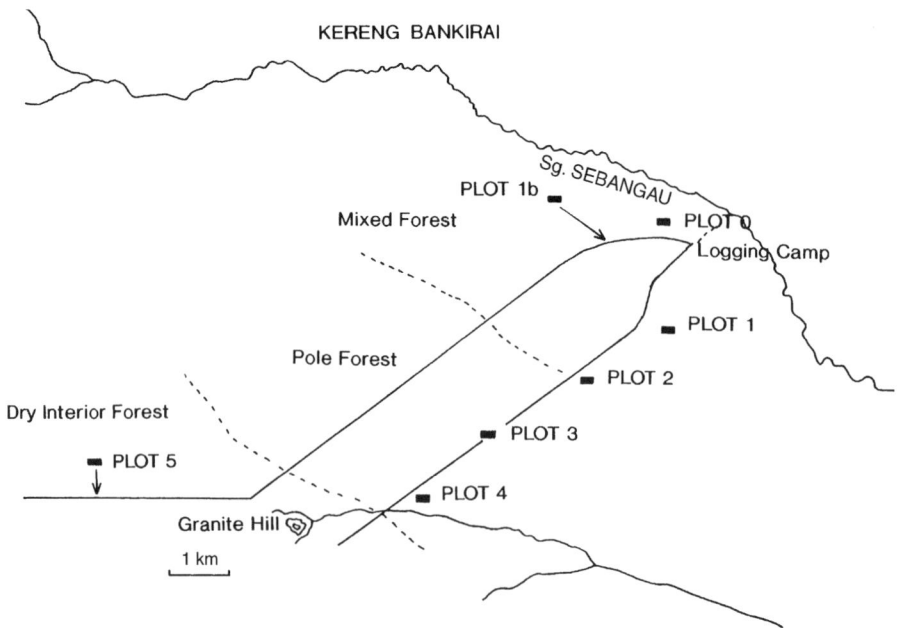

Figure 11.2 Transects and tree plots in the Sebangau river catchment, Kalimantan (access gained courtesy of the logging company Pt. Setia Alam Jaya)

in cross-section with the maximum depth 9–10 km from the river in the central part of a cupola (Figure 11.3). Subsequent data from the second transect revealed this to be incorrect. After a marked rise in surface elevation and depth over the first 5 km, the peat surface remained relatively flat from approximately 6–10 km from the river, whilst peat depth increased to 10 m. From 10 km onwards, however, rather than decreasing, both the surface elevation and the depth of peat continued to increase, reaching a maximum depth of 12.6 m at 16 km from the river with an overall surface elevation increase of 18 m. There was insufficient time available in the 1994 field programme to continue the transect beyond 16 km, although the peat-covered land-scape extends over the watershed into the catchment of the next river, Sg. Bulan (see

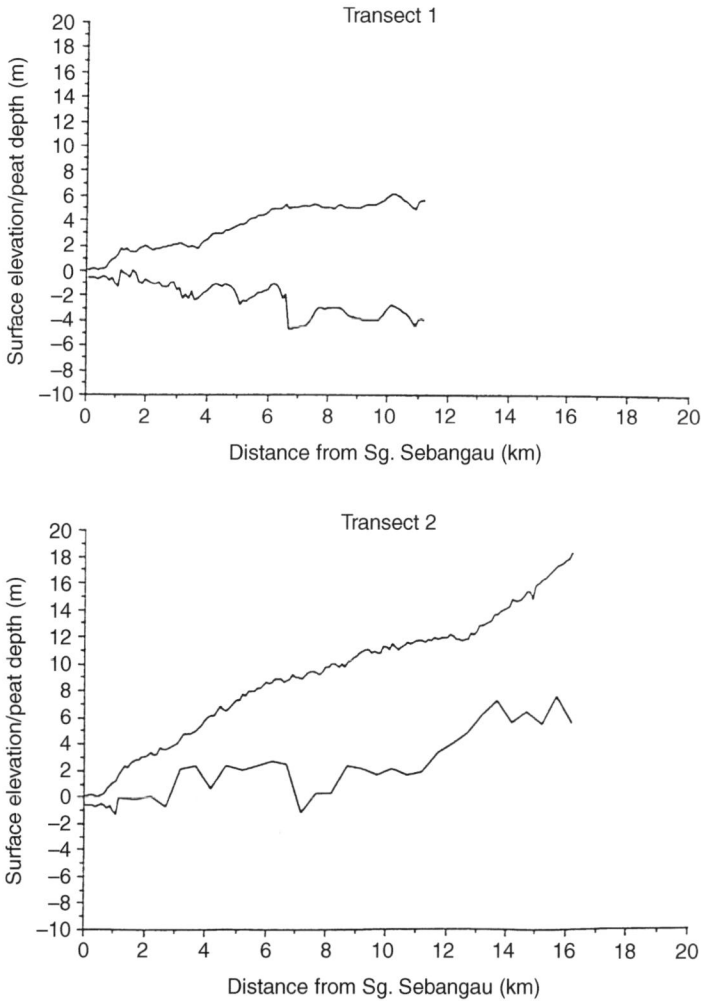

Figure 11.3 Surface elevation and peat depth along the transects established in 1993 (1) and 1994 (2)

Figure 11.1). A trial bore at 20 km from the river confirmed that the peat was still in excess of 5 m at that location.

Investigation of the stratigraphy of the peat revealed that it was woody and poorly humified, consisting mainly of the remains of tree roots and trunks. The peat was underlain by a very humified layer of compacted black, almost carbonised, organic material about 0.25 m thick with some mineral content, grading into clay. The peat stratigraphy conformed to this general description at all the sample locations from the river to 12 km into the forest; from 12 km onwards, however, the top 1.50 m of the profile was composed of drier, highly humified peat with little structure.

VEGETATION

In the upper catchment the edge of the peat swamp forest is about 1.5 km from the river, the original riverine forest of the flood zone having been removed by a combination of logging and burning. In the middle river catchment, however, away from intensive human disturbance, some forest still persists to the river bank. Following tree felling and/or burning, the riverine forest is replaced by species-poor sedge swamp with occasional bushes and stunted trees of *Ploiarium alternifolium* and *Combretocarpus rotundatus*. Secondary succession back to forest may be inhibited by the prolonged flooding to which this area is subject during the wet season, although frequent fires must also play a part in supressing tree regeneration.

The trees in the marginal peat swamp forest (maximum peat depth 6 m) up to a distance of approximately 5 km from the river are relatively tall (up to 40 m) and three canopy layers are discernible. Several of the canopy trees have stilt roots and a few have trunk buttresses. The forest floor is very irregular and consists of a mosaic of shallow depressions and mounds of varying sizes. The former produce a network of anastomosing channels that facilitate water runoff from the peat swamp during the wet season. Pneumatophores (breathing roots of the forest trees) are plentiful in the hollows. The elevated mounds are formed by aggregations of pneumatophores and stilting and arching tree roots which trap organic debris from fine tree roots and leaf litter. These mounds form the microhabitats for tree seed germination and establishment. Over 70 trees have been identified so far in study plots located within this marginal forest. Species of the upper canopy include *Gonystylus bancanus* and *Shorea* spp., in addition to *Combretocarpus rotundatus*, *Cratoxylon glaucum*, *Dactylocladus stenostachys* and several *Calophyllum* spp. Trees of the lower canopy layers include members of the genera *Eugenia*, *Diospyros* and *Garcinia*. The closed canopy ensures that light levels at the forest floor are low and, apart from a prolific growth of tree seedlings and saplings on the raised platforms, the ground vegetation is sparse and typical forest floor species are the sedge *Thoracostachyum bancanum* and the pandan *Pandanus andersonii*. Epiphytes, climbers and bryophytes are all poorly represented.

From 5 to 6 km from the river there is a gradual transition from the taller marginal forest to lower, denser "pole forest", on peat up to 10 m deep. In pole forest, the forest floor topography is extremely uneven, with deep pools between the hummocks; pneumatophores are very abundant. There are only two distinct canopy layers; the upper is open and dominated by *Combretocarpus rotundatus* at a maximum height of 25 m; the lower (10–15 m high) is also relatively open. Tree species diversity is less than in the marginal forest and genera such as *Calophyllum*, *Campnosperma*, *Diospyros*, *Palaquium* and *Tristania* predominate. Many trees have distorted trunks, chlorotic foliage and small, waxy, xeromorphic leaves. There are also fewer young trees than in the marginal forest. The more open canopy results in higher light levels reaching the forest floor, promoting a very dense, almost continuous under-growth of *Pandanus* spp. Insectivorous pitcher plants (*Nepenthes* spp.) are abundant.

Between 12 and 13 km from the river the pole forest is replaced abruptly by a less dense, taller, stratified forest in which most upper canopy trees exceed 35 m in height and some achieve 45 m. This interior tall forest, which is growing on peat between 9 and 13 m deep, has some superficial similarities with the marginal forest near to the river in that three canopy layers are discernible. Most of the upper canopy trees, however, achieve a greater height and girth than those of the edge forest and many have well-developed plank buttresses or stilt roots. The forest floor is undulating and although hummocks are less prominent than in either of the other two forest types, there are still raised platforms formed by tree root plates on which most tree growth and regeneration takes place. Between these raised areas there are shallow depressions from which occasional large-kneed pneumatophores protrude but there is not the prolific growth of smaller breathing roots which is such a significant feature of the pole forest and, to a lesser extent, the marginal forest. The main canopy species are *Agathis* sp., *Casuarina* sp., *Dactylocladus stenostachys*, *Dipterocarpus* spp., *Gonystylus bancanus*, *Koompassia malaccensis*, *Palaquium* spp., *Shorea* spp. and *Tristania grand-iflora*. Epiphytes, climbers (including several rattans) and lianas are more frequent than in the marginal or pole forests, but still do not form a major part of the forest structure; pandans are absent.

Additional diversity to the peat swamp forest is provided by granitic intrusions which form small hills above the influence of the peat. These support dry forest with a different range of tree species although they have some similarities with the marginal mixed swamp forest on shallow peat. Detailed studies of the species diversity of these small "islands" of dry forest have not yet been undertaken.

ANIMAL DIVERSITY

Faunal information was obtained from direct field observation, supplemented by live trapping of small mammals and various invertebrate capture methods. Preliminary surveys carried out during two dry seasons revealed a trend of decreasing species diversity in all animal groups along the transect from the marginal forest to the pole forest, followed by a subsequent increase again in the tall interior forest. This trend was particularly marked for birds and mammals.

Primates were most numerous in the marginal and interior forests and five species were recorded, all of which are protected under Indonesian Law; agile gibbon (*Hylobates agilis*), pig-tailed and long-tailed macaque (*Macaca nemestrina* and *M. fascicularis*), orang-utan (*Pongo pygmaeus*) and red leaf monkey (*Presbytis rubi-cunda*). Although most primates did not appear to utilise the pole forest, several fresh orang-utan sleeping platforms were observed in this forest type during the dry season of 1994.

A number of the other mammal species recorded also have statutory protection (e.g. leopard cat and marbled cat (*Felis bangalensis* and *F. marmorata*) and sun bear (*Helarctos malayanus*)), whilst six species (including the marbled cat) were found outside their previously known ranges in Borneo (Corbet and Hill, 1992).

Over the 2 years of observation, a total of 135 bird species were recorded. Of these, six are listed in the Red Data Book, constituting over 50% (6 out of 11) of the listed RDB species for the island of Borneo. They include wrinkled hornbill (*Aceros corrugatos*), helmeted hornbill (*Buceros vigil*) and Storm's stork (*Ciconia stormi*), which is a characteristic peat swamp species. As a whole the avian community is dominated by insectivores (63%), the main feeding guild being the foliage-gleaning insectivores. The lowest number of individuals and species were recorded in the pole forest, whilst the tall interior forest had a strikingly abundant and diverse community. Despite the limited time available for observation in this forest type, a number of species were only recorded from this habitat, including grey-breasted babbler (*Malacopteron albogulare*) and hook-billed bulbul (*Setornis criniger*), both of which are considered to be rare (Dutson et al., 1991).

A preliminary study of the fish of the forest pools and streams, and of the Sg. Sebangau and several tributaries, resulted in the collection of 34 species from 16 families, up to seven of which may be new species or subspecies. The fry of several economically important river fish were collected from forest pools, highlighting the importance of the peatland as a fish breeding ground.

HYDROLOGICAL AND MICROCLIMATIC MONITORING

Measurements of the peat water table were made using piezometer tubes inserted into the peat surface along the transects and in the permanent study plots. In addition, a permanent water-table recording device was set up in the pole forest at a distance of 6.3 km from the river. In the marginal forest the annual water-table fluctuation was between 0.75 and 1 m, reaching an average mimimum during the dry season of -0.5 m and a wet season maximum of $+0.5$ m relative to the mean peat surface level. During the wet season, the channels between the hummocks filled with rainwater leaving only the tops of the hummocks exposed. In the pole forest, water-table changes were determined, initially, over the important period between dry and wet seasons from September 1993 until January 1994. The lowest water-table level (relative to mean peat surface) of -0.44 m was recorded at the end of September 1993 whilst the highest level of $+0.06$ cm was obtained in mid-January 1994. A comparison with rainfall data for the same period shows that the water table fluctuated in response to the amount and frequency of rainfall events. There was a near immediate response of increased water table following high rainfall events followed by a much slower reduction in water level until the next period of rainfall. Frequent rainfall events during the wet season eventually produce surface flooding and runoff occurs into the small streams and rivers draining the catchment.

In the tall, interior forest the water table was always observed to be much lower at 1.0–1.5 m below the surface during the dry season, whilst even during the height of the wet season the water table was still 0.20–0.30 m below the mean peat surface.

Continuous monitoring of the forest microclimate was started in 1993 in areas of pristine and cleared peat swamp. Incident radiation, relative humidity, rainfall and air temperature were recorded in an open area close to the forest edge, whilst air and peat temperatures were determined in the pole forest 6.3 km from the river. Preliminary results show that in the first 12 months from September 1993 to August 1994 the total rainfall in this part of the Sg. Sebangau catchment was 2550 mm, which was 125% of the 4-year average for the nearest meteorological station at Palangka Raya (Table 11.3). Although the data available so far are not very extensive, they suggest that the peat swamp forest receives a higher rainfall than the deforested landscape around the town of Palangka Raya only some 20 km away.

Detailed daily and monthly mean minimum and maximum air temperature data are available for the 11 months from September 1993 to July 1994 for the two monitoring sites (open and forested). The monthly maximum and mean temperatures are consistently higher in the open than in the forest, whilst mean minimum temperatures are almost exactly the same. The role of the forest in moderating extremes of temperature is clear, with the annual mean maximum temperature reduced by almost 5°C (Table 11.4).

PEAT AND PEAT WATER CHEMISTRY

Samples of surface peat and surface peat water were collected from five of the permanent vegetation plots; in addition, samples of rainwater were collected outside the forest canopy. Chemical analysis of these reveals that this is an extremely acidic, nutrient-poor ecological system. Peat soils and surface waters have a very acidic pH, and low levels of total available and readily exchangeable plant mineral nutrient elements, particularly nitrogen, phosphorus, potassium, calcium and magnesium (Tables 11.5 and 11.6).

General trends can be identified in the changes in ionic content of the peat and peat water along the 10 km transect from the Sg. Sebangau to the pole forest on deep peat.

Table 11.3   Monthly rainfall (mm) recorded in the Sg. Sebangau catchment, Central Kalimantan, Indonesia, from September 1993 to August 1994 with the 4-year monthly averages (1989–93) from Palangka Raya for comparison

Year	Month	Base camp, Sg. Sebangau (monthly totals) 1993/94	Palangka Raya (4-year monthly averages) 1989–93
1993	Sep.	234.5	80.0
	Oct.	132.5	153.0
	Nov.	307.5	235.0
	Dec.	226.0	260.0
1994	Jan.	293.0	296.0
	Feb.	310.0	223.0
	Mar.	323.5	308.0
	Apr.	187.5	236.0
	May	169.0	42.0
	Jun.	305.5	136.0
	Jul.	61.0	68.0
	Aug.	0.0	91.0
Totals		2550.0	2128.0

Table 11.4   Air temperatures in the open and under the forest canopy in the upper Sg. Sebangau catchment for an 11-month period from September 1993 to July 1994 (values are monthly mean, maximum and minimum °C)

Year	Month	Open site			Forest floor		
		Mean	Max.	Min.	Mean	Max.	Min.
1993	Sep.	27.2	34.9	22.3	25.9	30.2	22.6
	Oct.	27.1	35.1	22.5	25.9	30.4	22.8
	Nov.	27.2	34.6	23.4	26.0	29.9	23.3
	Dec.	27.2	34.5	23.5	25.8	29.6	23.3
1994	Jan.	26.6	33.2	23.3	25.4	29.0	23.1
	Feb.	27.0	33.9	23.4	25.8	29.3	23.3
	Mar.	26.6	33.1	23.4	25.5	28.7	23.3
	Apr.	27.3	34.6	23.8	25.9	29.9	23.6
	May	27.6	34.5	23.8	26.0	29.6	23.6
	Jun.	27.1	33.7	23.5	25.5	29.0	23.4
	Jul.	26.9	35.2	21.5	25.1	29.2	21.6
	Mean	27.1	34.3	23.1	25.7	29.6	23.1

Although peat pH is low throughout, it increases slightly from 2.9 at the edge to 3.2 in the interior. Peat water pH is consistently higher than the peat soil pH by 0.2 to 1.0 unit. Peat water pH is highest towards the edge of the swamp and is more or less consistent elsewhere. Peat water electrical conductivity is very low throughout, ranging only between 44 and 54 $\mu S\,cm^{-1}$. It is highest at the edge and decreases into the peat swamp.

Table 11.5   Chemical analysis data for peat samples collected in the vegetation plots (see text for details) (values are ppm dry weight for all elements except N which is expressed as %). Figures used are means of five surface peat samples collected at random within each plot. (Exch. – total exchangeable; tot. – total available)

	Plot 0	Plot 1	Plot 2	Plot 3	Plot 4
Distance along transect from river (km)	1.3	2.5	6.3	7.9	9.4
pH	2.86	3.15	3.11	3.23	3.23
K tot.	185.0	144.5	167.1	158.9	146.5
K exch.	134.9	124.7	120.2	129.6	119.9
Ca tot.	44.9	46.2	51.9	96.1	95.0
Ca exch.	21.6	35.0	22.3	48.0	50.6
Na tot.	557.6	578.3	433.6	419.0	373.8
Na exch.	31.0	30.0	37.0	31.0	27.0
Mg tot.	177.4	152.7	166.7	225.3	190.8
Mg exch.	20.5	25.0	37.0	40.4	39.6
Fe tot.	58.6	70.7	48.0	48.2	32.5
Fe exch.	2.3	2.0	2.3	2.3	1.8
N tot.	1.8	1.0	1.9	1.4	0.8
N exch.	0.01	0.02	0.01	0.01	0.01
P tot.	278.3	272.3	373.2	340.0	339.0
P exch.	19.0	26.0	42.0	57.6	61.8

Table 11.6 Chemical analysis data for surface peat water samples collected in the vegetation plots (see text for details) (values are mg l^{-1}, conductivity is µS cm^{-1}). Figures used are means of 10 water samples collected within the plots; rainwater was collected outside the forest canopy. (n.d. = not detectable).

Distance along transect from river (km)	Plot 0 1.3	Plot 1 2.5	Plot 2 6.3	Plot 3 7.9	Plot 4 9.4	Rainwater collected outside the forest canopy
pH	3.88	3.44	3.44	3.46	3.59	5.88
Cond.	54	53	51	50	44	41
K	0.82	0.73	0.31	0.65	0.30	0.26
Ca	0.71	0.38	0.41	0.38	0.51	0.41
Na	2.63	2.83	2.59	2.97	2.68	3.62
Mg	0.12	0.06	0.06	0.09	0.07	0.20
Fe	0.20	0.19	0.16	0.15	0.13	0.01
NO$_3$	0.07	0.12	0.08	0.09	0.08	n.d.
PO$_4$	2.69	2.81	0.93	0.33	0.50	n.d.

The peat chemistry data show that exchangeable potassium, sodium and iron decrease from the edge to the centre as the peat depth increases; on the other hand, exchangeable calcium, magnesium and phosphorus show increases along the transect. The values for total available (soluble) ions indicate the degree to which each element within the peat may be available to plants for their mineral nutrition. In this respect most of the potassium is readily available, approximately half of the calcium, very little iron (3–5%) and variable amounts of magnesium (11–22%), phosphorus (7–18%) and nitrate-nitrogen (0.5–1.25%).

The concentration of all ions in the surface peat waters is very low. Sodium and phosphorus are highest, especially towards the edge of the forest, but whilst sodium remains relatively uniform throughout, phosphorus decreases from the edge to the centre. Potassium fluctuates, although highest in the marginal forest, whilst calcium is more or less constant throughout. Iron, magnesium and nitrate-nitrogen are consistently low. The concentrations of all elements in peat water are similar to those of rainwater which, however, has a higher pH (5.9), a lower content of iron, higher magnesium and sodium and no detectable nitrate-nitrogen or phosphorus.

TIMBER RESOURCES

The Sebangau catchment peat swamp forest is divided up into logging concessions. Logging activities are restricted to strips of forest 500 m wide either side of extraction railways and most of the timber removal takes place either close to the river in the marginal forest or beyond 14 km from the river in the tall, interior forest. For most of the distance in between, i.e. in the pole forest and transition zones, there is only minimal logging associated with the construction and maintenance of the extraction railways. The principal economic tree species are *Agathis* sp., *Calophyllum rhizophorum*, *Cratoxylon glaucum*, *Dactylocladus stenostachys*, *Dipterocarpus* spp., *Gonystylus bancanus* and *Shorea* spp.

In the marginal and tall interior forests, the selective removal of trees creates localised gaps in the forest canopy but more extensive destruction is associated with the construction of log skids and log storage areas beside the railway track. In these canopy gaps a dense pandan-dominated ground vegetation develops, except in the tall interior forest where the lower water table may inhibit the growth of these species; saplings of many forest trees proliferate together with other woody species, e.g. *Macaranga* spp., which are typical of secondary forest but are absent from the undisturbed forest. The regrowth of trees from cut stumps and the soil seed bank includes specimens of some commercial tree species, but studies of their abundance, growth rates and survival have not yet been undertaken.

NON-TIMBER RESOURCES

The non-timber products of the forests of the Sebangau catchment include latex, tree bark, rattan, medicinal plants, edible fungi, fish and firewood. Exploitation of the bark of the gemur tree (*Alseodaphne coriacea*) is a widespread activity in the Kalimantan peat swamp forests. Logging concession managers allow gemur collectors (usually migrant workers from Java) to enter the forest and to live and work there; access is provided along the railways in return for a percentage of the railhead selling price of the dried bark. Collectors travel great distances into the forest in search of gemur trees (in excess of 20 km). Once located, trees (even very small ones) are felled, and the bark is stripped and packed into baskets for transport to drying platforms in the forest. Java is the ultimate destination of the bark where it is processed for use as a cleansing agent in the recycling of industrial oils and as a constituent of anti-mosquito products (e.g. mosquito coils). In the Setia Alam Jaya logging concession in 1994, approximately 50 people, including wives and children, were involved in this activity. Gemur bark fetches a relatively high price and gemur collectors can earn a very good income by local standards.

It is not known what effect the widespread practice of gemur bark collection will ultimately have upon the survival of this species in the Kalimantan peat forests. Although regrowth from cut gemur stumps has been observed, this may be insufficient to ensure the sustainability of this species at the present rate of exploitation.

Collection of rattan and latex is also carried out in most of the Sg. Sebangau logging concessions. Rattans (principally *Calamus* spp.) are most abundant in the tall interior forest where they are gathered and transported to the river by rail. The tapping of latex from jelutong trees (*Dyera* spp.) is carried out largely by indigenous dayak people who live in the forest for several months at a time. A few local people also collect fungi and plants useful in the formulation of herbal medicines and tonics, but there does not appear to be any widespread exploitation of other potential non-timber forest products, for example gums, dyes, resins or tan barks.

DISCUSSION

NATURAL RESOURCE FUNCTIONS

Peat swamp forest is one of the largest components of the tropical rainforest mosaic in South-east Asia. It is perhaps surprising, therefore, that up until recently this ecosystem has received very little attention. This may be because these forested peat-lands have tended to fall between the two disciplines of peatland ecology and forest ecology. As peatlands globally acquire an increased scientific and socio-economic importance, arising from their role in the global carbon cycle and climatic change, more emphasis may be placed on investigations of tropical systems. Studies such as those described in this chapter highlight the natural resource functions which peat swamp forests perform and provide a scientific basis to promote a more effective approach to land use planning of large peat-covered catchments. The key points arising from our studies of the Sg. Sebangau peatland are considered below, followed by a discussion of the constraints to sustainable management of tropical peatland catchments.

BIODIVERSITY

The biological diversity of tropical peat swamp forests may be lower than that of humid tropical forests on mineral soils, but current observations of both animal and plant diversity indicate that they are an important and undervalued habitat for several rare and threatened species and worthy of more detailed investigation. For example, they appear to be one of the most important habitats for orang-utan in Borneo (Meijaard, pers. comm.) and may play a similar role in the conservation of other forest mammals. In addition, this study stresses their ornithological impor-tance; previously, little was known about the avian communities of the various forest sub-types, since work was concentrated on specialist peat swamp species (e.g. Dutson et al., 1991) or on comparative studies with the avifauna of dipterocarp forest (e.g. Gaither, 1994). It is now evident that many species have specific habitat preferences within the peat forest; for example, several notable bird species were recorded only from the tall interior forest. Unfortunately, this is the most productive forest type within this ecosystem and is subject to intensive logging which may ultimately compromise its conservation value.

In common with other tropical peatlands, the Sebangau catchment supports a zonation of different forest types which vary in their species complement and structure in response to changes in peat depth and hydrology. This is comparable to the catena of peat swamp forest types described by Anderson (1983) for northern Borneo. The sequence of forest types in the Sebangau catchment, however, differs in some important respects: in northern Borneo, the principal tree species, particularly of the middle forest sequence, is *Shorea albida*. This tree has not been recorded in the Sg. Sebangau forests and, significantly, its prominent role is not occupied by any other single species of tree. In addition, the central (interior) forests of Anderson's catenary sequence are dry (his *padang* type) and similar to heath forest (*kerangas*) in structure and species composition. The most depauperate forests in the Sebangau catchment (the pole forest and transition zones) are located on peat which maintains the highest water level, even during the dry season. The interior tall forest type is of particular interest since this very productive and species-diverse peatland forest type, which does not conform to any of Anderson's (1983) forest types, has not been described previously for the peat swamps of South-east Asia. This forest type requires more systematic recording of its plant and animal diversity.

ECOSYSTEM NUTRIENT DYNAMICS

In common with other tropical, domed peats for which chemical analysis data are available, the Sebangau peats are acidic and extremely nutrient poor; a constant but limited supply of elements from the atmosphere (from both rainfall and dryfall) is essential to maintain plant growth. Most of the nutrient capital of this ecosystem is stored in the current vegetation biomass and nutrient cycling must play a vital role in sustaining forest productivity.

Research carried out more than 30 years ago in Sarawak, northern Borneo, along a 25 km transect through pristine coastal peat swamp forest provided important reference information on the mineral content of surface peat (Anderson, 1964, 1983). Further information on peat and peat water chemistry is available from the studies of Neuzil et al. (1993) who compared the inorganic geochemistry of domed peatlands in Riau province (Sumatra) and West Kalimantan.

Comparison of the Sarawak data of Anderson with those from the Sebangau confirm the extreme nutrient-poor nature of the peat substrate in the latter and, in particular, the low levels of nitrogen, phosphorus and potassium. These differences may be explained in part by reference to the geographical locations of the two sites. The Sarawak site is sub-coastal, whilst the Sebangau study area is more than 150 km inland. Thus mineral inputs in precipitation derived from marine aerosols would be expected to be less at the interior site. The data provided by Neuzil et al. (1993) go some way to confirm that there may be chemical differences in surface peats from coastal and interior peatlands. They found that peat from coastal deposits was enriched in magnesium, calcium and sodium but low in silica, aluminium and iron compared to inland deposits. Differences in mineral inputs to the peat surface may also be attributed to differences in annual rainfall.

In view of the extreme nutrient impoverishment of tropical peatland systems, it is perhaps surprising that the deepest peats of the upper Sebangau catchment support a productive forest. The only differences observed so far to explain this are the greater degree of decomposition of the surface peat and the lower water table. It is likely that aerobic decomposition processes operate throughout the year and may be sufficient to provide an enhanced nutrient supply to the trees in this vegetation zone.

HYDROLOGY AND MICROCLIMATE

Ombrotrophic peatlands contain an accumulated reservoir of rainwater which sits above the mineral groundwater table. In the tropics, large peat-covered catchments are believed to reduce downstream flooding during the wet season by detaining rainwater and decreasing the rate of water movement into drainage systems. During the dry season, seepage from large peatland areas may continue to release water into the local rivers and in some areas can provide a source of water for irrigation and human consumption. The initial hydrological data from the Sg. Sebangau catchment emphasise the role that the peat plays as a water reservoir and buffer against severe events of either high or low rainfall. The mineral substrates underneath the peat contain a groundwater aquifer which is an important source of drinking water; however, it is not known what role the peatland plays in moderating mineral ground-water level fluctuations or supply either within or beyond the catchment.

Preliminary microclimatic data have confirmed the effect that forest canopies have in moderating extremes of temperature and influencing rainfall. Further work is required to evaluate whether peat swamp forests have a more significant role in this respect than other tropical forest types, likewise further studies are required to determine the part that large peat-covered catchments play in determining regional climates.

CARBON STORAGE

There is an urgent requirement for more accurate data on the extent of tropical peatlands, their depth and role in carbon flux. This will enable a much more accurate calculation of the total carbon stored in tropical peatlands than is presently available and also whether these systems function as carbon sinks or sources. Initial observations in the Sebangau catchment indicate that different parts of one peatland may have distinct roles. For example, the lower water table of the tall interior forest in the Sg. Sebangau catchment may be promoting a relatively rapid oxidative loss of peat in this area, a theory that is supported by evidence from the dating of these deep, interior high peats. Most present-day tropical peats started to accumulate around 4,500 years before present (BP) or later (Wilford, 1960; Anderson, 1964, 1983; Diemont and Supardi, 1987); those in the interior of Kalimantan originated over 9000 years BP (Sieffermann et al., 1988; Sieffermann, pers. comm.). Unexpectedly, however, peat collected from near to the surface of these interior peats is also very old and has been dated at between 5000 and 6000 years old (Sieffermann et al., 1988; Rieley et al., 1992). This suggests that either peat formation on these interior watersheds ceased several thousand years ago, whilst continuing to accumulate in coastal and riverine situations, or that the surface of this high peat has been lost more recently as a result of decomposition.

ECONOMIC VALUE

Peat swamp forests contain several commercial timber species, including some which are restricted to this forest type in South-east Asia, for example ramin (*Gonystylus bancanus*). The forests play a role in resource provision for local communities and in the supply of non-timber forest products some of which, for example gemur, are little known. This study in the Sg. Sebangau catchment has also shown that the peat swamp forests have a role as a breeding ground for several commercially important species of river fish. Less tangible economic benefits, such as the prevention of downstream flooding, require further investigation.

SUSTAINABLE MANAGEMENT

Ideally, large tropical peatland catchments should remain forested if their full range of natural resource functions are to be maintained: deforestation, drainage and agricultural conversion all bring about irreversible degradation. Even under forestry management there are several constraints to the long-term sustainability of the ecosystem. Some of these are considered below.

THE UNI-SECTORAL APPROACH TO LAND USE

In Indonesia, peat swamp areas are mainly designated for uni-sectoral forms of land use, for example, agriculture or forestry, and sectoral developers ignore or take for granted the direct or indirect effects that their activities have on other forms of land use. Experience obtained in temperate regions indicates that disturbance of peatland systems can have uncosted consequences on the functioning of the natural system (e.g. water supply and flood mitigation) and off-site impacts (e.g. water quality, disturbance of associated watersheds and natural drainage systems). Disruption of the normal hydrological functioning of tropical forested peatlands by the use of inappropriate timber extraction techniques could, for example, have consequential and detrimental effects on fish species which spawn in the forest pools. More serious medium- and long-term consequences of forest clearance and peat drainage include peat shrinkage, leading ultimately to loss of the entire peat profile and, as a consequence, exposure of potential acid sulphate soils. Effective land use planning must take full account of the range of environmental and socio-economic values that the peat swamp ecosystem performs. This requires the formulation of integrated management plans and the close co-operation of the various users (Silvius and Giesen, 1992).

INAPPROPRIATE TIMBER EXTRACTION METHODS

Logging concession conditions stipulated by the Indonesian Department of Forestry require logging companies and their contractors to follow a timber extraction cycle which is the same as that used in the dipterocarp forests on mineral soils, i.e. a 35-year logging cycle of commercial tree species. The logging companies are also required to undertake replanting programmes in those parts of the forest from which timber trees have been extracted. To our knowledge, however, there has not been any research into the growth, development and regeneration time of commercial peat swamp forest trees and many of the species used for replanting are inappropriate. On the Setia Alam concession, for example, several tree species, including *Acacia mangium* which is totally inappropriate for peat swamp, are grown in a tree nursery from seed to small sapling stage and then planted out; almost none survive. During 1993/94, 300 ha of selectively logged forest was planted with several thousand kapur nagur (*Calophyllum rhizophorum*) seedlings. This is a native peat swamp species, but the vast majority of seedlings failed to establish. Clearly, methodologies and techniques need to be developed for assisting forest regrowth and to ensure successful sylviculture.

LOGGING CONCESSION BOUNDARIES

These need to be redrawn for all peat swamp forest areas since the present boundaries make little ecological or hydrological sense. In the Sebangau catchment they bisect the peat dome and the river and disregard natural boundaries. They also fail to take into account the area within each concession which is occupied by the more productive forest types.

UNCONTROLLED FOREST EXPLOITATION

This includes the illegal logging activities referred to earlier in this chapter, the exploitation of non-timber forest products and the unlawful exploitation of forest animals. The latter is driven principally by a demand for young primates and birds for the pet trade; logging now provides easier access routes into previously undisturbed blocks of forest and the workers may be involved in wildlife exploitation. The removal of non-timber forest products is also unregulated and could rapidly lead to the disappearance of the very species upon which the collectors have become economically dependent. The Indonesian Department of Forestry has no responsibility for non-timber activities in State forests.

LACK OF PROTECTED STATUS

In Kalimantan only a small area of peat swamp forest has been set aside for wildlife conservation and most of the undeveloped forest is designated for timber extraction. Logging concession conditions stipulate that logging companies and their contractors set aside small areas of unlogged forest as "germ plasm" reserves to maintain biological and genetic diversity for the future. The latter are mostly far too small (e.g. 50 ha) to maintain forest biodiversity and may be located in parts of the forest which have been logged previously.

FIRE

Destruction of the forest canopy, human disturbance and settlement all increase the likelihood of forest fires. The rainfall of the whole Malesian region is very variable and there have been occasional and exceptional droughts in recent years. These dry conditions place the peat swamps and their forests at a high risk of fire damage. In 1974 it was estimated that the equivalent of a 1.9 cm layer of peat was being destroyed on average throughout Indonesia each year by burning (Subago and Driessen, 1974), but since then the pace of development has increased and the rate of loss is probably much greater. During 1994, Kalimantan experienced a protracted dry season and during September and October palls of smoke from forest fires affected the climate and visibility as far afield as Malaysia and Singapore. The consequences of fire for the natural peat swamp forest vegetation are poorly understood; there is no information on the influence of fire on tree growth, mortality or regeneration of commercially important tree species.

CONCLUSIONS

Tropical peat swamp forests are an undervalued tropical ecosystem which contribute greatly to global biodiversity and a large range of natural resource functions. Work being carried out by a joint Overseas Development Administration (ODA) and Indonesian Ministry of Forestry forest management project is promoting sustainable forest management practices in Indonesia. The aim is to ensure that the quantity of products removed from Indonesian forests is balanced by growth, thus ensuring a sustainable supply of forest products and other environmental and ecological benefits. The forest management project is proposing the management of large forest "landscape units" as a primary objective. For peat swamp forests the appropriate landscape unit is the entire hydrological catchment, encompassing the complete peatland dome and its associated drainage network; the ideal management objectives should give highest priority to the protection and maintenance of the various ecological processes and environmental benefits which the afforested catchment provides whilst allowing sustainable exploitation of timber and other products. This research programme on one peatland catchment is starting to elucidate some of these benefits. There are, however, many more requirements for research and monitoring before comprehensive proposals can be made for sustainable management of tropical peat swamp forests.

ACKNOWLEDGEMENTS

We would like to thank the various people and organisations who have made this research project possible and, in particular, the staff of the Faculty of Agriculture of the University of Palangka Raya. This research is Project 15 in the 1994–99 programme of the South-east Asian Rain Forest Research Committee supported by the Royal Society.

REFERENCES

Anderson, J.A.R. 1963. 'The flora of the peat swamp forests of Sarawak and Brunei including a catalogue of all recorded species of flowering plants, ferns and fern allies'. *Gardens Bulletin Singapore*, **20**, 131–228.

Anderson, J.A.R. 1964. 'The structure and development of the peat swamps of Sarawak and Brunei'. *Journal of Tropical Geography*, **18**, 7–16.

Anderson, J.A.R. 1983. 'The tropical peat swamps of Western Malesia'. In *Ecosystems of the World, Volume 43, Mires; Swamp, Bog, Fen and Moor* (ed. A.J.P. Gore), pp. 181–199. Elsevier, Amsterdam.

Andriesse, J.P. 1972. '*The Soils of West Sarawak*', Vol. 1. Department of Agriculture, Sarawak, East Malaysia.

Andriesse, J.P. 1974. '*Tropical Lowland Peats in South-east Asia*'. Communication 63, Royal Tropical Institute, Amsterdam.

Andriesse, J.P. 1988. '*Nature and Management of Tropical Peat Soils*'. FAO Soils Bulletin 59, Rome.

Anon. 1987. '*Malaysian Wetland Directory*'. Department of Wildlife and National Parks, Kuala Lumpur.

Boelter, D.H. 1964. 'Water storage characteristics of several peats *in situ*'. *Soil Science Society of America Proceedings*, **28**, 433–435.

Corbet, G.B. and Hill, J.E. 1992. '*The Mammals of the Indomalayan Region: A Systematic Review*'. Oxford University Press, Oxford.

Diemont, W.H. and Supardi, H. 1987. 'Accumulation of organic matter and inorganic constituents in a peat dome in Sumatra, Indonesia'. In *Proceedings of the International Symposium on Peat and Peatlands for Development*, Yogyakarta, Indonesia (unpublished).

Driessen, P.M. 1977. 'Peat soils'. In *Proceedings of the Soils and Rice Symposium*, pp. 763–779. IRRI, Los Banos, Philippines.

Driessen, P.M. and Rochimah, L. 1976. '*The Physical Properties of Lowland Peats from Kalimantan*'. Bulletin 3, Soil Research Institute, Bogor, Indonesia.

Driessen, P.M. and Soepraptohardjo, H. 1974. '*Soils for Agricultural Expansion in Indonesia*'. Bulletin 1, Soil Research Institute, Bogor, Indonesia.

Dutson, G., Wilkinson, R. and Sheldon, B. 1991. 'Hook-billed bulbul *Setornis criniger* and grey-breasted babbler *Malacopteran albogulare* at Barito Ulu, Kalimantan'. *Forktail*, **6**, 78–82.

FAO, 1974. '*Soil Map of the World 1, Legend*'. UNESCO, Paris.

Gaither, J.C. 1994. 'Understorey avifauna of a Bornean peat swamp forest. Is it depauperate?' *Wilson Bulletin*, **106**, 381–390.

Immirzi, P. and Maltby, E. 1992. '*The Global Status of Peatlands and their Role in the Carbon Cycle*'. Report No. 11, Wetland Ecosystems Research Group, University of Exeter, Exeter.

Janzen, D.H. 1974. 'Tropical blackwater rivers, animals, and mast fruiting by the Dipterocarpaceae'. *Biotropica*, **6**, 69–103.

Neuzil, S.G., Supardi, H., Blaine-Cecil, C., Kane, J.S. and Soedjono, K. 1993. 'Inorganic geochemistry of domed peat in Indonesia and its implication for the origin of mineral matter in coal'. *Geological Society of America, Special Paper*, **286**, 23–44.

Ng, F.S.P. and Low, C.M. 1982. '*Check List of Trees of the Malay Peninsula*'. Malaysian Forestry Department, Research Pamphlet No. 88.

Oraveinen, H., Crisologo, E.J., Abando, R., Herrera, N., Sankiaho, K. and Klemetti, V. 1992. 'Pre-feasibility study of the Philippines peat resources'. *Proceedings of the International Symposium on Tropical Peatland*. Kuching, Sarawak, Malaysia, pp. 49–56.

PHPA and AWB-Indonesia, 1990. '*Integrating Conservation and Land-Use Planning, Coastal Region of South Sumatra Indonesia*'. PHPA, Jakarta.

Prentice, C. and Parish, D. 1992. 'Conservation of peat swamp forest; a forgotten ecosystem'. *Proceedings of the International Conference on Tropical Biodiversity*, Kuala Lumpur, Malaysia, 1990, pp. 128–144.

RePPProt, 1990 '*A National Overview from the Regional Physical Planning Programme for Transmigration*'. UK Overseas Development Administration and Directorate BINA Programme. Ministry of Transmigration, Jakarta.

Rieley, J.O. and Page, S.E. 1997, '*Biodiversity and Sustainability of Tropical Peatlands*'. Sarara Publishing Ltd, Cardigan, UK.

Rieley, J.O., Sieffermann, R.G., Fournier, M. and Soubies, F. 1992. 'The peat swamp forests of Borneo: their origin, development, past and present vegetation and importance in regional and global environmental processes'. In *Proceedings of the 9th International Peat Congress*, Uppsala, Sweden, 1, Special edition of the *International Peat Journal*, 78–95.

Shier, C.W. 1985. 'Tropical peat resources – an overview'. In *Proceedings of the Symposium on Tropical Peat Resources – Prospects and Potential*. Kingston, Jamaica, pp. 29–46. International Peat Society, Helsinki.

Sieffermann, R.G., Fournier, M., Truitomo, S., Sadelman, M.T. and Semah, A.M. 1988. 'Velocity of tropical peat accumulation in Central Kalimantan province, Indonesia Borneo'. In '*Proceedings of the 8th International Peat Congress*'. Leningrad, 1, 90–98.

Silvius, M.J. 1984. '*Soils, Vegetation and Nature Conservation of the Berbak Game Reserve, Sumatra, Indonesia*'. RIN Contributions to Research on Management, Arnhem Research Institute for Nature Management.

Silvius, M.J. and Giesen, W. 1992. 'Towards integrated management of Sumatran peat swamp forests'. Paper presented at *Workshop on Sumatra Environment and Development, Seameo-Biotrop, Bogor, Indonesia*, 16–18 September 1992. Asian Wetland Bureau.

Sjarkowi, F. 1997. 'Economic valuation of biodiversity resources in the coastal zone of South Sumatra'. In *Biodiversity and Sustainability of Tropical Peatlands* (eds J.O. Rieley and S.E. Page), pp. 261–266. Sarara Publishing, Cardigan, UK.

Sorensen, K.W. 1993. 'Indonesian peat swamp forests and their role as a carbon sink'. *Chemosphere*, **27**, 1065–1082.

Subago and Driessen, P.M. 1974. '*The Ombrogenous Peats in Indonesia*'. Research Papers 1968–1974, 193–205. Indonesia/Netherlands Agricultural Co-operation, Amsterdam.

Wayi, B.M. and Freyne, D.F. 1992. 'The distribution, characterisation, utilisation and management of peat soils in Papua New Guinea'. *Proceedings of the International Symposium on Tropical Peatland*, pp. 28–32. Kuching, Sarawak, Malaysia.

Whitmore, T.C. 1984. '*Tropical Rain Forests of the Far East*'. Cambridge University Press, Cambridge.

Whitten, A.J., Damanik, S.J., Anwar, J. and Hisam, N. 1987. '*The Ecology of Sumatra*'. Gadjah Mada University Press, Yogyakarta, Indonesia.

Wilford, G.E. 1960. 'Radiocarbon age determinations of quaternary sediments in Brunei and N.E. Sarawak'. *British Borneo Geological Survey Annual Report, 1959*.

CHAPTER 12

Catchment Sustainability and River Biodiversity in Asia: A Case Study from Nepal

STEVE ORMEROD

and

INGRID JÜTTNER
Catchment Research Group, School of Biosciences,
Cardiff University, UK

INTRODUCTION

Although interest in river conservation is now growing substantially (Boon, 1992; Cairns and Lackey, 1992), the overwhelming emphasis has been in economically developed countries (Petts et al., 1989). Recently, however, concern about river conservation has extended to the less developed world (Barel et al., 1985; Whitten et al., 1987; Reinthal and Stiassny, 1991), particularly Asia (Sasekumar et al., 1994). Here, Dudgeon (1992) has argued that special conservation value arises from strongly seasonal flow regimes, and large habitat diversity over pronounced altitudinal and climatic gradients. Also, Asian rivers hold a particularly diverse fish fauna, and have extensive floodplains important for many specialised communities (e.g. Zakaria-Ismail, 1994). At the same time, there are suggestions of widespread threats to conservation from catchment degradation by deforestation, agriculture, flow regulation, and pollution (Dudgeon, 1992; Sasekumar et al., 1994). These features might be taken to indicate failures to manage river catchments sustainably (e.g. Bruntland, 1987).

In Nepal, river catchments have experienced large changes in the past few decades, including faster losses of forest than most other areas of Asia (Pereira, 1987; Gilmour, 1988; Thapa and Weber, 1990a, 1990b) for increased cultivation (Hrabovsky and Miyan, 1987). Continued change seems inevitable to meet the target of doubling agricultural production over the decade 1990–2000, even though virtually all land suitable for cultivation is already in use (His Majesty's Government of Nepal/IUCN,

---

1988). Such change presents a challenge to Nepal's globally important terrestrial ecosystems (Myers, 1988, 1990), but there will also be consequences for the rivers which drain the managed areas. So far, the consequences of environmental change for river biology and conservation in Nepal have seldom been considered except within a more general context of flooding and erosion (Ives and Messerli, 1989; Messerli and Ives, 1997; see below). Although the country has an enlightened approach to nature conservation, and recognises the international schedules which protect some aquatic species, river habitats do not currently receive strong emphasis in conservation planning (e.g. Heinen and Yonzon, 1994). This is despite their likely global importance – for example in holding more specialised species of river birds than anywhere else on earth (Buckton, 1998).

In this chapter, we review some of the first comprehensive data on the ecology, biodiversity and conservation value of rivers in Nepal (Rundle et al., 1993; Tyler and Ormerod, 1993; Brewin and Ormerod, 1994; Ormerod et al., 1994, 1997; Brewin et al., 1995; Wilkinson et al., 1995; Jüttner et al., 1996). The review is linked to concepts of sustainability through:

- suggesting that sustainable catchment management might provide an opportunity for simultaneously protecting river biodiversity and ensuring subsistence use as part of a wider conservation strategy; and
- asserting that river ecosystems provide important opportunities for monitoring whether sustainable catchment management is achieved.

THE SOURCES OF NEPAL'S DIVERSITY: GEOGRAPHICAL POSITION AND PHYSICO-CHEMICAL VARIATIONS

GEOGRAPHICAL POSITION AND PHYSICAL STRUCTURE

In part, the widely recognised biological diversity in Nepal reflects the country's biogeographical position on the sub-tropical borders of the Oriental and Palaearctic realms, where the ranges of many organisms overlap (e.g. Stainton, 1972; Inskipp, 1989). From a global perspective, Myers (1988, 1990) identified the region as one of the world's 18 major biodiversity "hotspots". So far, it is unknown whether the same richness is reflected in river biodiversity, partly because of the taxonomic difficulties that are so widespread outside the developed world, but also because comprehensive data are only now being assembled on diatoms, bryophytes, invertebrates, birds and fish.

The physical diversity of Nepal is an important additional source of ecological variation. The country's Himalayan mountains represent the largest altitudinal range on earth, with the land rising in a north-eastern direction from c. 100 m on the lowland floodplains (the Terai) to 7,000–8,848 m along the High Himalaya. Permanent rivers occupy around 5,000 m of this range, so that there are strong altitudinal variations in stream slope, size, geomorphology, temperature, catchment and riparian vegetation, monsoonal intensity and stream chemistry (Rundle et al., 1993; Ormerod et al., 1994).

High-altitude streams run either directly from glaciers or rise from springs, and

Table 12.1   Stream chemistry (mean values) in five regions of the Nepal Himalaya

	Simikot	Dunai	Annapurna	Langtang	Everest	Makalu
pH	7.8	8.1	7.8	7.3	7.3	7.4
Calcium (mg l^{-1})	15.7	19.3	19.7	6.3	3.8	4.3
Magnesium (mg l^{-1})	7.3	1.9	6.1	1.5	1.4	0.7
Sodium (mg l^{-1})	2.4	2.7	1.4	2.2	1.6	2.4
Silica (mg l^{-1})	3.4	5.2	2.9	4.6	4.0	4.5
Nitrate (mg l^{-1})	0.19	0.11	0.19	0.06	0.17	0.28
Conductivity ($\mu S\ cm^{-1}$)	139	128	163	89	37	41
No. of sites	24	20	22	22	14	31

have characteristics typical of other mountain regions (Ward, 1994). Glacial streams have low temperatures ($< 4°C$) and circadian fluctuations in discharge which relate to variations in air temperature; surface runoff is often turbid due to suspended rock "flour"; nutrient concentrations are low (Table 12.1), and ice acts as an important abrasive agent during both freezing and thawing cycles. By contrast, spring-fed streams have relatively constant discharge and temperature through the influence of groundwater; dense growths of aquatic bryophytes reflect flow stability, but also elevated concentrations of CO_2, which is used as a carbon source (Ward, 1994).

With downstream progression between 5000 and 1500 m OD, discharge volumes increase through steep and torrential cascades over bedrock, cobbles and large boulders (5–10 m diameter). Material of this size contributes to river bedload during extreme flood events between June and October, when over 90% of the 2000–3000 mm of annual rain falls during the monsoons. These may be intense events on saturated catchments with rivers already in flood (Starkel, 1972; Brusden et al., 1981; Gupta, 1983, 1988). Hillstream tributaries which rise in the lower Himalaya and Middle Hills have current velocities at these times which sometimes exceed 3–4 m s^{-1} (Brusden et al., 1981). Following flood events, sediment deposition occurs in previously eroded channels, and material is subsequently infilled between the large "superflood" boulders (Starkel, 1972) that are a frequent characteristic of Nepal's hillstreams. River terraces and shoals are also rearranged and add to habitat diversity in the river corridor.

As they leave the Middle Hills, Nepal's rivers open sometimes into widely braided channels (e.g. Karnali at Chisopani) with fans where bedload and suspended material have been deposited. Meanders through sandy alluvium (e.g. Narayni and Rapti at Meghauli) are common and, in their natural state, these rivers shift frequently over extensive floodplains. They hold a large network of semi-permanent channels, oxbows, larger lakes and other water-filled hollows, of varying age and size, which are important aquatic habitats in their own right.

CATCHMENT VEGETATION

Catchment and riparian vegetation change strongly over the altitudinal gradient, further influencing river function. Tributaries at the highest altitudes flow in unshaded channels through alpine tundra and boulderfields, with progressively lower altitudes influencing semi-natural vegetation and farming activity: in the valley of the Langtang Khola, for example, semi-natural vegetation at 3500–4000 m consists of dwarf shrubs, lichens, *Kobresia* and *Carex* sedges. Between 2500 and 3500 m, streams flow through degraded *Juniperus* scrub, with tussock, rosette and grass vegetation. Between 1200 and 2500 m, they are bordered by mixed evergreen and deciduous forests of *Quercus, Lithocarpus, Michelia, Betula, Alnus nepalensis, Rhododendron, Pinus* and *Tsuga* spp. (Miehe, 1990). At lower altitudes (300–1200 m), land is predominantly used for agriculture, though some degraded areas of sal (*Shorea robusta*) forest and scrub remain. Much of the lowlands of the Terai and inner Terai are now cultivated, but some tributaries still flow through remnant semi-natural forest in National Parks (e.g. Chitwan and Royal Bardia) where sal forms the climax vegetation. A range of other grass and tree species occur in succession as new floodbanks progress to riparian forest of silk cotton (*Bombax ceiba*), *Trewia nudiflora, Dalbergia sissoo* and *Acacia catechu*. In these forests, and in the extensive riparian grasslands of *Saccharum spontaneum, Narenga porphyrocoma, Themeda* spp., *Phragnites karka, Typha elephantina* and *Imperata cylindrica*, there are important habitats for Nepal's large mammals. The National Parks have been managed intentionally by fire and cutting for these mammals, but river erosion and river meandering following floods also have important effects in re-establishing grasslands.

RIVER HABITAT SURVEYS

These gross patterns in physical and riparian structure have been illustrated in Nepal using River Habitat Survey (RHS) methods developed in Britain by the National

Table 12.2   Principal components from a River Habitat Survey at 76 sites in Nepal

PC 1 (20%)	PC 2 (12%)	PC 3 (8%)	PC 4 (7%)
Water width (−)	Waterfalls (−)	Vegetated	Mature islands (+)
Bank width (−)	Overhanging	side bars (−)	Unvegetated
Woody debris (+)	boughs (−)	Cobbles (+)	mid-bars (+)
Depth (−)	Trees (−)	Unvegetated	Vegetated
Leaf litter (+)	Boulders (−)	point bars (+)	mid-bars (+)
Shade (+)	Gravel (+)	Slack flow (−)	Riffles (+)
Riffles (+)	Fallen trees (−)	Vegetated	Laminar flow (+)
Unvegetated	Shade (−)	mid-bars (−)	Riparian
side bars (−)	Bank height (−)	Waterfalls (−)	vegetation (+)
Bank height (−)	Bankside roots (−)	Sand (−)	Trees (−)
Underwater	Torrential flow (−)	Outfalls (−)	Unvegetated
roots (+)	Sand (+)	Boulders (−)	point-bars (+)
Laminar flow (−)	Silt (+)		Sand (+)
Bankside roots (+)	Debris dams (−)		Underwater
Fallen trees (+)	Exposed bedrock (−)		roots (+)
Boulders (−)			Torrential flow (+)
Pebbles (+)			Boulders (−)
Unvegetated			
mid-bars (−)			
Debris dams (+)			

Rivers Authority (now the Environment Agency) (Ormerod et al., 1997). The surveys show, for example, that downslope progression is accompanied by composite changes towards wider channels, deeper water, more laminar flow, fewer riffles, more unvegetated mid-channel bars, more unvegetated side bars, and fewer debris dams (Table 12.2). In addition to providing strong correlates with the richness and composition of river communities (Table 12.3), the RHS methodology appears to have the potential to detect structural changes in Nepali rivers which result from catchment use (Figure 12.1; Ormerod et al., 1997).

CHEMICAL VARIATIONS

In addition to changes in physical structure and catchment character, there are natural variations in stream chemistry both between altitudes and regions which are sufficiently large to be biologically important (Rundle et al., 1993; Ormerod et al., 1994). Streams in central and eastern areas (Makalu, Everest, Langtang) have lower ionic strength than in the west (Annapurna, Simikot, Dunai), probably reflecting geology (Table 12.1). Conductivity, sodium, chloride, silicate, potassium, calcium and nitrate all increase with downslope progression, probably as a result of fertiliser additions, catchment irrigation and increased weathering (Jenkins et al., 1995). These anthropogenic changes are also sufficient to influence biological communities, for example among diatoms (Jüttner et al., 1996).

BIOLOGICAL PATTERN

Reflecting trends in habitat structure, chemistry, riparian vegetation and climate, some of the strongest hydrobiological patterns in Nepal are downslope trends. For

Table 12.3 Correlation between measures of *alpha* and *beta* diversity in contrasting groups of organisms and the first four principal components from the River Habitat Survey

	PC1	PC2	PC3	PC4	Altitude
Diatoms					
Species richness	0.56***	0.10	−0.18	0.24*	0.37**
Ordination axis 1	0.32**	0.11	−0.43***	−0.03	−0.58***
Ordination axis 2	−0.60***	−0.04	0.06	−0.29	−0.51***
Invertebrates					
No. of families	0.03	−0.33**	−0.34**	−0.06	−0.13
Ordination axis 1	−0.59***	0.15	−0.27*	−0.11	−0.79***
Ordination axis 2	0.18	0.35**	0.16	−0.09	0.11
Bryophytes					
Species richness	0.44***	−0.10	−0.24*	0.05	0.36**
Birds					
Species richness	−0.62***	0.16	0.04	0.03	−0.69***
Ordination axis 1	0.54	0.07	−0.17	−0.05	0.42***
Ordination axis 2	0.11	0.18	−0.16	0.04	0.01
Altitude	0.74***	0.01	0.12	0.07	

Notes: Ordination scores from DECORANA were taken to represent turnover in community composition, and hence *beta* diversity. Taxon richnesses were taken as measures of *alpha* diversity

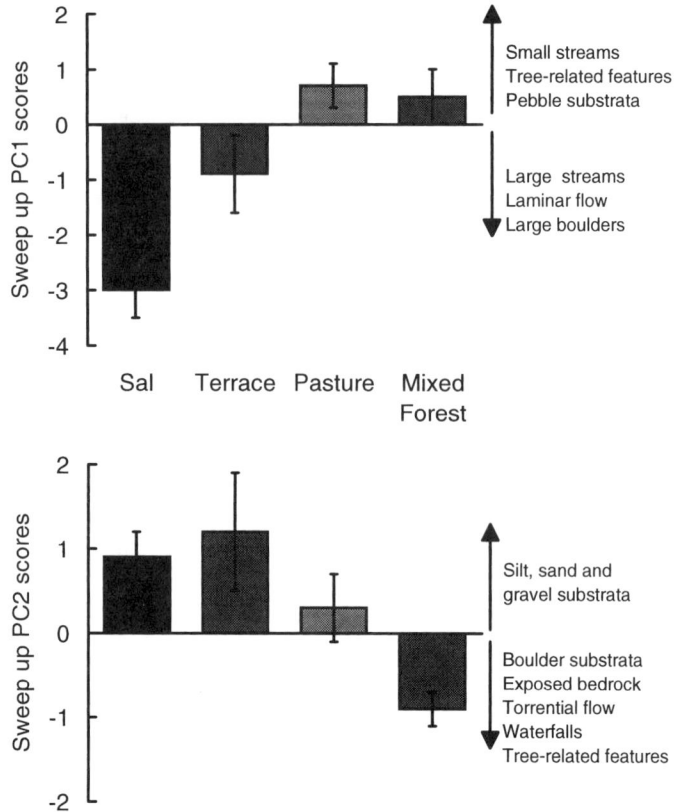

Figure 12.1 Variations in streams with different land use on principal components (Tables 12.2 and 12.3) derived from River Habitat Surveys

example, studies in the Langtang–Trishuli and Likhu Khola systems revealed a progressive increase with decreasing altitude in the number of species or families of all the taxa sampled (Figure 12.2; Ormerod et al., 1994). More recent studies have shown that this downslope pattern is variable seasonally, reflecting lower taxon richness in agricultural catchments during the summer monsoon, when densities fall drastically (Brewin et al., 1995). The altitude pattern is also variable regionally, so that other catchments show peaks in the richness of invertebrates and diatoms at middle altitudes (S. Ormerod et al., unpubl. data).

Stream communities change in taxonomic composition with altitude (Ormerod et al., 1994), and there are strong trends in functional structure. For example, primary producers in the hillstreams are seldom angiosperms, with bryophytes also scarce at the lowest or highest altitudes (Ormerod et al., 1994). Diatoms are abundant throughout, with communities dominated by prostrate, stalked and strongly attached forms in steeper streams (e.g. *Achnanthes minutissima, Fragilaria capucina, Cymbella affinis*), while those characteristic of lowland streams are often motile epipelic and episammic *Navicula* and *Nitzschia* spp. (Ormerod et al., 1994; Jüttner et al., 1996). Aquatic invertebrates in high-altitude streams are predominantly algal grazers, while

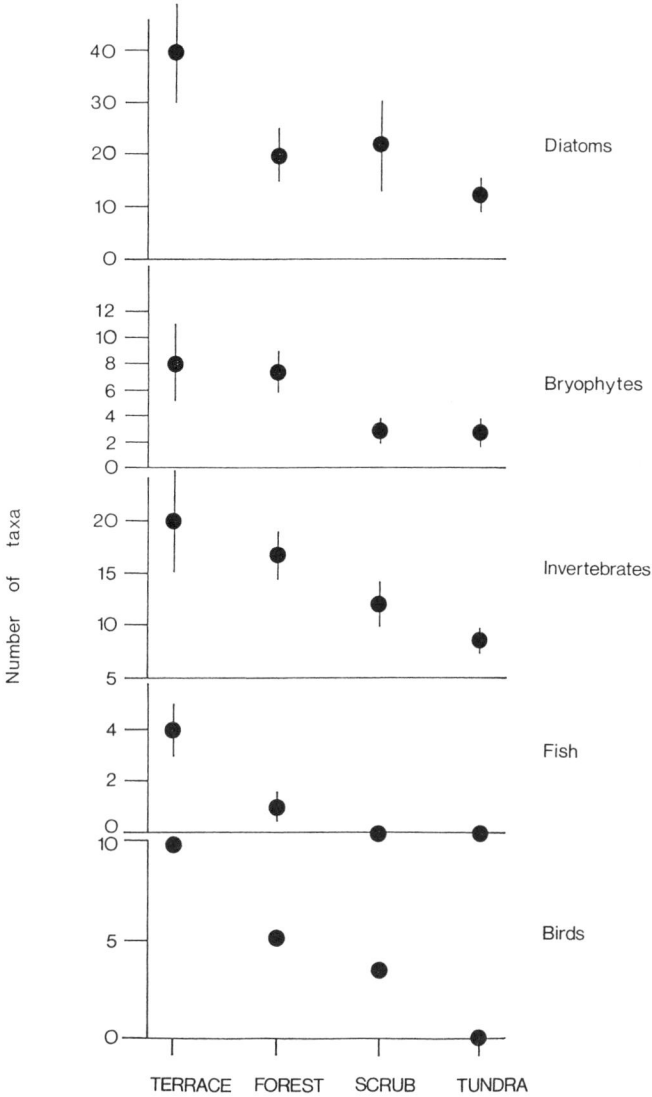

Figure 12.2   Changes in the number of species or families of aquatic groups in catchments of decreasing altitude from tundra, alpine scrub, forest and agricultural terracing in the Langtang and Likhu Khola valleys in Nepal

those in streams through forest or terracing are predominantly filter-feeders which take suspended material from the water column (Figure 12.3; Ormerod et al., 1994). This pattern is consistent with the inputs of particulate organic material from forests, while runoff from terraced catchments will be particularly rich in organic particles, maybe even as living seston.

For river habitat structure to be important ecologically, the activities or distributions of organisms should be associated with certain habitat features. In Nepali

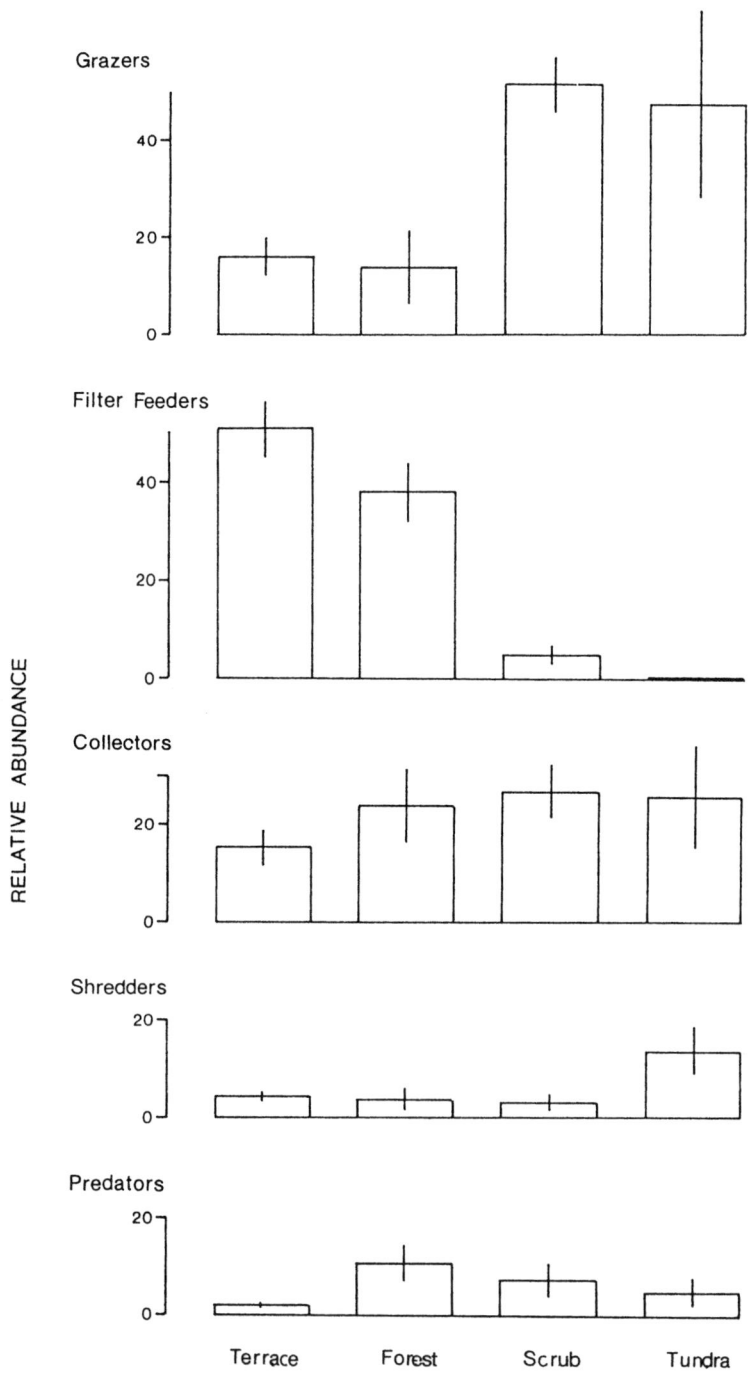

Figure 12.3   Changes in the functional feeding guilds of the invertebrates in the catchments of Figure 12.2

streams, this is the case in birds, diatoms and invertebrates (Tyler and Ormerod, 1993; Brewin et al., 1995; Jüttner et al., 1996), and probably also fish (Bassett, 1993). Species of inscctivorous birds along hillstreams provide the most conspicuous examples of habitat use, which varies between species in complex ways (Figure 12.4; Tyler and Ormerod, 1993). Some feed dominantly by flycatching over rocks and boulders (e.g. plumbeous redstart), some are ground gleaners on rocks (e.g. white-capped river chat) or shingle (spotted forktail), while others are purely aquatic feeders in riffles (brown dipper) or on the splash zone and partially submerged surface of large boulders (little forktail). Such specific patterns of niche use reflect the diverse community of insectivorous birds along Himalayan streams (Buckton, 1998).

In the Likhu Khola and Langtang rivers, some invertebrate groups occur dominantly in riffles (e.g. Baetidae, Hydropsychidae, Simuliidae, Ceratopogonidae), while others occur dominantly in structurally complex margins (e.g. Corixidae, Limnephilidae, Ephemeridae, Perlidae, Anisoptera). Taxa associated with margins or pools

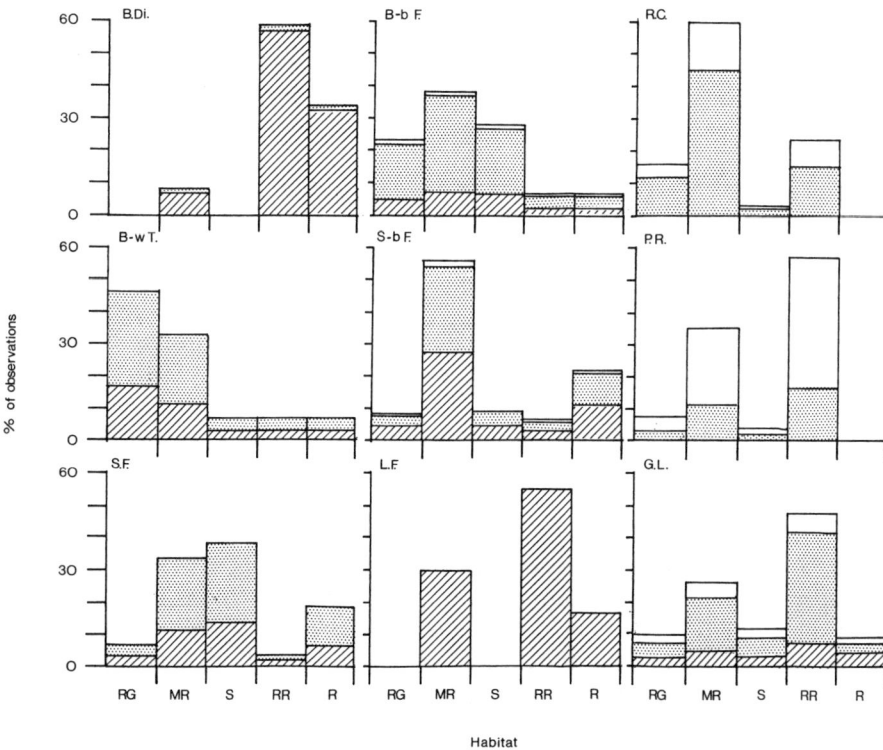

Figure 12.4   The occurrence of birds in different habitats in seven tributaries of the Likhu Khola, Nepal. Species codes are: B.Di = Brown Dipper; B-b F. = Black-backed Forktail; R.C. = River Chat; B-w.t = Blue-Whistling thrush; S-b F. = Slaty-backed Forktail; P.R. = Plumbeous Redstart; S.F. = Spotted Forktail; L.F. = Little Forktail; G.L. = Grey Wagtail. Habitat codes: RG = riparian ground; MR = marginal rocks; S = shingle; RR = mid-river rocks; R = river. Open bars = aerial foraging; stipple = ground foraging; hatch = water foraging

make an increasing contribution to invertebrate communities at lower altitudes (Brewin et al., 1995; Wilkinson et al., 1995), but sandy habitats, which are sometimes common in streams bordered by terracing, have low invertebrate density and diversity. By contrast, deposited leaf litter or other naturally organic substrata hold a diverse fauna (Brewin et al., 1995). Among diatoms, habitat differentiation involves inter-specific variations in abundance between pools and riffles (Jüttner et al., 1996), but can also involve micro-scale variations where diatoms are epiphytic on bryophytes (Rothfritz et al., 1997).

In the absence of complete knowledge of the distribution and taxonomy of invertebrates, bryophytes, or other groups, we cannot yet appreciate fully the status of many Nepali river organisms. There are already indications, however, that some groups are scarce, and even globally threatened. They include several fish species from the Cyprinidae, Sisoridae, Amblycipitidae, Bagaridae and Anguillidae (Shrestha, 1990). The fish-eating crocidilian, the gharial *Gavialus qangeticus*, has been the subject of a captive breeding and release programme, although numbers are still said to be in decline, perhaps due to pollution. The Gangetic dolphin *Platanista qangetica* is also vulnerable, and feeds in the remaining areas of the lower Karnali river with large micro-habitat complexity and abundant fish (Smith, 1993). In all these cases, knowledge of ecology is patchy, preventing the informed basis for sound management.

CATCHMENT-SCALE CHANGE

These biological data from upland and lowland rivers are an indication that complex river structures in Nepal are important in the habitat distribution of several taxa, while taxonomic richness, community composition and trophic function are marked by strong altitudinal trends. Catchment changes specific to each altitudinal zone will thus have the potential to affect distinct communities or ecosystem types. However, most of Nepal's riverine species are concentrated in streams of the Middle Hills, and perhaps also in the hitherto unexplored but varied aquatic habitats of the Terai. Previous work in terrestrial ecosystems has shown similar patterns (Hunter and Yonzon, 1993). Since the Terai and Middle Hills also hold most of Nepal's human population (>90%; Jha, 1992), it is here that anthropogenic change and conservation impacts might be greatest.

There are few data which illustrate unequivocally the effects of human activity on Nepali river catchments. Of the potential threats to Asian rivers from pollution, impoundment and deforestation identified by Dudgeon (1992), pronounced organic enrichment is clearly apparent in the Kathmandu valley (e.g. Jüttner et al., 1996), and probably occurs around other settlements. There has been speculation also about effects by impounds on migratory fish or dolphins in the Sun Kosi, Trisuli and Karnali rivers (Shrestha, 1990; Smith, 1993); the globally endangered mahseers (*Tor putitora* and *T. tor*; Nautiyal, 1989) are among the species potentially affected, but to an unquantified degree (Shrestha, 1986).

The third of Dudgeon's (1992) "threats", from catchment degradation, potentially involves the whole of Nepal's populated area. This idea relates to the "Himalayan degradation" scenario of Eckholm (1975), in which tree clearance and agriculture affect soil erosion, landslips, nutrient regimes and flooding, with potentially far-reaching influences throughout the Ganges, Indus and Brahmaputra basins. In reality, however, this has proved a difficult scenario to validate, with diffuse and complex changes simultaneously involving hydrology, sediment dynamics, stream habitat structure and stream chemistry (Table 12.4). Since most of the Terai and Middle Hills are now under cultivation, quantitative comparisons between semi-natural and perturbed catchments are restricted. Furthermore, the natural effects of seasonal rain, to which deforested areas are assumed to be sensitive (Dudgeon, 1992), are themselves poorly understood, with data subject to large uncertainty or conflict at different geographical scales (Thompson and Warburton, 1988). Intense monsoonal rain can cause pronounced erosion even in uncultivated areas (Sharma, 1987; Ives and Messerli, 1989), while landslips and erosion might be expected anyway in Nepal due to continued orogeny and seismic activity (Ramsay, 1987; Ives and Pitt, 1988; Ives and Messerli, 1989). Across Nepali rivers as a whole, estimated annual sediment

Table 12.4   Potential changes in the catchments of Nepalese rivers as a result of human activity with an indication of the scale at which consequences occur (C = catchment, R = riparian, A = aquatic).

1. Catchment conversion to agricultural/fuelwood exploitation
Tree clearance (C,R)
Vegetation change (C,R,A)
Soil loss and exhaustion (C)
Livestock trampling (C)
Altered organic inputs (R,A)
Abstraction for irrigation (C,A)
Altered hydrology (C,A)
Pesticide use (C)
Village development (C)
Habitat restructuring (C,R,A)
Chemical change (A)
Temperature change (A)
Altered light regime (A)
Sediment mobilisation (C,A,R)
Altered sediment dynamics (A)
Fertiliser use (C)
Susceptibility to natural events (C,A,R)

2. Urbanisation
Altered hydrology (C)
Industrial pollution (A)
Domestic pollution (A)

3. Water resource development
Inundation (C,R)
Altered temperature regime (A)
Altered hydrology (C,A)
Altered chemistry (A)
Altered sediment dynamics (A)
Physical barriers (C,A)

4. Exploitation, disturbance, persecution, introductions
Populations of plants, fish, birds, reptiles, mammals (R,A)

losses approach those expected from mountain uplift of $5\,mm\,yr^{-1}$ (Ramsay, 1987). At smaller scales, such as erosion plots, runoff and sediment losses from bare land or degraded forest exceed values from undegraded forest or grassland (e.g. $30\,l\,s^{-1}$ and $100\,g\,m^{-2}$ during events of 30–40 mm over hours to days or 2–$5\,l\,s^{-1}$ and 2–$20\,g\,m^{-2}$; Gardner and Jenkins, 1995). However, bare land occupies only a small proportion of catchments, and with much agricultural land under crops during the monsoon period grassland development might even stabilise slopes rather than degrade them (Smadja, 1992). While landslides can number 5–$10\,km^{-2}\,yr^{-1}$ during monsoons, and rarely occur in forest areas, most are small slumps ($<5\,m$ long) in terraces which have low connectivity to streams or rivers (Gardner and Jenkins, 1995). Thus, while the effects of monsoonal events on runoff and sediment flux can be locally pronounced, there is still major uncertainty in assessing how effects scale up across whole catchments or regions.

Clear relationships between land use, river structure and stream biology have also proved elusive. Several large-scale surveys (Rundle et al., 1993; Ormerod et al., 1994, 1997; Brewin et al., 1995; Jenkins et al., 1995) have shown how geomorphology, stream chemistry and biotic communities differ systematically between streams in forest and agricultural areas; however, conclusions about cause–effect links have been confounded because comparisons also involved difference in altitude. Studies involving catchment comparisons at similar altitudes, for reasons given above, involve limited replication. They are, nevertheless, instructive, providing some of the only data from which assessments are possible.

Table 12.5   Selected differences in river biota between catchments in different land use in the Likhu Khola, Nepal

	Afforested	Intermediate	Agricultural	All streams
Macroinvertebrates				
Family richness	15(6)	20(6)	19(4)	18(5)
Relative abundance (%) of:				
Ephemeroptera	36.9(3.8)	47.4(22.1)	39.3(12.4)	41.5(14.1)
Plecoptera	3.8(3.3)	0.7(0.5)	0.8(1.3)	1.6(2.3)
Trichoptera	39.8(14.7)	38.1(14.0)	42.6(7.8)	40(11.4)
Coleoptera	10.5(6.3)	2.0(1.7)	9.9(7.8)	6.9(6.4)
Diptera	6.4(3.1)	7.6(3.6)	4.2(1.1)	6.1(5.4)
Fish				
Species richness	6	5(1)	6(1)	6(1)
Winter biomass (g $100\,m^{-2}$)	152	266(157)	1212(98)	563(517)
Summer biomass	685	398(170)	1187(371)	709(435)
Winter density (n $100\,m^{-2}$)	84	63(37)	243(19)	126(94)
Summer density	267	348(71)	460(76)	372(94)
Riparian birds				
Species richness	6(1)	5(1)	3(0)	4(2)
Abundance (n km^{-1})	11.5(1.4)	9.0(0.7)	1.5(0)	6.5(4.8)

Note: richness = number of species or families

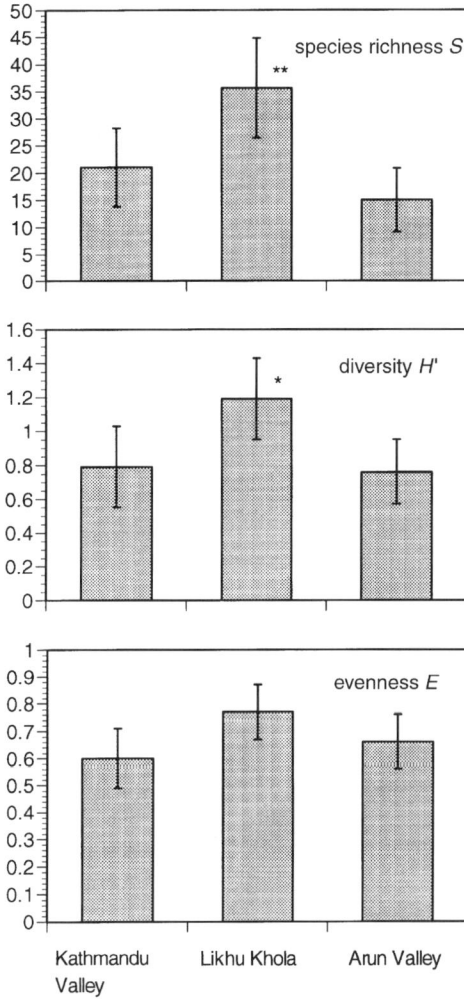

Figure 12.5   Mean species richness, diversity H', and evenness from diatom communities in groups of five replicate sites of rivers receiving agricultural runoff (Likhu Khola) and sewage (Kathmandu) by comparison with reference rivers (Arun Valley) at comparable altitude

In the Likhu Khola, adjacent sub-catchments of $2-5\,km^2$ at similar altitudes (600–800 m) have intensities of terrace cultivation varying from almost complete cover (91–98%) through intermediates (70–80%), to catchments with substantial forest, albeit secondary growth (60%). Inter-catchment comparisons here showed large increases in fish density and biomass in terraced streams by comparison with forest, due mostly to the air-breathing *Opicephalus gachua* (snakehead or murrel; Ormerod et al., 1994). Invertebrate communities showed only modest differences (Table 12.5), but included increased density of filter-feeding hydropsychid caddis from $500\,m^{-2}$ in forest streams to *c.* $2300\,m^{-2}$ in terraced streams; this is a typical response to moderate organic enrichment. The unusually speciose guild of Himalayan river passerines can be markedly reduced locally in richness and density along terraced streams due particularly to big reductions among two chats (plumbeous redstart; white-capped river chat) and four forktails, all of them insectivores (Tyler and Ormerod, 1993). Since aquatic food abundances were not reduced in the same streams, changes in habitat complexity or food available from riparian sources may be responsible. Terrestrial invertebrates are, indeed, scarcer along the Likhu Khola's terraced streams, contributing less than 1% of drifting invertebrates, contrasting with around 10–13% of organisms drifting in the tree-lined streams (Brewin and Ormerod, 1994). These negative effects on birds do not appear to be widespread, however, across the Himalaya (S. Manel and S. J. Ormerod, unpubl. data).

Diatoms have revealed particularly interesting and marked differences between stream types. Communities were more species rich and diverse in the terraced streams of the Likhu Khola by comparison with either unpolluted reference streams (Arun Valley) at similar altitudes, or organically polluted streams in the Kathmandu valley (Figure 12.5; Jüttner et al., 1996). Variations in community composition were particularly strong, with recognisable indicators of water quality present where streams were affected either by sewage (e.g. *Gomphonema parvulum, Navicula atomus, Nitzschia palea*) or agricultural runoff (e.g. *Navicula shroeteri*). These patterns matched pollution effects shown by stream chemistry, and show the strengths of using globally widespread, identifiable organisms such as diatoms as biological indicators in Nepal (Jüttner et al., 1996). They show also, however, that diversity can increase as a result of catchment disturbance, so that changes in community composition are more meaningful indicators of catchment change than diversity alone.

OPTIONS FOR CONSERVATION AND SUSTAINABILITY

These discussions illustrate the biological features of Nepali rivers which underpin their potentially large value to conservation: they hold habitat types which vary over a large range of altitude, and which are utilised in particular ways by river organisms; they hold communities of several groups which contribute to *alpha* diversity at individual locations, and promote *beta* diversity by changing between locations; they illustrate changing trophic and functional characters; they hold representative species, but also species which are scarce and in some cases globally endangered. At the same time, the catchments and aquatic systems of Nepali rivers face an array of potential changes which might impact on both conservation interest, and on subsistence use (Table 12.4). Changes of this type have been indicated throughout the tropics and sub-tropics of Asia and East Africa (Barel et al., 1985; Whitten et al., 1987; Reinthal and Stiassny, 1991; Dudgeon, 1992; Dudgeon et al., 1994: Sasekumar et al., 1994), with many authors arguing that catchment protection and conservation measures are urgently required. But what are the realistic options?

From a predominantly Western perspective, Boon (1992) has suggested that river conservation has five principal strategies. These are:

1. Preservation of pristine systems.
2. Limitation of further change in catchments which have undergone only partial change.

3. Mitigation of potential damage in catchments where catchment and riparian change proceeds.
4. Restoration of detrimentally affected systems.
5. Abandonment of rivers damaged beyond the current limits of restoration.

With the exceptions of the High Himalaya, probably no areas in Nepal are wholly pristine; some National Parks are managed intentionally for important terrestrial organisms, and other protected areas have been affected by human activity over decades to centuries (e.g. Miehe, 1990 on Langtang). Pure preservation of pristine river catchments is thus unlikely. By contrast, the National Parks limit further developments in the Annapurna Conservation Area (2,600 km^2), Langtang (1,710 km^2), Chitwan (1,432 km^2) and other parks, each of 106–3555 km^2, which in total cover a diffuse 10% of Nepal's land surface (Heinen and Yonzon, 1994). The upper reaches and tributary systems of some upland (e.g. Langtang–Khola) and lowland rivers (e.g. Reu) rise in these designated areas, often away from densely populated centres, so that their catchments, floodplains and riparian zones are maintained in a semi-natural state. In these cases, river conservation probably benefits opportunistically from efforts aimed at terrestrial conservation. Much will depend on how effectively park regulations governing forest protection, tourism, agriculture or industrial developments are enforced in the future (e.g. Yonzon and Hunter, 1991). Should there be any expansion in protected space, as some conservationists have advocated (e.g. Myers, 1988, 1990; Hunter and Yonzon, 1993; Heinen and Yonzon, 1994), opportunity might be taken to include important rivers. Even with increased cover, however, protected areas in Nepal will remain scattered, so that safeguards to whole catchments will not be achieved.

Several methods of mitigating the effects of productive catchment use on tropical rivers are available for Nepal. They reflect the need for soil or catchment conservation, and include the maintenance of protective tree or grass cover on hillslopes; adequate control of livestock; stabilisation of road embankments using grasses; buffer strips of vegetation along streams; more efficient use of fuelwood (currently 98% of Nepal's energy supply); and restrictions to subsistence activities on marginal, sensitive or currently uncultivated land (Pereira, 1987). Some of these strategies have been adopted (e.g. Fleming, 1983), but require expanded uptake; this will depend on public involvement to overcome the complex range of socio-economic, cultural, organisational and individual behavioural factors which affect catchment use (Bruntland, 1987; see Figure 3 in Thapa and Weber, 1990b).

The restoration of rivers affected by past activity will be limited by finance, but can occur through these same strategies.

In reality, the application of any of these measures in Nepal purely for river conservation is economically impractical and politically unlikely. This is the case also in the developed world: in the UK, for example, the natural or semi-natural character of most river catchments has been removed over centuries. Even though per capita GNP which is over 30 times that in Nepal might be expected to provide funds for environmental benefits, river conservation in the UK is still seen as only one among a wide array of sometimes conflicting catchment uses. A more likely practical option for river conservation in Nepal in the medium term will be a strategy which recognises the following issues simultaneously:

- That representative parts of important semi-natural rivers and their catchments can be maintained opportunistically in existing protected areas for terrestrial organisms (e.g. Hunter and Yonzon, 1993).
- That, outside protected space, rivers and river catchments have value as "natural capital", able to simultaneously support aquatic biodiversity and satisfy subsistence needs: both these uses will benefit if rivers and their catchments are maintained and managed, rather than if they are degraded (Turner, 1993).
- That in the wider environment, productive uses of land must continue, but in ways that are sustainable (Bruntland, 1987; McNeely, 1990; Schweitzer, 1992; Holdgate, 1994).
- That the integrated management of whole catchments (e.g. Newson, 1992; Burbridge, 1994) can provide a strategic framework for achieving the above, while minimising conflicts where they might arise.
- That, in order to facilitate integrated management, there is a need to develop an appropriate framework in political, legislative, regulatory, technical, educational and ethical terms.

These views strongly parallel those of other workers in the tropics (e.g. Dudgeon et al., 1994), and are close to mainstream international perspectives on the protection of biodiversity (Heywood, 1995). Care is needed, however, over definitions. For example, contrastingly "hard" and "soft" concepts of sustainability place strongly differing emphasis on the values of resources in semi-natural ecosystems; "softer" perspectives suggest that human ingenuity will allow the substitution of naturally available resources, while technological solutions will solve problems of catchment management where they arise (Turner, 1993; Bartelmus, 1994; Table 12.6). Thus, the Bruntland Commission (Bruntland, 1987) defined sustainability in a way that made no explicit mention of the protection of biodiversity, although cognisant of the importance of damage to ecosystems, and providing important leadership in recognition of the precautionary principle. This same definition has since been taken up by

Table 12.6 Contrasting and simplified attributes of extreme forms of "weak" and "strong" sustainability

Weak sustainability	Strong sustainability
Requires only that a total capital stock is sustained	Requires human impact < global carrying capacity
Perfect substitution between natural and anthropogenic capital	Perfect interchange between natural and anthropogenic capital
Technical innovation offsets natural resource depreciation	No depreciation on natural capital (i.e. harvesting rate < regeneration rate)
No critical natural capital	Soils, forest, catchments, rivers represent natural capital that should be maintained and restored
Environmental quality can decline	Growth, decoupled from impacts on environmental quality

many national governments (e.g. Department of the Environment, 1993). Nature conservationists might prefer the earlier IUCN view (IUCN, 1980), that sustainable management allows productive use while ensuring "the maintenance of essential ecological processes and life support systems, the preservation of genetic diversity, and the sustainable utilisation of species and ecosystems", a view subsequently echoed in the Rio Convention on Biological Diversity (Heywood, 1995, p. 15). These fundamental contrasts reflect widely differing notions of sustainability, with the potential for ethical conflict, confusion and misinformation.

Other writers also recognise that the attainment of sustainability is beset with difficulties. There are socio-political barriers to implementation (e.g. Barbier, 1987; O'Riordan, 1993) and a pressing need to incorporate concepts of sustainability into legislature (Moran, 1994), but neither of these needs will be aided by the conceptual weakness outlined above. To some conservationists, there is inadequate emphasis on protecting wilderness against encroachment in view of the drive for sustainable "use" and "development" (Noss, 1991; Holdgate, 1994). Equally, to some scientists, there are difficulties in understanding catchment ecosystems in Nepal sufficiently well to know what level of use will be sustainable (cf. Ludwig et al., 1993; see above). Also, as with many other "common" resources, river catchments are potentially subject to external costs not only through failures of economic markets to value them appropriately (Heywood, 1995) but also because of impacts from external environmental factors outside catchment control; examples would be transboundary air pollution and the diffuse effects of climate change. Perhaps above all in Nepal, there is the major force of a population which will double in the next 20–30 years, and hence will increase resource requirements to twice current levels unless per capita consumption declines (cf. O'Riordan, 1993). These problems are liable to affect many other countries which have biological richness equal to Nepal but also similar economic poverty.

ARE RIVER SYSTEMS INDICATORS OF CATCHMENT SUSTAINABILITY?

In the face of such socio-political difficulties and current limits to understanding river catchments, how are we to judge whether sustainable catchment management is being achieved? Rivers and their biodiversity can provide an important lead. Already, there is recognition throughout the developed world that rivers integrate processes in whole catchments, and their biota are regularly monitored to detect change at this scale (e.g. Calow and Petts, 1994; Norris et al., 1995). They are regarded as indicators of ecosystem health, degradation and integrity (e.g. Karr, 1993; Cairns et al., 1993), and can provide the targets towards which sustainable catchment management might be directed (Bartelmus, 1994). So far, however, these principles have not been exploited in the less developed world, which is unfortunate because of the cost-effective and advantageous methods available. Our own work, reviewed briefly here, shows that there is scope for such biological surveillance in Nepal (e.g. Jüttner et al. 1996; Ormerod et al., 1997), where natural biological patterns are increasingly understood, and anthropogenic perturbations increasingly recognisable. There is now a need for the development of a monitoring strategy that will enlarge and develop the baseline already in existence, while allowing also the subsequent measurements that can detect change (e.g. Dudgeon et al., 1994). Almost certainly, this will require an organisation to carry out surveillance, to train its exponents, and to advise bodies already managing catchments for soil and watershed conservation (Dr Sudobh Sharma, pers. comm.). Without doubt, Nepal is not alone in the less developed world in all these respects, and there may be important and transportable lessons for other tropical and sub-tropical countries from this brief case study.

ACKNOWLEDGEMENTS

Our thanks are due to many individuals and organisations who have debated these issues, or who have provided the opportunities for the work on which the review is based. They include Hem Sagar Bharal of Bird Conservation Nepal; Dr Sudobh Sharma of Kathmandu University and LEAD Nepal; Dr Ajaxa Dixit of the Nepal Water Conservation Foundation; staff of the Department of National Parks and Wildlife Conservation and the Central Soil Science Division of His Majesty's Government of Nepal; the staff of the Catchment Research Group in the University of Wales; Dr Rita Gardner (Royal Geographical Society); Dr Alan Jenkins and Dick Johnson (UK Institute of Hydrology); Dr Phil Boon; Dr David Dudgeon; the Overseas Development Administration, and the Darwin Initiative for the Survival of Species. The views expressed are our own and implicate none of these others.

REFERENCES

Barbier, E.E. 1987. 'The concept of sustainable economic development'. *Environmental Conservation,* **14**, 101–110.

Barel, C.N.D., Dorit, R. and Greenwood, D.H. 1985. 'Destruction of fisheries in Africa's Lakes'. *Nature,* **315**, 19–20.

Bartelmus, P. 1994. 'Environment, Growth and Development'. Routledge, New York.

Bassett, M. 1993. 'The distribution of fish in streams of the Likhu Khola, Nepal'. Unpublished MSc Thesis, University of Wales, Cardiff.

Boon, P.J. 1992. 'Essential elements in the case for river conservation'. In *River Conservation and Management* (eds P.J. Boon, P. Calow and G.E. Petts), pp. 11–34. Wiley, Chichester.

Brewin, P.A. and Ormerod, S.J. 1994. 'The drift of macroinvertebrates in streams of the Nepalese Himalaya'. *Freshwater Biology,* **32**, 573–584.

Brewin, P.A., Newman, T. and Ormerod, S.J. 1995. 'Patterns of macroinvertebate distribution in relation to altitude, habitat structure and land use in streams of the Nepalese Himalaya'. *Archiv für Hydrobiologie,* **135**, 79–100.

Brunsden, D., Jones, D.K.C., Martin, R.P., Doorkamp, J.C. 1981. 'The geomorphological character of part of the lower Himalaya of eastern Nepal'. *Zeitschrift für Geomorphologie,* **37**, 25–72.

Bruntland, G.H. 1987. '*Our Common Future*'. Report of the World Commission on Environment and Development, Oxford University Press, Oxford.

Buckton, S.T. 1998. Spatio-temporal patterns in the distribution and ecology of river birds. Unpublished PhD thesis, Cardiff University.

Burbridge, P.R. 1994. 'Integrated planning and management of freshwater habitats, including wetlands'. *Hydrobiologia,* **285**, 311–322.

Cairns, J. Jr, McCormack, P.V. and Niederlehner, B.R. 1993. 'A proposed framework for developing indicators of ecosystem health'. *Hydrobiologia, 263*, 1–44.

Cairns, M.A. and Lackey, R.T. 1992. 'Biodiversity and management of natural resources: the issues'. *Fisheries, 17*, 6–10.

Calow, P. and Petts, G. 1994. '*The Rivers Handbook, Volume 2*'. Blackwell Scientific Publications, Oxford.

Department of the Environment, 1993. '*UK Strategy for Sustainable Development*'. Consultation Paper, Department of the Environment, London.

Dudgeon, D. 1992. 'Endangered ecosystems: a review of the conservation status of tropical Asian rivers'. *Hydrobiologia, 248*, 167–191.

Dudgeon, D., Arthington, A.H., Change, W.Y.B., Davies, J., Humphrey, C.L., Pearson, R.G. and Lam, P.K.S. 1994. 'Conservation and management of tropical Asian and Australian waters: problems, solutions and prospects'. *Mitteilungen Internationale Verienigung Limnologie, 24*, 369–386.

Eckholm, E. 1975. 'The deterioration of mountain environments'. *Science, 189*, 764–770.

Fleming, W.M. 1983. 'Pewa Tal catchment management programme: benefits and costs of forestry and soil conservation in Nepal'. In *Forest and Watershed Development and Conservation in Asia and the Pacific* (ed. L.S. Hamilton). Westview Press, Boulder, Colorado.

Gardner, R. and Jenkins A. 1995. 'Land use, soil conservation and water resource manaqement in the Nepal Middle Hills'. Unpublished report to the Overseas Development Administration, Royal Geographical Society and Institute of Hydrology, Wallingford.

Gilmour, D.A. 1988. 'Not seeing the wood for the trees: a reappraisal of the deforestation crisis in two hill districts of Nepal'. *Mountain Research and Development, 8*, 343–350.

Gupta, A. 1983. 'High magnitude floods and stream channel response'. *Special Publications of the International Association of Sedimentology, 6*, 219–227.

Gupta, A. 1988. 'Large floods as geomorphic events in the humid tropics. In *Flood Geomorphology* (ed. by V.R. Baker, R.C. Kochel and P.C. Patton). Wiley, Chichester.

Heinen, J.T. and Yonzon, P.B. 1994. 'A review of conservation issues and programs in Nepal: from a single species focus towards biodiversity protection'. *Mountain Research and Development, 14*, 6176.

Heywood, V.H. 1995. '*Global Biodiversity Assessment*'. Cambridge University Press, Cambridge.

His Majesty's Government of Nepal/International Union for the Conservation of Nature, 1988. '*Building on Success: The National Conservation Strategy for Nepal*'. Kathmandu, Nepal.

Holdgate, M.W. 1994. 'Ecology, development and global policy'. *Journal of Applied Ecology, 31*, 201–211.

Hrabovsky, J.P. and Miyan, K. 1987. 'Populations growth and land use in Nepal: the Great Turnabout'. *Mountain Research and Development, 7*, 264–270.

Hunter, M.L. and Yonzon, P. 1993. 'Altitudinal distributions of birds, mammals, people, forests and parks in Nepal'. *Conservation Biology, 7*, 420–423.

Inskipp, C. 1989. '*Nepal's Forest Birds: Their Status and Conservation*'. International Council for Bird Preservation, Cambridge.

IUCN, 1980. '*World Conservation Strategy*'. IUCN–UNEP–WWF, Gland, Switzerland.

Ives, J.D. and Messerli, B. 1989. '*The Himalayan Dilemma: Reconciling Development and Conservation*'. Routledge, New York.

Ives, J.D. and Pitt, D.C. 1988. '*Deforestation: Social Dynamics in Watershed and Mountain Ecosystems*'. Routledge, New York.

Jenkins, A., Sloan, W.T. and Cosby, B.J. 1995. 'Stream chemistry in the Middle Hills and high mountains of the Himalayas, Nepal'. *Journal of Hydrology, 166*, 61–79.

Jha, P.K. 1992. *Environment and Man in Nepal*. Craftsman Press, Bangkok.

Jüttner, I., Rothfritz, H. and Ormerod, S.J. 1996. 'Diatoms as indicators of river quality in the Nepalese Middle Hills with consideration of effects by habitat-specific sampling'. *Freshwater Biology, 36*, 101–112.

Karr, J.B. 1993. 'Defining and assessing ecological integrity: beyond water quality'. *Environmental Toxicology and Chemistry, 12*, 1251–1531.

Ludwig, D., Hilborn, R. and Walters, C. 1993. 'Uncertainty, resource exploitation and conservation: lessons from history'. *Science*, **260**, 17–36.

McNeely, J.A. 1990. 'How conservation strategies contribute to sustainable development'. *Environmental Conservation*, **17**, 9–13.

Messerli, B. and Ives, J.D. 1997. '*Mountains of the World: A Global Priority*'. Parthenon, New York.

Miehe, G. 1990. '*Langtang Himal: Flora und Vegetation als Klimazeiqer und Zeugen in Himalaya*'. Dissertation Botanicae Band 58 in der Gebründer Borntraeger Verlagsbuchhandlung, Berlin, Germany.

Moran, V. 1994. 'A different perspective on sustainability'. *Ecological Applications*, **4**, 405–406.

Myers, N. 1988. 'Threatened biotas: "hot spots" in tropical forests'. *Environmentalist*, **8**, 187–208.

Myers, N. 1990. 'The biodiversity challenge: expanded hot-spot analysis'. *Environmentalist*, **10**, 243–256.

Nautiyal, P. 1989. 'Mahseer conservation problems and prospects'. *Journal of the Bombay Natural History Society*, **86**, 32–36.

Newson, M. 1992. '*Land, Water and Development: River Basin Systems and their Sustainable Management*'. Routledge, London.

Norris, R.H., Hart, B.T., Finlayson, M. and Norris, K.R. 1995. 'Use of biota to assess water quality'. *Australian Journal of Ecology*, **20**, 1–227.

Noss, R.F. 1991. 'Sustainability and wilderness'. *Conservation Biology*, **5**, 120–122.

O'Riordan, T. 1993. 'The politics of sustainability'. In *Sustainable Environmental Economics and Management: Principles and Practice* (ed. R. Kerry Turner), pp. 37–69. Belhaven, London.

Ormerod, S.J., Rundle, S.D., Wilkinson, S.M., Daly, G.P., Dale, K.M. and Jüttner, I. 1994. 'Altitudinal trends in the diatoms, bryophytes, macroinvertebrates and fish of a Nepalese River System'. Freshwater Biology, **31**, 309–322.

Ormerod, S.J., Baral, H.S., Brewin, P.A., Buckton, S.T., Jüttner, I., Rothfritz, R. and Suren, A.M. 1997. 'River habitat surveys and biodiversity in the Nepal Himalaya'. In *Freshwater Quality: Defining the Indefinable* (ed. P.J. Boon and D.L. Howell), pp. 241–250. HMSO, Edinburgh.

Pereira, H.C. (1987). '*Policy and Practice in the Management of Tropical Watersheds*'. Westview Press, Boulder.

Petts, G.E., Moller, H. and Roux, A.L. 1989. '*Historical Change of Large Alluvial Rivers: Western Europe*'. Wiley, Chichester.

Ramsay, W.J.H. 1987. '*Deforestation and Erosion in the Nepalese Himalaya: Is the Link Myth or Reality?*'. IAHS Publication 167, 239–250.

Reinthal, P.N. and Stiassny, M.L.J. 1991. 'The freshwater fishes of Madagascar: a study of an endangered fauna with recommendations for a conservation strategy'. *Conservation Biology*, **5**, 231–243.

Rothfritz, H., Jüttner, I., Suren, A.M. and Ormerod, S.J. 1997. Epiphytic and epilithic diatom communities along environmental gradients in the Nepalese Himalaya: implications for the assessment of biodiversity and water quality. *Archiv für Hydrobiologie*, **138**, 465–482.

Rundle S.D., Jenkins A. and Ormerod S.J. 1993. 'Macroinvertebrate communities in the middle hills and Himalaya of Nepal'. *Freshwater Biology*, **30**, 169–180.

Sasekumar, A., Marshall, N. and Macintosh, D.J. 1994. 'Ecology and conservation of Southeast Asian marine and freshwater environments including wetlands'. *Hydrobiologia*, **285**, 1–325.

Schweitzer, J. 1992. 'Conserving biodiversity in developing countries'. *Fisheries*, **17**, 35–38.

Sharma, C.K. 1987. 'The problem of sediment load in the development of water resources in Nepal'. *Mountain Research and Development*, **7**, 316–318.

Shrestha, T.K. 1986. 'Spawning ecology and behaviour of Himalayan mahseer *Tor putitoria* (Ham) in the Himalayan waters of Nepal'. In *First Asian Forum* (eds J.L. Maclean, L.B. Dixon and L.V. Hoslillos), pp. 689–992. Asian Fisheries Society, Phillipines.

Shrestha, T.K. 1990. 'Rare fishes of Himalayan waters of Nepal'. *Journal of Fish Biology*, **37** (Suppl. A), 213–216.

Smadja, J. 1992. 'Studies of climatic and human impacts and their relationship on a mountain slope above Salme in the Himalayan Middle Mountains, Nepal'. *Mountain Research and Development*, **12**, 1–28.

Smith, B.D. 1993. 1990 'Status and conservation of the Ganges river dolphin *Platanista gangetica* in the Karnali River, Nepal'. *Biological Conservation*, **66**, 159–169.

Stainton, J.D.A. (1972). '*Forests of Nepal*.' John Murray, London.

Starkel, S. 1972. 'The role of catastrophic rainfall in the shaping of relief of the lower Himalaya (Darjeeling Hills)'. *Geographica Polonica*, **21**, 103–147.

Thapa, G.P. and Weber, K.E. 1990a. 'Actors and factors in deforestation in "Tropical Asia"'. *Environmental Conservation*, **17**, 19–27.

Thapa, G.P. and Weber, K.E. 1990b. '*Managing Mountain Watersheds: The Upper Pokhara Valley, Nepal*'. Studies in Regional Environmental Planning, No. 22. Asian Institute of Technology, Bangkok.

Thompson, M. and Warburton, M. (1988). 'Uncertainty on a Himalayan scale'. In *Deforestation: Social Dynamics in Watersheds and Mountain Ecosystems* (eds J. Ives and D.C. Pitt). Routledge, London.

Turner, R.K. 1993. 'Sustainability: principles and practice'. In *Sustainable Environmental Economics and Management: Principles and Practice* (ed. R. Kerry Turner), pp. 3–36. Belhaven, London.

Tyler, S.J. and Ormerod, S.J. 1993. 'The ecology of river birds in Nepal'. *The Forktail,* **9**, 59–82.

Ward, J.V. 1994. 'Ecology of headwater streams at high altitudes'. *Freshwater Biology,* **31**, 277–294.

Whitten, A.J., Bishop, K.D., Nash, S.V. and Clayton, L. 1987. 'One or more extinctions from Sulawesi, Indonesia?' *Conservation Biology*, **1**, 42–48.

Wilkinson, S.M., Brewin, P.A., Rundle, S.D. and Ormerod, S.J. 1995. 'Observations on the behaviour of a *Dineutus* whirligig beetle (Gyrinidae) in a Nepalese hillstream'. *The Entomologist*, **114**, 131–137.

Yonzon, P.B. and Hunter, M.L. 1991. 'Cheese, tourists and red pandas in the Nepal Himalaya'. *Conservation Biology*, **5**, 196–202.

Zakiria-Ismail, M. 1994. 'Zoogeography and biodiversity of the freshwater fishes of Southwest Asia'. *Hydrobiologia*, **285**, 41–48.

CHAPTER 13

Conservation of Inland Deltas: A Case Study of the Gash Delta, Sudan

JOHN KIRKBY

and

PHIL O'KEEFE

Department of Geography, University of Northumbria at Newcastle, UK

INTRODUCTION: THE PHYSICAL SETTING OF THE GASH DELTA

The Gash river, with a catchment of $21\,000\,km^2$, rises near the coast in Eritrea, reaching Sudan near the town of Kassala (Figure 13.1). Here, heavily loaded with silt and sand eroded from the hills of Eritrea, the Gash starts to deposit its sediment forming a huge inland delta. Some 150 km to the north of Kassala town and on the outer fringe of the delta, the river disappears in the wooded Gash Die (Figure 13.2), depositing about 5.5 million tonnes of sediment each year in its channels and on its floodplain. In this way alluvium is continually added to the delta and the general elevation of the delta is raised.

For 9 months of the year, the Sudanese section of the Gash is dry, but from early June to late September, the river floods. These flood waters are the basis of a complex land use system supporting some 150,000 people living on the delta. This system is now under stress from rising populations of people and animals, recurrent droughts, over-use of groundwater, deforestation, drifting sand dunes, soil erosion and the disintegration of water management systems. Malnutrition, disease and starvation are evidence of the stress on people and may be linked to environmental degradation. In parts of the delta, 50% of children under five suffer from malnutrition. Many animals have died during recent droughts, as in 1990 when local rainfall was only 50 mm rather than the average of 150 mm.

The Sustainable Management of Tropical Catchments. Edited by David Harper and Tony Brown.
© 1998 John Wiley & Sons Ltd.

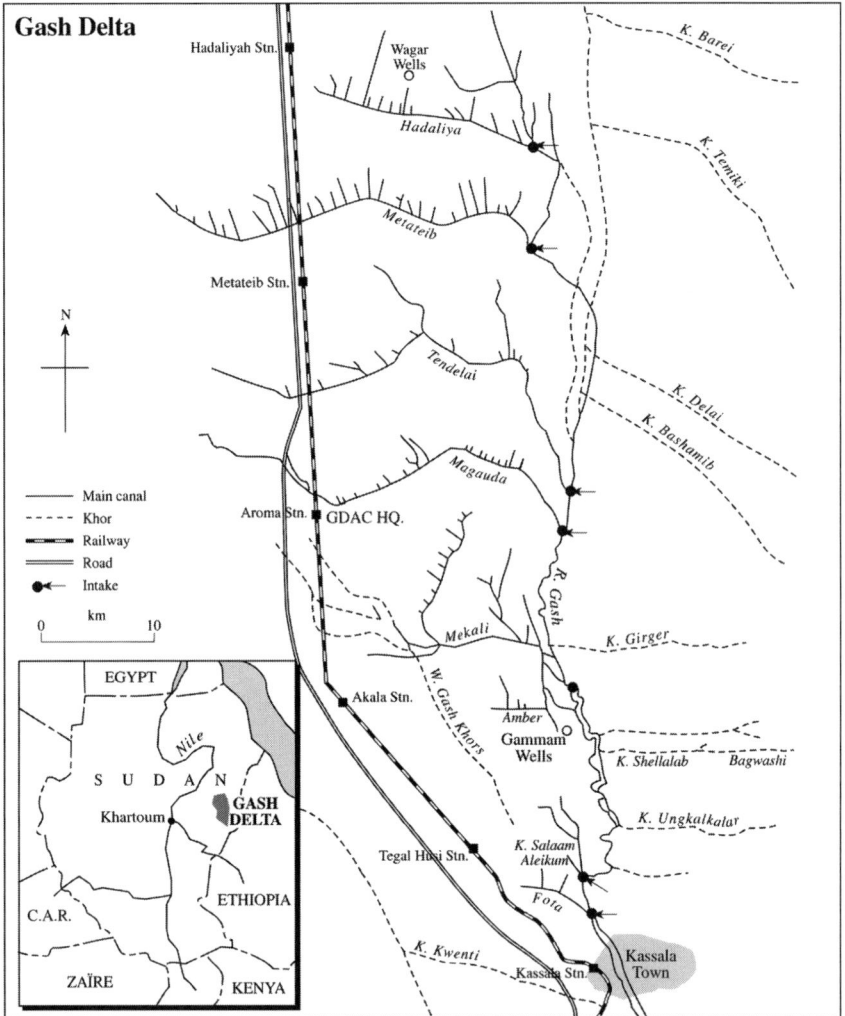

Figure 13.1   Location of Gash catchment and geography of the delta

HUMAN GROUPS ON THE DELTA

Several groups of people make use of the Gash delta, some living there permanently, and others temporarily, during the dry season. Their different lifestyles and the demands that they make on the delta's resources create a variety of stresses on the ecosystem. There are also significant differences between the abilities of the different groups to respond to environmental degradation and to the effects of modernisation.

The 100 000 Hadendowa, part of the Beja group, have lived here longest. They regard themselves as the traditional owners of the area, though formally land rights now belong to the Gash Delta Agricultural Corporation (GDAC). Though initially

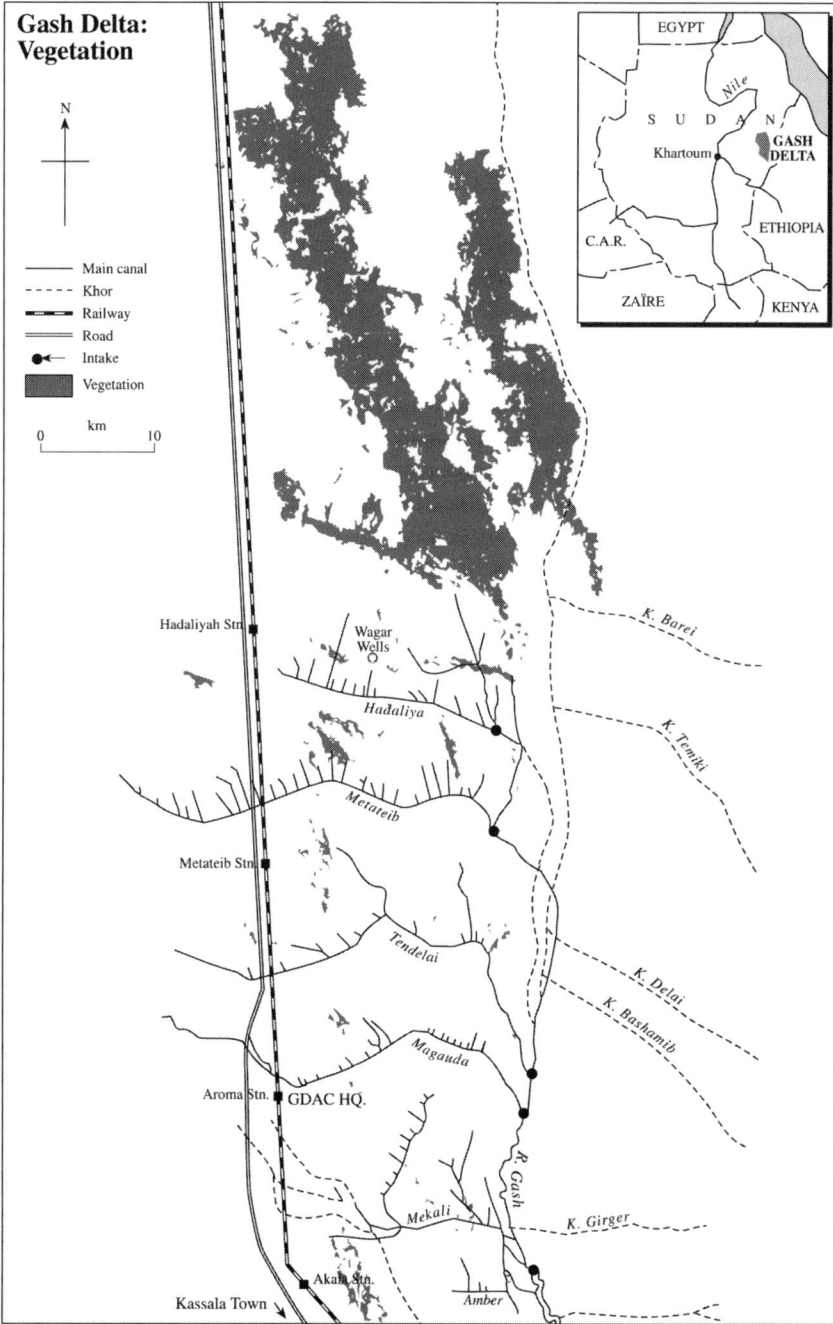

Figure 13.2   Location of wooded vegetation (Gash Die) at the natural termination of the delta

the Hadendowa appear to have been pastoralists with little involvement in agriculture, from 1927 the Government of the Sudan started a programme of sedenterising the Hadendowa (Daly, 1986) and they now fall into two groups: first the pastoral semi-nomadic Hadendowa who gain a living from animals but with a limited amount of agriculture, and second the sedentary agricultural Hadenowa who are farmers but keep some animals. Herds are taken away from the Gash delta during the wet season by the men who find grazing and browse over a considerable area of eastern Sudan, for example in the Qoz to the north of the Gash delta (Ibrahim, 1987). During the dry season their herds feed on the biomass of the delta. The semi-nomadic Hadendowa let their rights to irrigated land to some 25,000 sharecroppers (about 5,000 families) who are mainly Fellata.

The Fellata are descendants of the West African Bantu migrants who arrived in the early years of the twentieth century. The then Government of the Sudan encouraged them up to the 1920s to settle in the Gash delta as a source of labour (Daly, 1986). They are mainly agriculturalists but are also involved in trading, lorry driving, money lending, as mechanics and in a range of other non-agricultural activities. As a group, the Fellata have responded more positively to the opportunities offered by the modernisation process and are thus more able to adapt to the effects of environmental degradation.

The 20,000 Rashida, originally from the Arabian peninsula, are mainly pastoralists concentrating on camel herding. They arrived during the early years of the nineteenth century. The Rashida migrate over large distances, using the Gash delta as a dry season base for grazing and browsing their animals. Because the Rashida are not involved in agriculture, they are not directly at risk from the present threats to crop production. Nevertheless, the decline in biomass production on the delta undermines the sustainability of their economy. A number of other groups live on the delta, including refugees from Eritrea who have not yet been repatriated.

In 1982 it was estimated that 40% of household income on the delta was gained from farming, 20% from animal sales and 40% from other activities such as woodcutting, casual labouring, shepherding and varied services. Farming and pastoralism of course also contribute to the satisfaction of subsistence needs.

LAND USE AND PRODUCTION SYSTEMS

IRRIGATION SYSTEMS

Until the early years of the twentieth century, the Gash river was largely uncontrolled and liable to switch from its present eastern course into the Western Gash, as the build-up of sediment caused avulsions. In 1907 (Grabham, no date) embankments were constructed south of Kassala to permanently close off the Western Gash. Since then the Gash has been restricted to its eastern course, but with continuing troublesome local changes, associated with sedimentation of the river bed and erosion of the sandy banks. If the Gash were to return to its former western course, the present irrigation system would break down completely.

Contemporary accounts, such as the diary of R.V. Savile, an officer in the Sudan Political Service (Savile, 1902), indicate that trees were then much more widespread and a wide range of game, including lion, was being hunted on the delta. Clearly the delta was less intensively used at that time. Aerial photographs show that the amount of tree cover has significantly reduced since 1980, particularly along the Eastern Gash. Much of the reduction of tree cover can be explained by increased demand on land resources linked to rising population, but the development of the present flush irrigation system (see below) between 1923 and 1937 accounts for the removal of large amounts of tree cover.

It is not known when agriculture started on the delta but it was certainly before the nineteenth century (Richards, 1950). Cotton was being grown here for the British market at the time of the American Civil War, to replace the disrupted supplies from the southern USA. By the early years of the twentieth century, small canals had been dug in the southern part of the delta to control the Gash floods. Daly (1986) noted that some cotton was being produced on the delta in the early twentieth century. One important advantage for cotton production on the Gash delta was that the use of water was not restricted by the Nile Waters Agreement which severely limited the Sudanese use of water from the Nile.

In 1922 the Kassala Cotton Company initiated the irrigation scheme which is still in use, with some extension during the 1950s. No suitable site could be found in the Sudanese section of the Gash for a dam to create a store for perennial irrigation, and therefore a flush irrigation scheme was developed. Flood water is diverted through eight canals, several of which follow the courses of major Gash distributaries. From these canals, which are about 10 m wide, water is diverted into smaller canals leading to *misgas* – large flat-floored areas. These *misgas* are up to 4 km wide and 15 km long and run north from the main canals. To allow better control of water application, the *misgas* are further divided into units of about 30 ha (Figure 13.3).

Silt makes up about 2.6–5.4% of the volume of Gash water and management of this silt is a critical problem for both the irrigation engineers and farmers. The need to control this silt has influenced the management of the Gash, the design of the canal system and methods of water application to fields. Because the silt load is particularly heavy during the rising flood, the diversion of water to the canals is delayed until the water is less silty. As water engineers were aware that silt could quickly block the canals, they were designed with gradients between 1 in 2,000 and 1 in 3,000. On these gradients, water flows fast enough to carry most of the sediment but not so fast as to erode the unlined bed and banks of the canals. When water is released from the Gash, in order to minimise deposition, it is essential that it is not ponded in the canals but is allowed to flow directly into the *misgas*. Even with the careful alignment of canals and continuous release of water to the *misgas* there are problems of siltation and erosion, necessitating dredging and canal bank reconstruction. At present, water escapes from unrepaired gaps in the northernmost canals so that the northern *misgas* can no longer be adequately irrigated. It has also been necessary to redesign and change the locations of a number of canal offtake structures as the Gash has changed its course and silted its bed. At the Salaam Aleikum offtake, spur dikes were reconstructed in the early 1990s in an attempt to stabilise the course of the Gash. Such river structures are extremely expensive.

Water, diverted from the main canals, spreads through the *misgas* and smaller basins, guided by low embankments. In the basins the water deposits its loads of silt

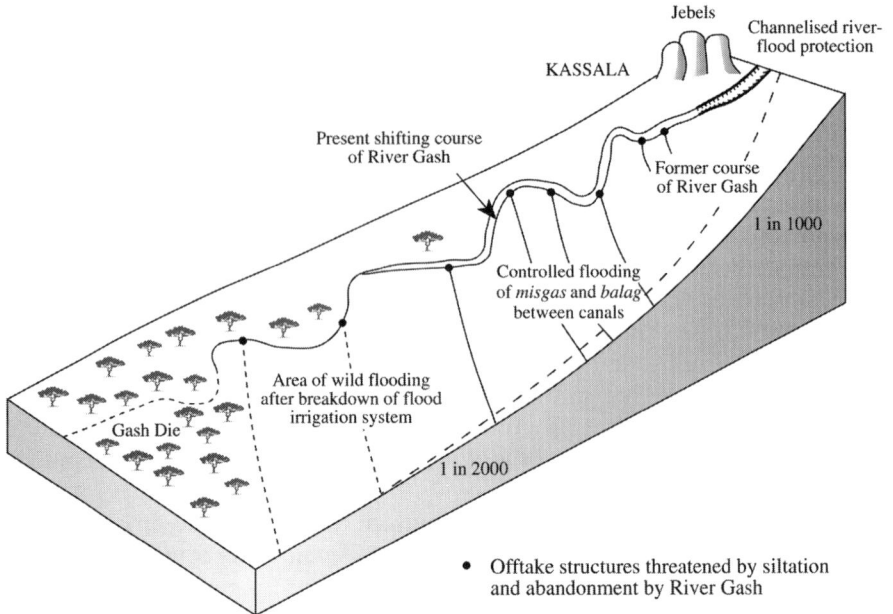

Figure 13.3   Diagrammatic representation of water management on the Gash delta

and infiltrates the soil, creating a store of soil moisture for the farming season. Ideally, depending on the texture of the soil and whether sufficient water is available, water is left standing for between 20 and 30 days. Since the availability of water is unpredictable, it is difficult to ensure that the period of submergence is optimal. Normally, if the quantity of water allows, there is a second period of irrigation. The layer of silt reduces permeability, and infiltration rates depend on local soil conditions so that there is local under-irrigation and elsewhere, local over-irrigation even within individual units. Overall the efficiency of irrigation is probably about 50%. Most of the *misgas* are used for annual field crops which take water from shallow depths. Deep percolating water is not wasted since this recharges the wells on the delta and trees elsewhere on the delta are able to benefit from the water at greater depths.

Unfortunately, the deposition of silt, which averages about $3 \, cm \, m^{-1}$ depth of water, reduces the rate of percolation and standing water is also lost through evaporation. Potential evapotranspiration is about $3,000 \, mm \, yr^{-1}$ on the Gash delta. Deposition of silt is greatest at the entrance to basins, up to 30 cm. This leads to two further problems: first the percolation rate here is particularly slow and second the period of submergence is shorter near the basin entrance. Weeds, particularly *Striga*, quickly establish on the emerged areas. Some of the local rainfall of between 100 and 200 mm may arrive before the river flood, allowing weeds to establish even before the irrigation season. In an attempt to control weed development, one of the more serious practical difficulties of farming on the Gash delta, soil preparation is started as soon as possible after the land dries. Farmers prefer to use tractors and disc ploughs to prepare seed-beds, but there are few tractors and weeds may have established before later users have disced their land. The GDAC further attempts to control weeds by flooding *misgas* in rotation, one year in three. Flood water is also used to recharge *hafirs* – large artificial ponds, which are used for watering stock during the dry season.

During the early years of the scheme it was possible to irrigate about 33,000 ha in any one year. From 1934 to 1939 between 28,000 and 36,000 ha were irrigated each year (Daly, 1991) but the disintegration of the irrigation system has not reduced the maximum irrigable area. In 1989, 15,000 ha were irrigated, in 1990, 16,000 ha, and in 1991, 25,000 ha. Priority in irrigation is given to up to 16,500 ha of *misgas*; any surplus water is used to irrigate forests. The variation in the amount irrigated reflects the size of the Gash flood and the effectiveness of management by the irrigation engineers. It is clear that these variations in the area irrigated can greatly affect the production of food. This is particularly problematic when a poor Gash flood coincides with poor rains locally as happened in 1985 and 1990.

Land in the *misgas* which are to be irrigated in any one year is divided by lottery before planting begins. The average notional entitlement is about 2 ha, but rarely is as much as this actually allocated. Entitlements vary greatly and some members of the group, particularly the sheiks and their immediate families, are given much larger holdings. For many farmers, particularly sharecroppers, who give half of the crop to the title holder, the allocation is sufficient to satisfy little more than subsistence needs. Although sharecropping is technically illegal in Sudan, 40% of the 11,600 farmers are sharecroppers who farm the traditional allocations of the Hadendowa pastoralists. Few of the farmers have livestock as an additional source of livelihood. Because allocations of land vary in location each year, farmers live in temporary settlements near their field during the cropping season. Eighty per cent of the tenancies are held by the Hadendowa who let out half of their land to sharecroppers, mainly the Fellata, who themselves hold 14% of the tenancies. The remainder of the tenancies are held by Sudanese Arabs.

From the 1920s to the 1980s the dominant crop in the *misgas* was cotton. The GDAC until recently demanded that half of the farmers' crops were cotton but the food grain *dura* (sorghum) now dominates though some cotton is still grown. The grain from the *dura* is used as human food and the stalks as valuable dry season fodder for camels. Farmers also grow a wide range of vegetables for home use and for local sale. Oilseeds are becoming important as cash crops. When the main crop has been harvested there may be enough moisture for a follow-on crop such as pumpkin or for vegetables. Animals may graze the crop residues at the end of the harvest.

In addition to the flood waters of the Gash, groundwater is also used. Two sets of wells tap this groundwater. Hadendowa tribespeople use 13 large wells on the main delta for watering their cattle and for household uses. These wells are recharged during river floods. A second group of wells, in the Kassala area at the apex of the delta and along the Eastern Gash, have long been used for perennial irrigation to produce vegetables. Overpumping of the Kassala wells, using modern diesel pumps, has led to a serious fall in the water table from about 8 m to about 35 m. Clearly this rate of abstraction is not sustainable and ultimately perennial irrigation must be reduced.

Each year there is enough water to irrigate about 5% of the whole delta. Table 13.1 shows the present extent of different land cover on the Gash delta.

Table 13.1   Land cover on the Gash delta. Source: HVA International

Section	Area (ha)
Total area of the delta	300,000
Misgas	50,000
Maximum area of *misgas* irrigated in one year	16,500
Total area of forest	50,000
Closed and open savanna (mainly in the north)	30,000
Mixed *balag* (floodable) forest	10,000
Riverine forest (along Eastern Gash)	10,000

Because the rainfall of the area is about 150 mm, the natural vegetation of the plain surrounding the delta is semi-desert or, at best, open woodland. For at least 30 years (Lebon, 1965) the plain has been suffering desertification, mainly through over-grazing. Consequently, and in addition to the effects of rising populations of people and animals, demands on the resources of the delta are increasing. Until the 1920s the damper areas of the delta were covered by lush forest and much of the rest by open woodland. Local Hadendowa tribesmen have traditionally used this woodland resource to graze and browse their animals.

ANIMAL HUSBANDRY

About half a million animals are on the delta. Camels and cattle, making up about one-third of the total, are the most desired animals. The other two-thirds are sheep and goats. More prosperous households also own donkeys, used for carrying water, fuelwood and crops. Camels, the most prestigious animals, are owned by the Hadendowa and Rashida. Unlike cattle, they browse on the move and thus are unlikely to over-consume the biomass. Camels' feet are soft and do not disturb the soil making it less prone to wind erosion. Because they are able to survive for long periods without water, they are better able than cattle to use the browse of the dry Gash Die. Cattle and sheep tend to graze intensively in small areas and consequently can expose large areas of soil to wind erosion; their feet cut up the surface soil much more than do those of camels. Goats are omnivorous and better able to survive in very degraded areas so they are more valuable than sheep or cattle during periods of drought. However, goats, through the thoroughness of their browsing, may severely deplete the biomass.

FOREST AND WOODLAND

About 50,000 ha of the Gash delta are now under forest or woodland. Open savanna covers 30,000 ha, particularly in the Gash Die in the north of the delta where agriculture is not possible due to lack of water. The formerly extensive area of *Acacia* savanna woodland has been steadily reduced during the past century and natural regeneration is limited by the effects of grazing.

Balag (irrigated forest), covering 10,000 ha, has a much fuller tree canopy than open savanna. Most *balags* are located on the north side of the individual *misgas* and are supplied with any spare water after the irrigation of the *misgas*. Much of the area of *balags* is in the northern part of the delta, using the surplus Gash water that has not been fed onto *misgas*. Because the northern canals are now in disrepair, much of the *balag* there is poorly irrigated and over considerable areas trees are dying. Although *balag* has traditionally been used for browse and grazing, attempts are being made to enclose sections of them to protect against over-use; as might be expected, these attempts at exclosure are resented by the Hadendowa. The Hadendowa deliberately set fire to parts of the *balag* in hunting wild pigs; and charcoal burners exploit them both legally and illegally to supply fuel for urban markets as far away as Khartoum. In an attempt to control deforestation, the GADC has, since 1982, attempted (unsuccessfully) to ban the cutting of greenwood in all forest areas.

Some 10,000 ha of *Tamarix* and *Prosopis* forest remain along the Eastern Gash, almost half of it on islands within the river. These areas of forest are being progressively reduced and, in consequence, the banks of the Gash are destabilised. This has allowed widening of the river. The consequent loss of hydraulic efficiency has led to locally increased sedimentation of the river bed. Bank destabilisation and increased sedimentation make changes of the course more likely, leading to bypassing of the canal intake structures. Silting of the river bed raises the height of the river flood beyond the design level of canal intakes, making it necessary to rebuild these structures.

From the 1920s to the 1950s, the area of forest was seriously reduced by the extension of agriculture. Paradoxically, one species of tree, *Prosopis*, originally introduced as a drought-tolerant source of fodder, has been spreading invasively onto agricultural land. FAO forestry experts are divided in their views on the desirability of planting *Prosopis*. The continuing depletion of forest and woodland through the effects of grazing and cutting is almost certainly more significant than the spread of agriculture. An estimated 80% of the dry season fodder is presently obtained from the various forests and woodlands.

Clearly the woodlands and forests are essential to the functioning of the pastoral system, yet the pastoral system threatens their very survival. There is no tradition of tree planting or woodland management among the local Hadendowa because resources have hitherto been adequate without such activities. During the last generation, however, wood resources have been overstretched; it has been suggested, for example, that the rate of exploitation is possibly 10 times the present sustainable yield of trees on the Gash delta. Unless effective conservation policies are implemented, the only areas of forest or woodland that are likely to remain in the long term are the parts of the Gash Die which lack surface water and thus do not allow cattle to be kept. If boreholes were to be sunk in the Gash Die, extensive deforestation would almost certainly follow as seedlings were grazed out.

SUSTAINABILITY OF THE LAND USE SYSTEM

For a number of reasons, the food production system of the Gash Delta is now under stress linked to environmental degradation, increasing to potentially catastrophic levels. Components of this environmental degradation include the removal and reduction of vegetation cover and the local invasion of settlements, fields and canals by small sand dunes. Such events appear to be evidence of desertification.

It is also clear that the irrigation system is suffering progressive breakdown. In other words the environment is deteriorating and the major component of the man-made production system is collapsing. If an adequate response is to be made to the apparent threat, it is important that the rate at which the degradation is occurring is verified and that the full range of proximate and ultimate causes is understood.

Environmental degradation here is best interpreted on two time scales, each with its distinctive momentum: long-term, secular decline and short-term, contingent events. Year-by-year fluctuations, for example in climate, particularly local rainfall, the size of the Gash flood, breaks in canal banks and changes in the level of the water table may cause improvement or deterioration in crop and livestock productivity and in the conditions of forests. These effects must be disentangled from the long-term and cumulative effects of, for example, overgrazing, overbrowsing, increased cutting of trees for fuelwood and charcoal and the collapse of the irrigation system. Short-term changes may trigger disaster but take place in environments created by the more significant long-term trends. Nevertheless, the shorter- and longer-term changes interact, with the possibility that stress conditions in a series of bad years may finally kill off already stressed vegetation, and that overuse of the environment in the short term may force it to cross a threshold of degradation. Similarly, long-term pressures on the environment reduce the resilience of vegetation to short-term and intense stress.

Evidence of long-term changes is complex and in some ways contradictory. One indicator of increasing environmental degradation is said to be *haboobs*, duststorms, which are widely believed to be increasing in frequency. If vegetation is removed, it is argued that the soil is more prone to wind erosion. In fact the evidence for an increase in *haboobs* is questionable: R.V. Savile's diary (Savile, 1902) lists numerous *haboobs* at that time. Lebon's map of land misuse in Sudan (Lebon, 1965) shows the Gash delta as being clearly within the area of "overstocking on semi arid grazing by camel-owning tribes". This map was based on his extensive field experience in Sudan and is presumably a reflection of demonstrable environmental degradation. It would appear, then, that some aspects of desertification are not of recent origin. One the other hand, the aerial photographic evidence for deforestation, and measurements of the rate of destruction of trees, compared with the calculated sustainable yield, both suggest that the rate of environmental destruction is relatively fast. Similarly, the continuing deterioration of the irrigation system is unquestionable and present trends are potentially catastrophic.

Whether the health and nutritional status of the people can be taken as evidence of long-term environmental degradation is very uncertain, since few data are available for earlier periods. In any case, the causes of malnutrition and disease are not simply environmental: social factors such as the place in society of women and children, the age of marriage and family size may explain high levels of disease, poor health and poor nutrition. The available evidence suggests that some components of the environment have been deteriorating for a considerable time and others from more recently. In most cases the trend is towards deterioration.

The environmental decline of the Gash delta could be interpreted simply as a classic case of Malthusian population pressure on resources leading to cumulative and self-reinforcing environmental collapse. The arrival of the Rashida and Fellata early in the twentieth century, certainly led to an increase in human population and, in the case of the Rashida, animal population. The implication of such an interpretation is that little can be done to save the Gash delta and create a sustainable future in either the short or longer term. A reduction in population could be achieved only in the long term and would entail revolutionary social and cultural change which are unlikely in the present political regime. Alternatively, emphasis might be placed on controlling the activities of pastoralists, for example, reducing the number of their animals and controlling their browsing and grazing. Such attempts have been very unpopular, in particular the attempts at exclusion. It must be accepted that pastoralism is virtually the only possible production system in much of the delta and that pastoralism is an effective way of converting biomass into a product that can be used by humans. Methods of meeting the needs of pastoralists and integrating them with the agricultural system are considered below.

Breakdown of the 70-year-old irrigation system contributes significantly to stresses in both the agricultural and pastoral systems and thus to the sustainability of the delta production system as a whole. As the opportunities for agriculture are reduced, people need to depend more on pastoralism, placing more stress on trees and other plants. The possibility of managed irrigation of the *balag* is also reduced. Ultimately, if the irrigation system breaks down, the possibilities of production will be determined by the vagaries of nature. It is clear that the decision to construct the Gash irrigation system was a Faustian bargain, as is any irrigation system. Hydraulic environments do not fossilise in a situation advantageous to mankind: the sediment carried by the stream, while importing fertile soil, silts the irrigation system and causes the river to switch course. Continuous investment is needed to maintain the advantages of irrigation (increased yields, reduced risk of crop failure, increased control on planting time).

A further set of environmental stress factors relate to the nature of irrigation. Wittfogel (1957) developed the notion of hydraulic totalitarianism: irrigation schemes of any size depend on co-ordinated actions. Whoever controls the co-ordination is in a position of enormous power. During the period of control by the Kassala Cotton Company and until the 1980s, irrigation has been managed mainly for commercial monoculture with limited scope for household production of food crops. Consequently, land users have needed to look to pastoralism for subsistence. Similarly, the small size of the bulk of allocations causes farmers to seek additional income from the woodland, for example in cutting wood for charcoal.

In the short term, full rehabilitation of the whole irrigation scheme is a pressing need, but little money is available. In truth, the sustainability of the irrigation scheme, while in proximate terms environmental, is in a more fundamental sense social and political. In 1988, the Development Assistance Committee and Organisation for Economic Cooperation and Development (DAC/OECD, 1988) investigated the circumstances affecting the sustainability of development projects worldwide. The adjustment of activities to the circumstances of the physical environment were one factor only. Other factors were the availability of financial support, management capacities, the amount of local co-operation, suitable technology, political and economic stability and favourable macro-economic policies. Undoubtedly, the technical capacity to manage re-establishment of the project exists in Sudan, but other circumstances are much less favourable. In particular, economic stability and the ability to provide long-term financial support are lacking. This is partly due to long-term political and economic problems of Sudan and partly due to the with-drawal of international aid following the Sudanese Government's support for Iraq in the Gulf War. The rise of Islamic fundamentalism and its strong presence in the Government of Sudan make an early resumption of aid unlikely. Local participation has hitherto been limited by the feudal social relations and the top-down delivery focus of GDAC. A recent addition to the feudal relationships has been the appearance of four large companies as entitlement holders. It would be surprising if they did not receive preferential treatment for water supply.

ACTION TO IMPROVE SUSTAINABILITY

Since 1987 the Netherlands government has been supporting the development and rehabilitation of the Gash delta as part of its support for development in the Sudan generally (IOV, 1992). In this area, it is working with GDAC. At present, however, the Netherlands is almost alone among western governments in giving aid to Sudan. Netherlands aid for the Gash delta has supported action in a number of linked activities. Clearly, the production system will break down if irrigation is not maintained, therefore the limited money available is used to attempt to stabilise the river's course near the canal intakes and to improve water and silt management in canals and *misgas*. Planting of trees on canal margins should limit sand blowing in to the canals. In the agricultural sector, farmers have been helped in experiments on alternative crops, weed control, soil preparation and planting techniques. Agrono-mists have been investigating the relationship between soil quality, water manage-ment and crop yields and the possibility of growing a wider range of crops. In assessing progress it was decided that greater emphasis should be placed on the farmers' own experiments and on spreading information on successful methods used by the most go-ahead farmers, rather than on research station activities alone.

Trees have a wide range of uses for farmers, and so their unnecessary removal, for example in preparing *misgas* using machinery, has to be part of the education process. Uses are:

- Fodder for animals (grass and leaves)
- Shade trees for meetings
- Firewood
- Charcoal for household use and for sale
- Fruits, seeds, pods as household food
- Poles for building
- Timber for household use and for sale
- Wood for making farming equipment
- Raw materials for thatch

- Raw materials for tent making
- Traditional medicines
- Food flavourings
- Hunting
- Environmental protection against wind erosion and desiccation

Foresters are encouraging and providing resources to support farmers in replacing trees that have been lost from the *misgas* and other parts of the environment since the 1920s. One of the promising methods for increasing tree cover is through agro-silvo-pastoralism. Crops dominate while the trees are small, and fodder and other products can be taken from trees later in the 15-year rotation. Trees have grown rapidly on the experimental plots and intercropping of vegetables has been successful. Fast-growing introduced trees such as *Eucalyptus* perform well in research station experiments but on-farm experiments have shown that they are unlikely to survive droughts such as would occur if the irrigation system deteriorated further. Indigenous trees such as *Acacia* in similar conditions have shown that they can tolerate drought. Outside the *misgas*, shelter belts have been planted in and near villages and along canals to attempt to reduce the blowing of sand but also to give a range of other benefits. Tree nurseries are being developed to support agroforestry. Both men and women are working on separate communal woodlots near villages. Forest guards attempt to protect the remaining areas of *balag* against grazing and cutting for charcoal making.

These investments in irrigation, agriculture and forestry should start to restore the deteriorating physical environment of the delta and to improve food security. A full restoration of the environment is a distant prospect and realisable only if it achieves support from the bulk of the people. Such support depends on their perceiving real and direct benefits to their incomes and household security. Further, such support is likely to be achieved only if people have a sense of ownership and full participation in decision making and allocation of benefits. This ideal of participation and empowerment will be difficult to achieve in the present feudal framework of the Hadendowa and Rashida socio-politico-economic systems.

In its aid policy, the Netherlands Government pays particular attention to improving the position of the disadvantaged members of society, particularly women. Among the Hadendowa, men and women live almost completely separate lives. For most of the time Hadendowa women are confined to the home and unable to participate in economic activities. The challenge is to reduce the enforced seclusion of the Hadendowa women and to allow them to take a more active part in economic life. In comparison, Fellata women have much more freedom, are involved in agriculture and able to earn income from marketing the vegetables they grow. Woodlots have been provided for groups of Hadendowa women. This has been helpful in several ways: women earn an income from selling wood products, they have recently started to grow vegetables and bush crops in their woodlots and women now have more freedom of movement away from the home. Women are being encouraged to create vegetable gardens near their homes. This helps to improve nutrition and allows the possibility of earning extra income.

Figure 13.4 summarises the changes to the land use system that it is hoped to achieve during the next few years and compares the resulting landscape with the

Present land use pattern

Intended land use pattern

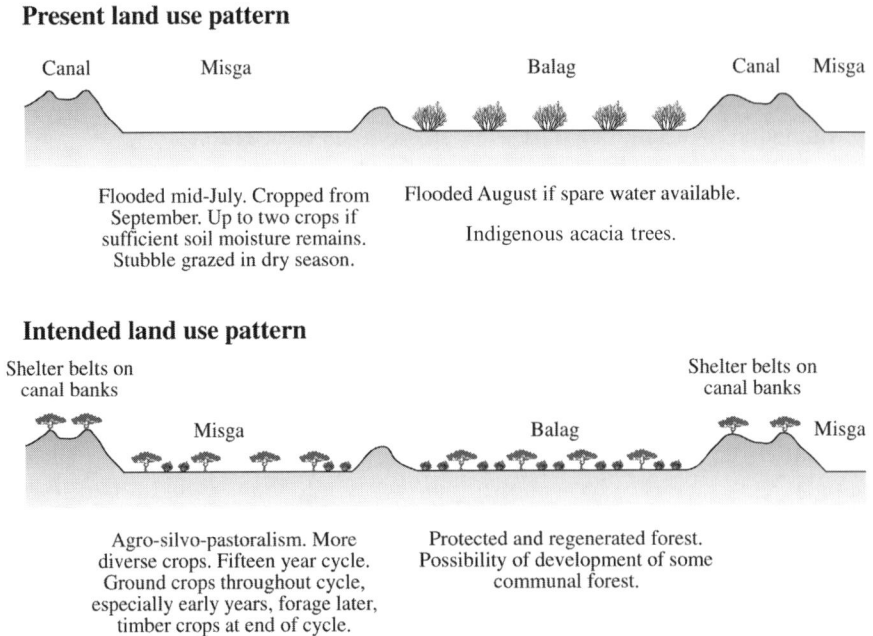

Figure 13.4 Present land use pattern compared with the future, pattern necessary for sustainability.

present system. It is unlikely that money will be available to rehabilitate the irrigation system in the near future so that the challenge is to maintain as much of the system as possible, and to make the most efficient use of the water that is available. The key to successful environmental management is agro-silvo-pastoralism. Evidence from on-farm experiments and schemes involving women indicates that this approach can work, but significant changes in the social system and power relations will be needed.

REFERENCES

The material in this chapter is based on field investigations in Kassala Province during 1992, internal reports of HVA International (Agricultural Development Specialists), internal reports of Gash Delta Agricultural Corporation, and internal papers of ETC (UK) Consultants in Development Programmes, 117 Norfolk Street, North Shields, Tyne and Wear. The authors work with ETC (UK) and would welcome comments on and discussion of the paper.

Bascomb, J.B. 1990. Border pastoralism in eastern Sudan. *Geographical Review,* **8**(4), 416–430.
DAC/OECD, 1988. *Sustainability of Development Programmes: A Compendium of Donor Experience.* Paris.
Daly, M.W. 1986. *Empire on the Nile.* Cambridge University Press, Cambridge.
Daly, M.W. 1991. *Imperial Sudan: The Anglo-Egyptian Condominium 1934–1956.* Cambridge University Press, Cambridge.

Grabham, G.W. no date. *Water Supplies on the Gash Delta*. Report D29, held in Sudan Archive, Durham University.

Ibrahim, F.N. 1987. Ecology and land use change in the semi-arid zone of the Sudan. In *Lands at Risk in the Third World* (eds P.D. Little, M.M. Horowitz and E.E. Nyerges), pp. 213–229. Westview Press, Boulder and London.

IOV (Operations Review Unit), 1992. *The Sector Programme for Rural Development*. Netherlands Ministry of Foreign Affairs, The Hague.

Lebon, J.H.G. 1965. *Land Use in Sudan*. Geographical Publications, Bude, Cornwall.

Richards, C.H. 1950. *The Gash Delta*. Ministry of Agriculture Bulletin Number 3, Khartoum.

Savile, R.V. 1902. Diary (unpublished), held in the Sudan Archive, Durham University.

Wittfogel, K.A. 1957. *Oriental Despotism: A Comparative Study of Total Power*. Yale University Press, New Haven.

CHAPTER 14

The Importance of People in the Management of Tropical Catchments

CLARE MADGE

Department of Geography, University of Leicester, UK

INTRODUCTION

This paper explores the importance of indigenous people to the management of tropical catchments and the contribution that an ethnographic approach can make to understanding agroforestry issues and management policies in such areas. Although recent research has stressed the utility of an ethnographic approach for aiding rural management, such an approach has not yet markedly filtered through to government and NGO management practices (Long and Long, 1992; Parkin and Croll, 1992; Hobart, 1993; Pottier 1993). To illustrate the benefits of an ethnographic approach, this paper concentrates on a case study of one Jola community in The Gambia, West Africa. The paper explores a facet of agroforestry that, according to Falconer (1990), has hitherto received only limited research attention: the role that collected products (local plant and animal species gathered, fished and hunted, from fallow, semi-cultivated, forest and river areas) play in rural African households. According to Ogle and Grivetti (1985) collected products have often remained "invisible" in agricultural and forestry management policies and are assigned a low status and value compared to commercial trees. This case study reveals the vital role that collected products play in a rural Gambian community.

In particular, the study investigates the interrelated subsistence, commercial and socio-cultural roles of collected products and highlights gender differences in the use of these products. It demonstrates that an ethnographic approach can provide rich and detailed sources of data from which to understand the multiple roles of collected products, and the meanings associated with their use. Moreover, since these roles are interpreted from within the cultural framework of the studied population, the research also potentially provides a base from which to develop sustainable management policies.

The Sustainable Management of Tropical Catchments. Edited by David Harper and Tony Brown.
© 1998 John Wiley & Sons Ltd.

The discussion is divided into five sections. After a brief introduction to the case study area and methodology, the second section examines the various subsistence, commercial and socio-cultural roles of collected products. In the third section the use of collected products is placed in historical perspective and explanations sought for their increased significance. In response to this marked expansion, indigenous management techniques have been employed to regulate environmental resources; these are examined in the fourth section. In the fifth section the discussion speculates on whether such management techniques can be built upon by formal planning mechanisms, leading to an assessment of the applicability of an ethnographic approach for tropical catchment management.

CASE STUDY AND METHODOLOGY

The Gambia is a small country situated in the westernmost part of Africa, identified by the World Bank (1988) as one of the poorest countries in the world (Figure 14.1). The Gambian economy is based on subsistence agriculture, the most important food crops cultivated being rice, maize and millet grown on small family farms. The groundnut is the main cash crop, representing 88% of exports in 1991 (Europa Publications, 1992). The tourist industry has recently expanded in The Gambia, and now accounts for 11% of GDP and involves over 100,000 tourist visits per year, over half of whom originate from Britain. The estimated population of the country was 930,249 in 1993, and comprises an amalgamation of 14 ethnic groups. The Jola are an acephalous group of sedentary wet rice agriculturalists, consisting of 8% of the total population (Linares, 1985). The Gambia is situated in the sub-Saharan tropical climate zone, the natural vegetation being mostly Guinea Savanna or Guinea Woodland Savanna. Although the country is not covered uniformly by dense forest, it is nevertheless very rich in tree life, containing over 150 species of woody plants (Percival, 1968). During the last 20 years over one-third of The Gambia's woodlands have been destroyed or drastically altered, the fallow interval in the

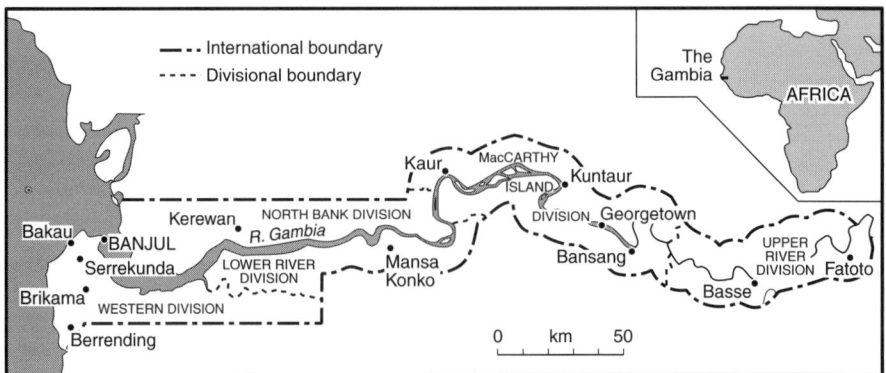

Figure 14.1   The Gambia showing the location of Berrending village (south-west corner)

traditional bush-fallow system being reduced from 30 years to 3 years (Department of Wildlife Conservation, 1979).

One village, Berrending, was selected for detailed study (Figure 14.1). The physical organisation of the village was based upon divisions according to ward, household, sinkiro (the consumption unit) and dabada (the production unit). The village consisted of six wards of approximately 650 individuals in 40 households and showed a complex socio-political organisation. The household was usually inhabited by one extended patrifilial group consisting of up to four generations of related males. In the village farming predominated during the rainy season and collecting in the dry season. Domestic work was undertaken all year round, usually by young women. Collected products were procured from fallow, semi-cultivated, river and forest areas (Figure 14.2).

Eight households (one-fifth of all village households) were selected for detailed ethnographic study; they were chosen after completing a village survey which collected data on household size, ethnicity, marital status, wealth and education levels. Quantitative and qualitative data were collected through surveys, structured interviews and participant observation. Two surveys were undertaken each week from March 1988 to February 1989. The first collected information on all the plant and animal species that the 38 individuals in the eight households had collected during the past week, the use made of these products, whom they shared/sold the products with/to, whom they went collecting with and the amount (local measurements) of the products collected. The second survey collected information on the source of any income earned on a weekly basis. Informal and structured individual and group interviews were undertaken in conjunction with participatory research to determine in-depth information on values, uses and beliefs associated with collected products. By combining these methods over a one year period it was possible to attempt to interpret information from within the cultural framework of the studied population.

THE ROLE OF COLLECTED PRODUCTS: SUBSISTENCE

During the year of the survey, plants accounted for 83% of the total collections and animals for the remaining 17%. Figure 14.3 summarises the different subsistence uses of the collections as categorised by the villagers. When the total data were disaggregated according to gender, women collected 56% of the total and men 44%. Collecting for subsistence was clearly important to both women and men. However, the use made of the collected product varied according to gender (Figure 14.4). Women predominated in the collection of fuelwood and food sauces while the men concentrated on collecting structural materials, beverages and food snacks. Most previous research suggests that women collect more food than men. However, this work shows that food snacks accounted for over one-quarter of the men's total collection. These snacks were mostly collected by young men and boys since they searched for food snacks to supplement their main food intake, and sometimes even to replace it. This observation is noteworthy since not only is the role of boys and young men in food procurement often overlooked but food snacks are also often not

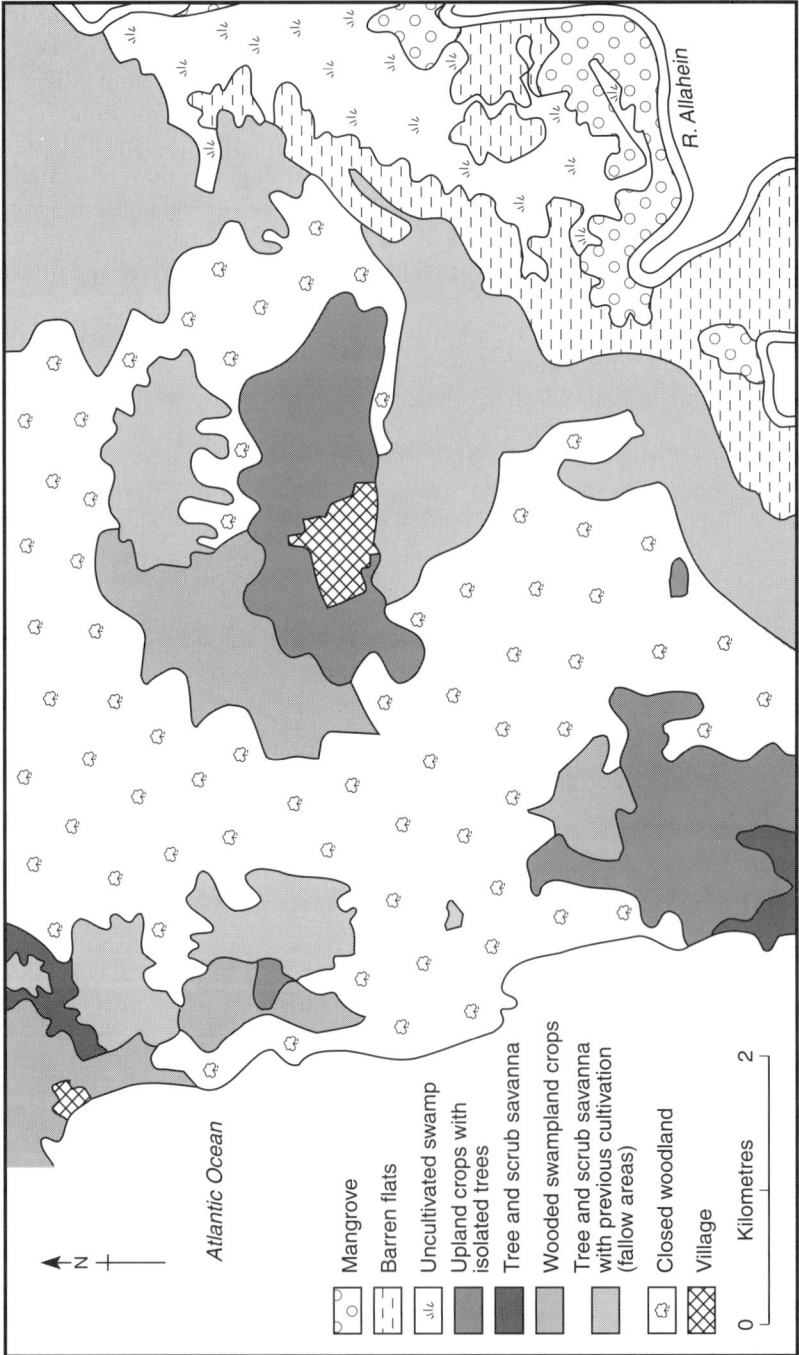

Figure 14.2 Berrending village, showing vegetation and collecting zones

Mangrove
Barren flats
Uncultivated swamp
Upland crops with isolated trees
Tree and scrub savanna
Wooded swampland crops
Tree and scrub savanna with previous cultivation (fallow areas)
Closed woodland
Village

R. Allahein

Atlantic Ocean

N

0   2
Kilometres

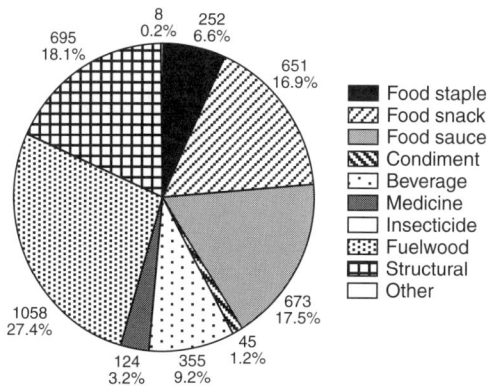

Figure 14.3   Subsistence use of collected products (total and percentage shown). Note there is some overlap, for example bush teas are used both as beverage and medicine

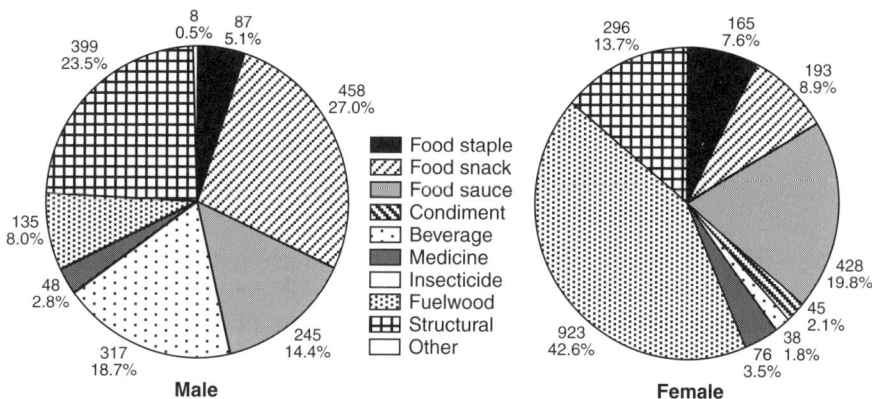

Figure 14.4   Subsistence use by gender

accounted for in food surveys although they clearly play an important nutritional role to rural individuals.

Additionally, beliefs concerning appropriate gender roles were significant in explaining the gendered use of collected products for subsistence. For example, in this research village women were expected to collect fuelwood for cooking, which represented slightly under half (43%) of their total collections. Similarly, the gender division of labour meant that men tapped palm trees for wine, accounting for their relatively high proportion of beverages (19%). Men were also the main collectors of structural materials for building construction, such as poles, timbers, fences and leaves for roofing. Women also collected structural materials, but usually for handicrafts rather than for building construction, such as materials for weaving mats and food containers. While there were important gender differences in the products collected, there were also differences between women in products collected for

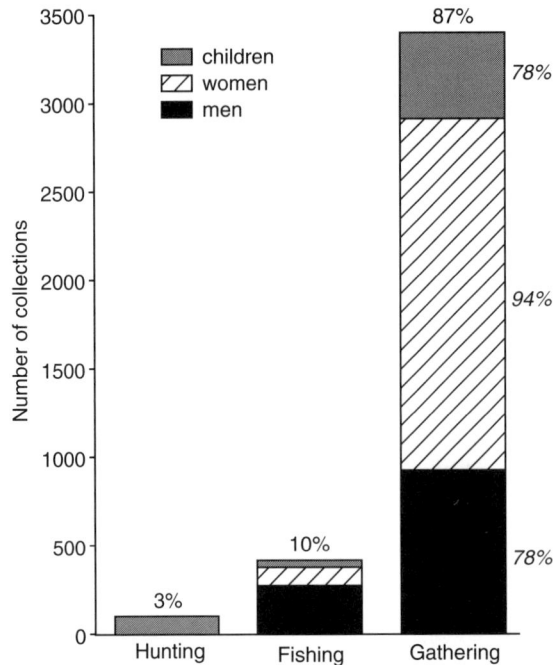

Figure 14.5   Method of collecting, by gender

subsistence. What a woman collected, when, how much and the use made of the product depended on five major factors: ethnicity, wealth, household responsibilities, indigenous technical knowledge and motivation (Madge, 1993).

A further gender component to collection for subsistence may be illustrated when considering the method of collection. Figure 14.5 shows that when total numbers of collections in the village were subdivided by gender according to method of collection (gathering, hunting or fishing), gathering was significantly more important than hunting or fishing. Gathering represented 87% of the total collections compared to 10% for fishing and 3% for hunting. All gender groups concentrated on gathering, representing 94% of women's total collections and 78% of the men's and children's total collections. Overemphasis on hunting in the literature has often obscured the vital role of gathering but in this case study both women and men displayed highly specialised knowledge concerning the gathering of collected products, such as when, where and how to collect them. Women in particular knew how to process, cook, preserve and store such products (Madge, 1994).

THE ROLE OF COLLECTED PRODUCTS: COMMERCIAL

Figure 14.6 illustrates the diverse income-earning strategy adopted by the villagers. It clearly demonstrates the vital role of income from the sale of collected products,

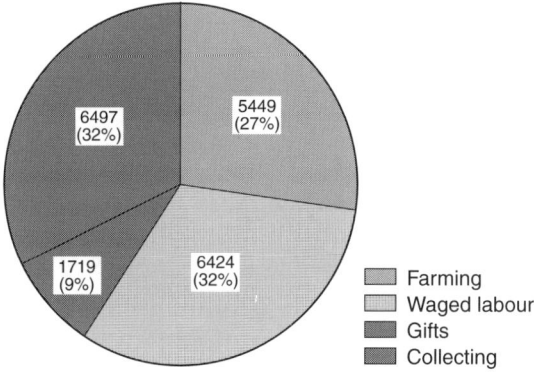

Figure 14.6   Sources of income, total cash and percentages

providing one-third of the total income (20,089 dalasi, D) earned by the individuals during the survey period (12 D = £1 sterling at the time of fieldwork). The importance of income from the sale of collected products has also been recorded by Lagemann (1977) who noted that indigenous tree crops provided one-quarter to one-half the total cash income earned during one season in eastern Nigeria. These results suggest that the Food and Agriculture Organisation's (1984, p.36) assertion that collected products are of "minor economic importance and are not likely to become major sources of cash" is misleading. Additionally, when the information from the Gambian case study was disaggregated by gender it became apparent that the men in the village gained 61% (12,180D) and women 39% (7,909D) of the total income earned during the survey period. However, women and men earned money from different sources (Figure 14.7). Male income was predominately earned through wage

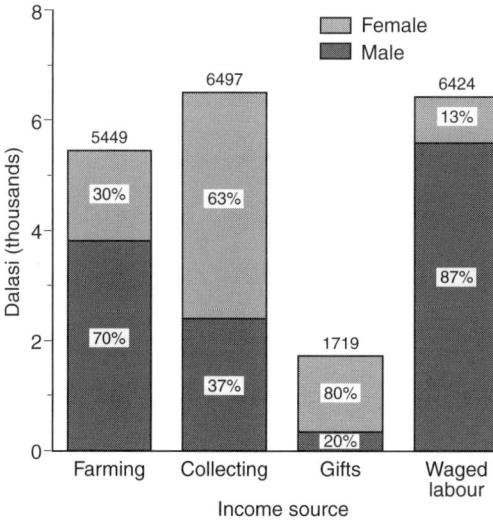

Figure 14.7   Sources of income by gender, total cash and percentages

labour and sale of agricultural produce, while female income was obtained through gifts from other household members and through the sale of collected products. In commercial terms, the sale of collected products was therefore especially important to women, accounting for slightly over half of their total earnings (52%) but less than one-quarter of male total earnings (20%). The income earned from the sale of collected products also varied among the women according to ethnicity, wealth and household size. Furthermore, in a survey undertaken in the local town, Brikama, it was revealed that over 50% of market stalls sold one or more collected products and twice as many female traders sold collected products as male traders. The commercial role of collected products is potentially as important to urban women as to their rural counterparts.

THE ROLE OF COLLECTED PRODUCTS: SOCIO-CULTURAL

Collected products played a prominent social role in the case study village and were used to invest in long-term social relationships. They were used to regenerate and create links both between and within households. This was particularly important in times of economic and environmental stress, wherein individuals with whom one had previously shared collected products could be called upon to give aid, in a patron–client relationship. Collecting in the village was also an important social activity, especially for young boys. Who decided what to collect, where to collect it and with whom to go was a measure of authority and a means by which young boys displayed their prowess.

The symbolic and cultural values that village individuals placed on the forest and collected products were also important to Jola society. Collected products acted as "cultural cementers" and played a vital function in upholding the Jola "way of life". Village individuals commonly stated: "We are Jola. We are people of the forest". The Jola could be described as having a "forest culture" (Shiva, 1988). Collected products also played a prominent role in village ceremonies and were used as cultural artefacts. For example, palm wine consumption was important in marriage and funeral ceremonies in the past and special forest groves played an important symbolic role during ceremonies, both acting as a site for communication with the ancestral spirits, and as a sacred site for rights of passage and women's fertility association meetings.

All use made of collected products was, moreover, mediated through cultural beliefs and understandings of the environment, trees and people. The Jola had an ambivalent attitude towards the forest (n'ehramba), both fearing and respecting it. The fear was translated in the existence of "evil" forests and trees which harboured devils, witches and "spirits". The spatial extent of collection was mediated by these cultural values as evil areas could only be entered by specialist professionals, such as herbalists and hunters. However, the forest was also respected; a commonly stated phrase in the village was that "all life comes from the forest" which was linked to the healing (medicinal) power of trees. Such "endogenous" beliefs about the environment were held in parallel with "exogenous" beliefs which diffused from the urban centres, extension services, education system and contact with the tourist areas.

HISTORICAL PERSPECTIVES ON COLLECTING

Collecting was clearly important to the villagers during 1988–89 but in order to make any generalisations it is important to place the use of collected products in historical context. Archaeological evidence from Linares (1971) suggests that the Jola have a long history of use of collected products. Over a century ago Mitchinson (1881) stated that the Jola were people who: "collect palm wine and to a smaller extent plant rice... The numerous minor creeks which intersect the country abound with crocodiles and fish, both affording plentiful supply of food for the people". Mungo Park (1799) also indicated that the Jola were involved in the trade of collected products, in particular wax, wine and whitewash. However, the villagers also unaminously stressed that the frequency of collecting had increased significantly during the last 10 to 15 years owing to interrelated environmental, economic and social factors.

During the last 15 years there has been a 30% decline in national average rainfall in The Gambia which has reduced rice production. Income was needed by rural Gambians to purchase rice that was hitherto cultivated; this income was obtained from the sale of collected products. Local alterations in the micro-environment were exacerbated by national and international economic and political developments which also resulted in increased collection. In 1986 The Gambia embarked on an Economic Recovery Programme (ERP) encouraged by the International Monetary Fund and World Bank to reduce The Gambia's balance of payment deficit which had arisen during the 1970s owing to accelerated public expenditure. This resulted in two important changes to the Gambian rural economy. Firstly, it reduced subsidies for groundnut production and, secondly, it increased commodity prices for basic household requirements (rice, soap, sugar). Rural Gambians found that they not only had less income owing to the reduced price obtained for the sale of groundnuts, but also the cost of goods had increased. To overcome these difficulties, the collecting economy became diversified and commoditised. Products previously collected for subsistence, such as oysters, fencing, timbers and fuelwood, were collected for sale.

These wider economic changes were mediated at the local level through gender relations. According to the women of the village, the increased significance of collecting during the last two decades was due to the decline of the palm oil industry which in the past was the main Jola dry season activity. There had been a strict gender division of labour in palm oil production, the men collecting palm kernels while the women processed the oil. The men then gained control of the oil to sell and the women the kernels which needed further processing before sale. When the international market price of kernels fell in the late 1970s, the women said that they were no longer willing to perform the heavier work in the processing and gain the smaller remuneration. The village women became dissatisfied with the returns to their labour and contested the basis of the conjugal contract, namely the right of husbands to control their wives' labour power. The women refused to work in the family unit with their husbands and started processing palm oil on their own, purchasing the palm kernels from migrant men. This resulted in a general decline in palm oil processing in the village for both women and men. A new income source was required to replace that previously supplied by the sale of palm oil, which was provided by the commoditisation of collected products. The relationship between the diversification and commoditisation of the collecting economy and economic, environmental and socio-cultural change can thus be established. This change has also been paralleled by the employment of indigenous management strategies to regulate environmental resources.

MANAGEMENT OF ENVIRONMENTAL RESOURCES

Dynamic and overlapping mechanisms of indigenous management have been implemented to encourage sustainable use of environmental resources in the case study village. The people were acutely aware of the process of environmental change resulting from commoditisation of the collecting economy. One elderly village woman stated that: "Now people must look after the forest more than in the past because they have more use for the trees". Another said: "We do not like to cut our trees down. We are not fools but what can we do? We need money to buy rice. Poverty has killed the forest". Individuals were well aware of their increased reliance on collected products and were actively managing these resources. Three main management strategies were identified by the villagers.

Firstly, conservative use was made of forest resources. Fuelwood was not usually obtained from food trees, thus ensuring that the food supply was not affected. The use of fruit was also regulated; village children stated that they would only collect ripe fruit, leaving unripe fruit for others and would not collect more than they were capable of eating. Fruit was collected by hand from branches which were never cut to recover the fruit.

Secondly, there was also selective maintenance and promotion of particular plant species. Certain trees were left intact as fields were cleared to cultivate rice or groundnuts, a process also noted by Asamoah (1985) in Ghana, Leach (1991) in Sierra Leone and Umeh (1985) in Nigeria. In the case study village these trees were used for food, medicine, shade or for providing monetary income. Large trees were also often left since it was considered that they prevented soil erosion. Other tree species were lopped at certain times of the year to stimulate plant growth. Through the manipulation of the life-cycle habit of the trees, there was the active creation of a suitable environment for the production of useful tree species. Shepherd (1992) suggests that the preservation of valued plant species in managed land is an increasingly common response to the decline in availability of forest and uncultivated bush in much of West Africa.

Thirdly, there was also management through taboos and religious sanction. Certain areas of the forest were seen as "evil forest" and certain trees seen as "evil trees". Most individuals would not even enter the "evil forest" and would never consider cutting down an "evil tree". Other areas of the forest were also considered sacred and were not cleared. Also, hunters had an indigenous management system whereby strict self-regulatory control was placed both on the particular animal species that could be killed and the timing of the kill. For example, between May and November female antelopes could not be killed for during this period they were pregnant, lactating or tending their young. Also, it was firmly believed that if a hunter slaughtered too many animals he would perish himself. This acted as an indigenous regulation on the number, species and timing of animals killed.

In addition to environmental management at the local level, the villagers also became organised politically to prevent forest destruction. Individuals frequently advised one another not to cut down young trees whereas in the past they would not do so. In 1976 the villagers grouped together to prevent further forest destruction from outsiders felling palm trees. However, indigenous management of the forest is not an original concept introduced in the 1990s. Dawe (1922) noted that the Jola tapped their palm wine in an environmentally responsible manner so that "the growing part of the palm is not injured in any way". Also, in 1948 in Kerewan, Gamble (1949) noted that measures were taken to control *Icacina senegalensis* by placing a prohibition order on cutting these bushes as farms were being cleared or on picking fruit until a day set by the alkali (village headman). Management techniques, like other survival strategies, have fluctuated throughout history and become adapted to prevailing environmental conditions. Management practices vary over both time and space; local people thoughtfully respond to economic, political and environmental change. The important question for the future is whether formal management policies can build upon these local management strategies.

CONCLUSIONS

In 1989 in the case study village one young man commenced an orchard project, growing oranges, mangoes and indigenous tree species. However, his mother insisted that he pulled out all the indigenous tree seedlings. Investigation into the reasoning behind this request revels three overlapping explanations and highlights the complexities of decision-making regarding environmental management.

Firstly, the request was motivated by practical reasons. Some indigenous medicinal species would be harmful if consumed by animals or children.

Secondly, there was a social explanation – it was considered bad luck for men under 35 years of age to plant trees. This belief was linked to the social power structure of the village. Positions of power were invested in the elders in traditional Jola society, which were reliant on social control based on the commercial revenue gained from the sale of oil palm produce. Control of the use of trees by elder members of Jola society was thus crucial for continued political power. Therefore a young man planting trees was a threat to the established social order, and was thus discouraged.

A third explanation revealed that cultural beliefs of the Jola about people's relationship with the environment were also important in understanding the situation. The Jola do not see themselves as separate from the forest, but as part of it, both fearing and respecting it. As such, they do not see themselves as agents or replenishers of the forest, nor are they its "masters", but they see themselves as part of, intimately connected to, the "natural world". To plant indigenous "wild" species was thus to act as a "god", to act against the social "order" between people and trees, to subvert commonly accepted ecological power relations. Moreover, "wild" forest trees were considered to be publicly owned, held in custodial guardianship by the ancestors. In contrast, tree seedlings cultivated specifically by an individual were considered privately owned. To cultivate indigenous seedings was thus to "act against" Jola beliefs about their relationship with trees and with the ancestors; it was a challenge to the custodial role of the ancestors and the traditionally held property rites regarding trees and their use. The planting of indigenous seedlings was thus a "dangerous" cultural act.

This example highlights the difficulties that formal management policy may have in building upon indigenous management strategies. Local people and outside agents (extension officers, NGO workers, academics) are not necessary in opposition but they may have immediate difficulty in comprehending each other's cultural under-standings of the environment. This is where an ethnographic approach can make a valuable contribution in attempting to understand local people's understandings of their environment and working within those frameworks of understanding. Indeed, Arce and Long (1992) demonstrated that management policies lacking such a sensitive approach are unlikely to succeed. Moreover, this example demonstrates that cultural interpretations of the environment cannot be viewed in isolation but must be linked to an appreciation of ecological power relations.

This chapter demonstrates that an ethnographic approach can provide invaluable information regarding agroforestry issues and can also highlight some of the problems that may be encountered in attempts to manage tropical catchment areas. "Management" may occur at several (sometimes competing and sometimes comple-mentary) scales using varying "systems" of knowledge (indigenous, urban, Western), but an ethnographically informed approach can provide invaluable data to map the "mindscape" of environmental management which is bound to be an essential component of any successful sustainable management policy for tropical catchment areas.

ACKNOWLEDGEMENTS

The author would like to thank all the villagers of Berrending in The Gambia and, in particular, Ansu Badjie and Mba'Sally Badjie for their friendship, help and willing co-operation. She is also grateful to Mr F.M.S. Janneh for his help in completing the town market survey. The Economic and Social Research Council funded the research and their support is gratefully acknowledged. Also thanks to Andrew Millington for his helpful comments.

This work is based upon material originally published as 'Ethnography and agroforestry research: a case study from the Gambia', in *Agroforestry Systems, 32*, 127–146, 1995, reproduced by kind permission of Kluwer Academic Publishers, The Netherlands.

REFERENCES

Arce, A. and Long, N. 1992. 'The dynamics of knowledge: the interface between bureaucrats and peasants'. In *Battlefields of Knowledge: the Interlocking of Theory and Practice in Social Research and Development* (eds N. Long and A. Long), pp. 211–246. Routledge, London.

Asamoah, R.K.F. 1985. 'Uses of fallow trees and farm practices in Ho Forest Districts (Ghana)', Unpublished PhD Thesis, Institute of Renewable Natural Resources, University of Science and Technology, Kumasi, Ghana.

Dawe, M.T. 1922. '*List of Plants Collected in The Gambia*'. Government Printer, Bathurst, The Gambia.

Department of Wildlife Conservation, 1979. '*Fifty Trees of Abuko Nature Reserve*'. Conservation Factsheet No. 9. Government Printer, Banjul, The Gambia.

Europa Publications, 1992. '*Africa South of the Sahara*'. Europa, London.

Falconer, J. 1990. '*The Major Significance of Minor Forest Products: The Local Use and Value of Forests in the West African Humid Forest Zone*'. FAO, Rome.

Food and Agriculture Organisation, 1984. '*Government Consultation on the Role of Women in Food Production and Food Security*', 10–30th July 1984, Harare, Zimbabwe. FAO, Rome.

Gamble, D.P. 1949. '*Contributions to a Socio-Economic Survey of The Gambia*'. Colonial Office, London.

Hobart, M. (ed.) 1993. '*An Anthropological Critique of Development*'. Routledge, London.

Lagemann, J. 1977. '*Traditional African Farming Systems in Eastern Nigeria*'. Weltforum Verlag, Munich.

Leach, M. 1991. 'Endangered environments: understanding natural resource management in the West African forest zone'. *Institute of Development Studies Bulletin*, **22**, 17–24.

Linares, O. 1971. 'Shell middens of Lower Casamance and problems of Diola protohistory', *West African Journal of Archaeology*, **1**, 23–54.

Linares, O. 1985. 'Cash crops and gender constructs: The Jola of Senegal', *Ethnology*, **24**, 83–94.

Long, N. and Long, A. (eds) 1992. '*Battlefields of Knowledge: the Interlocking of Theory and Practice in Social Research and Development*'. Routledge, London.

Madge, C. 1993. '*Medicine, Money and Masquerades: Gender, Collecting and "Development" in The Gambia*'. Social Sciences Faculty Paper No. G93/1, University of Leicester.

Madge, C. 1994. 'Collected food and domestic knowledge in The Gambia, West Africa'. *Geographical Journal*, **160**, 280–294.

Mitchinson, A.W. 1881. '*The Expiring Continent; A Narrative of Travel in the Senegambia with Observations on Native Character, the Present Condition and Future Prospects of Africa and Colonialism*'. Allen, London.

Ogle, B.M. and Grivetti, L.E. 1985. 'Legacy of the chameleon: edible wild plants in the kingdom of Swaziland, Southern Africa. A cultural, ecological and nutritional study'. *Ecology of Food and Nutrition*, **16**, 198–208; **17**, 1–64.

Park, M. 1799. '*Travels in the Interior Parts of Africa: Performed under the Direction and Patronage of the African Association in the Years of 1795, 1796 and 1797*'. Bryan Edwards, London.

Parkin, D. and Croll, E. (eds) 1992. '*Bush Base: Forest Farm: Culture, Environment and Development*'. Routledge, London.

Percival, D.A. 1968. '*The Common Trees and Shrubs of The Gambia*'. Government Printer, Bathurst, The Gambia.

Pottier, J. (ed.) 1993. '*Practising Development: Social Science Perspectives*'. Routledge, London.

Shepherd, G. 1992. '*Managing Africa's Tropical Dry Forests*'. Overseas Development Institute, London.

Shiva, V. 1988. '*Staying Alive. Women, Ecology and Development*'. Zed Books, London.

Umeh, L. 1985. 'Management of agroforestry systems in the Lowland Humid Tropics: A study of some village communities in Southern Nigeria'. Unpublished PhD Thesis, Department of Forest Resource Management, University of Ibadan, Nigeria.

World Bank, 1988. '*World Development Report*'. Oxford University Press for the World Bank, London.

CHAPTER 15

Hydrological and Ecological Considerations in the Management of a Catchment Controlled by a Reservoir Cascade: The Tana River, Kenya

NIC PACINI
Ministry of Environment, Rome, Italy

DAVID HARPER
Department of Biology, University of Leicester, UK

KENNETH MAVUTI
Department of Zoology, University of Nairobi, Kenya

INTRODUCTION

The Tana is the largest river in Kenya. The total catchment area is approximately 95,131 km^2 (17% of the country; Figure 15.1). With 800 km^2 surface at peak flood, the region of the Tana delta is among the 20 major African floodplains (Thompson, 1996). Its total length is around 1,000 km, rising in the Mt Kenya (5,199 m a.s.l.) and the Nyandarua (Mt Kinangop, 3,908 m a.s.l.) mountains and flowing to Kipini on the Indian Ocean. The catchment contains some of the most important resources of the country, which are:

- Intensively farmed agricultural land (Murang'a, Kirinyaga and Embu Districts; Chapters 2–5).
- Major hydro-electric power plants: the "Seven Forks" Development Scheme.
- Unique habitats internationally recognised for their beauty and their importance for the survival of endemic species of animals and plants and indigenous human populations (Mount Kenya National Park, the Aberdares National Park, Mwea National Reserve, Kora and Meru National Parks, the Tana River National Primate Reserve, the Tana river floodplain and the Tana Delta).

The Sustainable Management of Tropical Catchments. Edited by David Harper and Tony Brown.

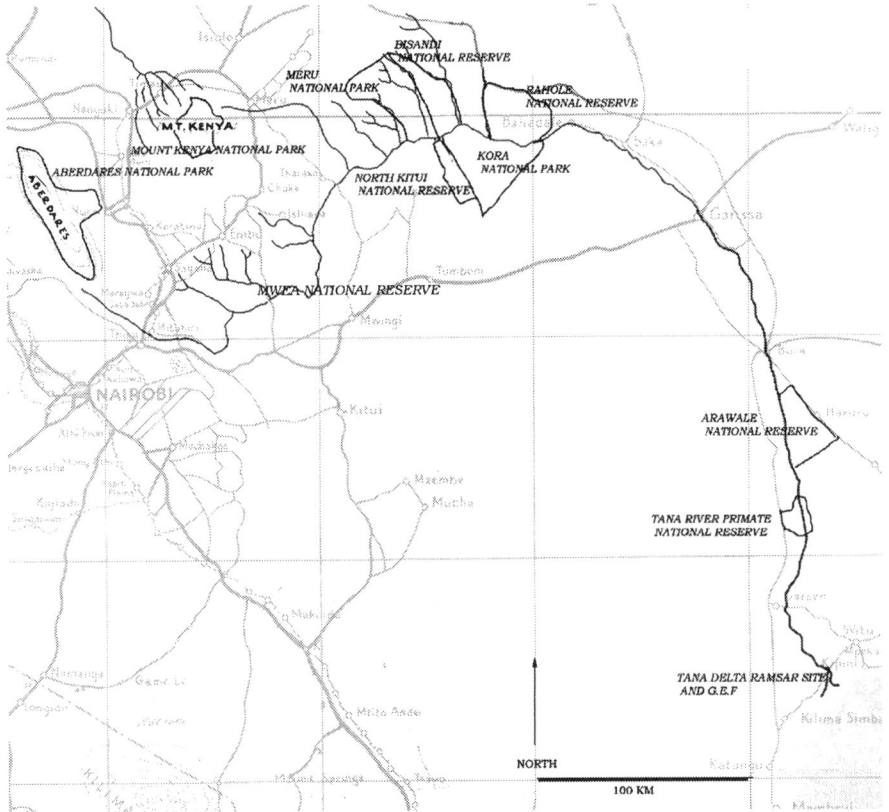

Figure 15.1   The Tana catchment in Kenya, showing the course of the river and the national parks and national reserves which are associated with and dependent upon it

Annual average rainfall figures in the upper catchment range between 2400 mm on the mountains to some 700 mm at Masinga Dam wall (1,056 m a.s.l.). These highlands contribute more than 50% of the total runoff of the whole Tana basin, which converges to power the turbines of Masinga Dam at the top of the "Seven Forks" reservoir cascade. This is a hydro-electric power generation scheme which currently consists of five reservoirs, potentially rising to seven, which supply some 75% of the country's electricity.

Below Masinga Dam the Tana river receives the waters of the Thiba (at Kamburu Dam), then, below the three lower dams, the tributaries Ena, the Mutonga and the Kazita. After this, there is no permanent tributary down to the river mouth at Kipini (600 km downstream). The region below the confluence of the Kazita has an average annual Penman evaporation exceeding 2200 mm, an average annual rainfall between 300 and 600 mm and is part of the Sahel zone according to the definition of Grove (1978). The coastal region of the Tana catchment contains an extensive delta stretching from Mto Tana, the original river mouth, to Kipini, the actual exit after channel diversion works carried out by the Malindi administration at the end of last century. Here a stable easterly wind brings moisture to a thin strip of the Eastern Africa coastal region with 900 mm average rainfall per year at the river mouth. The delta constitutes a complex and species-rich environment with a gradual succession of aquatic habitats between the freshwater and the Indian Ocean (Njuguna, 1992).

In 1974 the Tana River Development Authority (TRDA) was created with the specific statutory duty to co-ordinate hydro-electric development in the upper basin with intensive irrigated agriculture in the lower part and general economic development throughout the catchment.

The Seven Forks Development Scheme (Figure 15.2, Table 15.1) started with the construction of a small reservoir at Kindaruma in 1968. Pre-investment studies gathered little information about the area and it was not until 1975 and then 1977 that some more extended investigations were carried out to give a background for the assessment of development impacts (Odingo, 1979). By then siltation had significantly diminished the storage capacity of the first dams and planners had become progressively more concerned about soil erosion control and human activities in the catchment (see Chapters 2–4).

Sediment deposition in the reservoirs had been underestimated at the beginning of the project due to the over-optimistic attitude of reservoir designers: "In the absence of reliable sediment transport data and sediment yield estimates from tropical areas, designers have had to use data from temperate regions" (White, 1990). In Kindaruma, sediment accumulation caused a progressive drop in power generation from the

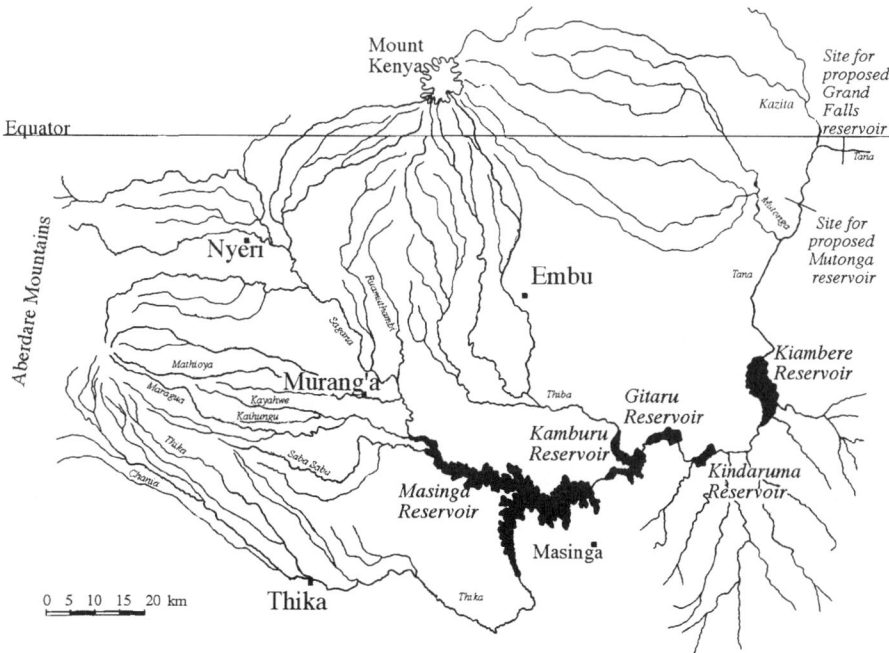

Figure 15.2 The "Seven Forks" reservoir cascade, showing the five existing dams and the two proposed new ones

Table 15.1 The Seven Forks reservoir cascade ranged by altitude. Storage capacity and generating power are estimates obtained at time of closure of the dams

Reservoir	Year	Altitude (m a.s.l.)	Surface (km^2)	Storage volume (10^6 m^3)	Generating capacity (MW)	Approximate cost (US$ × 10^6)
Masinga	1981	1056.5	125.0	1560	40	100
Kamburu	1975	1006	15.0	146	84	2
Gitaru	1978	924	3.1	20	145	5
Kindaruma	1968	780	2.6	18	140	1
Kiambere	1988	700	25.0	585	140	312

planned 140 MW down to 22 MW (Rowntree, 1990). Today Kindaruma is only 13 m deep. The threat of rapid siltation and the need for higher control over water availability for power generation prompted developers to construct a high dam at Masinga, which impounded a large reservoir. The high cost per MW of power generated was partly due to the need to rebuild a bridge and a major road connection, partly due to high environmental and land opportunity costs as the land submerged in the upper Tana catchment was relatively fertile and productive, and partly due to the relatively low fall for the Masinga turbines compared to the lower dams.

The TARDA (the name was changed after the inclusion of the Athi river basin) considered, however, that the increased control over river flows enhanced the efficiency of operation of the lower reservoirs to a total benefit of at least 70 MW. After the closing of the dam, turbines at Gitaru realised higher productivity reaching 216 MW. Additional benefits of Masinga Dam were linked to the prospects of further exploitation of the Chania sub-basin above the reservoir, for Nairobi's water supply. The presence of a large upstream "buffer" reservoir allowed water abstractions without negative consequences for downstream power generation.

A UNEP study published in 1977 warned of the possible detrimental effects of building the reservoir but went without official response by TARDA. On the contrary, several other sites below Kiambere were indicated as suitable for the potential creation of new reservoirs. These future dams (one or two sites are under consideration) were intended to have the specific function of controlling dry season discharge for the management of downstream irrigation schemes.

ECOLOGY OF THE SEVEN FORKS RESERVOIR CASCADE

Masinga Reservoir (Figure 15.3, Table 15.2) is the direct recipient of water and sediment coming from the upper catchment and its dendritic shape enables the land–water linkages to be understood by spatial comparisons. Sediment yield from the catchment, associated with riverine currents within the reservoir, contributes to a reduction of light availability at the western end. Large sediment deposits at the inlets of the major rivers – Tana and Saba Saba – have created deltaic swampy areas. The filling of the reservoir was completed at the end of 1981. During pre-impoundment studies (Ward et al., 1976), computer simulation indicated an annual theoretical water level fluctuation of 5 m. In reality decrease in water level by as much as 10 m has

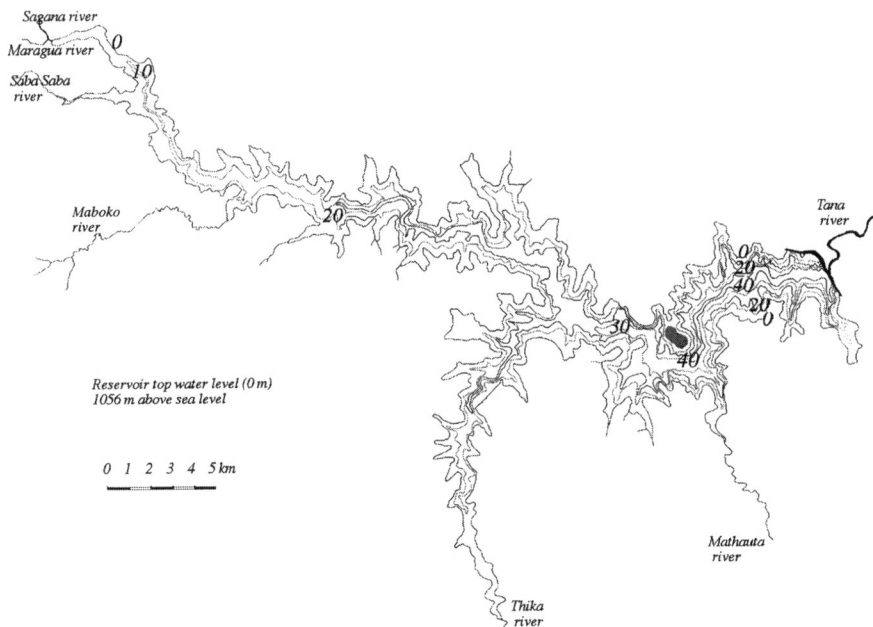

Figure 15.3 The morphometry of Masinga Reservoir

occurred at least three times since filling (1984, 1988 and 1992) leaving up to 50% of the surface area alternately exposed and flooded. The primary consequence of this hydraulic regime is the absence of littoral vegetation. *Cyperus papyrus*, which colonises most lentic habitats in these regions rarely survives water fluctuations above 1.5 m (Muthuri, 1985); the absolute limit is considered to be 2.5 m (John, 1986). *Polygonum senegalense*, the only true aquatic plant in the reservoir, which can sustain water level variations of up to 5 m, grows at a low density on the limited shallow

Table 15.2   Physical limnology of Masinga Reservoir

Characteristic	Quantity
Latitude	$0°47' - 1°00'$ S
Longitude	$37°16' - 37°36'$ E
Altitude (m a.s.l.)	1056.5
Drainage area (km^2)	7335
Surface area	125
Drainage/surface ratio	58.7
Volume ($10^6 m^3$)	1460
Reservoir length (km)	42
Mean depth (m)	13.8
Maximum depth (m)	48
Lake shoreline (km)	285
Shoreline development ratio	7.2
Hydraulic residence time	3 months – 2 years

shores of the reservoir and plays a marginal role in the formation of a habitat for fish and invertebrates. On land, dry bushland vegetation dominated by *Acacia* and *Commiphora* species reach to the reservoir upper water edge. Not withstanding the high shoreline development (285 km in length), the interaction between the land and the aquatic habitat (lateral connectivity) is reduced.

Masinga is a bottom-withdrawal, warm monomictic reservoir. During the rainy seasons the reservoir experiences maximal stratification. Cool river floods traverse the entire reservoir from the inlet to the turbines as a deep (35–40 m) underflow. The density stratification of water masses is enhanced by the high water temperatures prevailing during the whole year (ranging between 19.5 and 27.1°C) and the high loads of suspended sediment (with concentrations reaching $0.8\,g\,l^{-1}$) transported by hypolimnetic currents. Strong density stratification confines the high phosphorus loading to the lower depths, restricting the development of algal biomass in the epilimnion. The rainy season is also the period of maximum flushing, as the retention time is of the order of one week; suspended sediments and the nutrients bound to their surface sink at the inlet and leave the reservoir through a hypolimnetic outlet without affecting the isolated epilimnion.

Circulation occurs typically between June and July as wind strength increases, in concert with the arrival of the monsoon, and air temperatures decrease. Cooling is due to strong winds which tend to clear the skies of clouds after the April–May rains, increasing irradiance and causing high evaporation. High irradiance coupled with the mixing of the entire water column, which renews the supply of nutrients to the epilimnion, should be conducive to high primary production. This is, however, inhibited by high turbidity, which occurs following the recirculation of sediment-laden hypolimnetic water – the Secchi disc transparency in May is about 1.5 m while in July it may drop down to 0.1 m (Pacini et al., 1993).

Simultaneous light and phosphorus limitation can be deduced from Figure 15.4 which shows a peak in phytoplankton biomass midway between the reservoir inlet, which is characterised by high light limitation, and the reservoir outlet, which is characterised by high water transparency but nutrient limitation associated with the progressive longitudinal decrease in total phosphorus (Pacini, 1994). Thus at all major climatic periods through the year, the reservoir productivity is either light- or phosphorus-limited. Despite high sediment and phosphorus loading (Chapter 5), rates of primary production are low when compared to those of natural tropical lakes.

Figure 15.5 shows how maximal primary production in the main basin of the reservoir is dependent upon climatic instability (shown as run of wind and rainfall pattern) causing mixed water column conditions. This factor overrules total phosphorus inputs as a result of erosion in the catchment which determine the concentrations of total phosphorus in the two rivers bringing the largest amount of the flow to the reservoir (the Tana and the Saba Saba) and the concentration of total phosphorus in the reservoir itself. The difference in concentration within the reservoir is dependent on the stability of the water column. The stability conditions can be inferred from the lower plot showing the succession of wind and temperature during the same period.

Figure 15.6 compares the nitrogen/phosphorus ratio, as an expression of potential limitation by nitrogen or phosphorus, in the rivers of the catchment and in Masinga Reservoir through the year. The data do not allow a seasonal pattern to be followed in

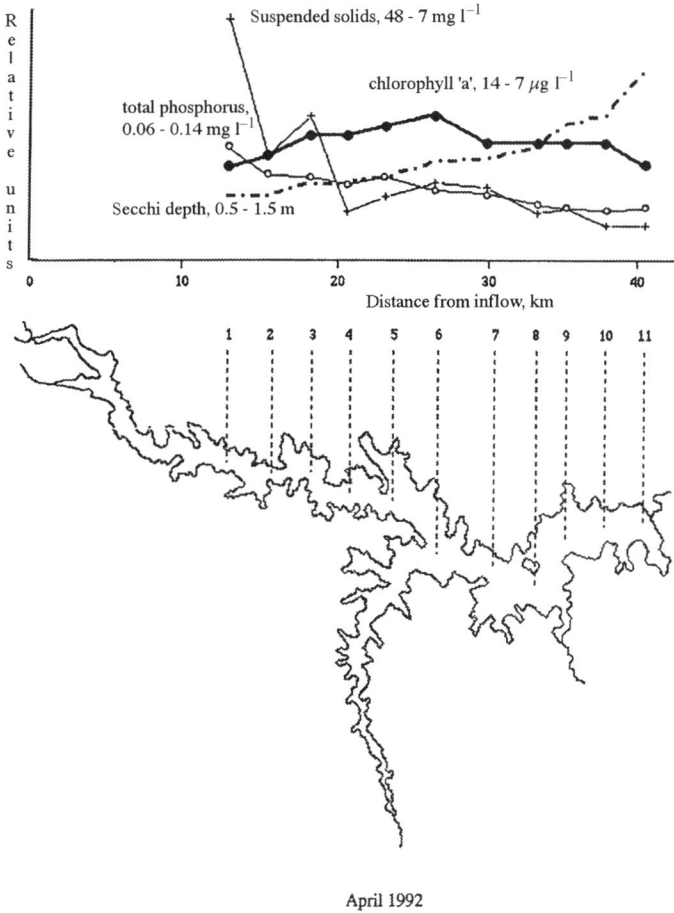

Figure 15.4 Light and phosphorus limitation in Masinga Reservoir: chlorophyll biomass is optimal in mid-reservoir because of reduced transparency caused by silt at the western half and reduced phosphorus caused by sedimentation at the eastern half

detail, but indicate that the ratio usually is higher in the reservoir than in the rivers. In the rivers N/P varies between 6 and 67 while the range is 20–260 for the reservoir. In rivers an increase of nitrogen transport during the rainy seasons is indicated by higher N/P in November and in March–June. (This variation in the N/P ratio is likely to be an underestimate as a result of the poor detection of nitrogen transport in rivers: rapid transport occurs in the first runoff flush of storms due to the solubility of nitrate and ammonia.) Intuitively it would be expected that a similar pattern would appear in the variation of N/P ratios in the reservoir. The opposite actually occurs – lowest values of N/P in the reservoir coincide with the rainy seasons (in March–April 1992, November 1992 and March–June 1993) when the N/P ratio in rivers is supposed to be higher. Nutrient ratios in the reservoir are not a simple reflection of nutrient inflows. Independent autochthonous processes are able to modify the relationship between the

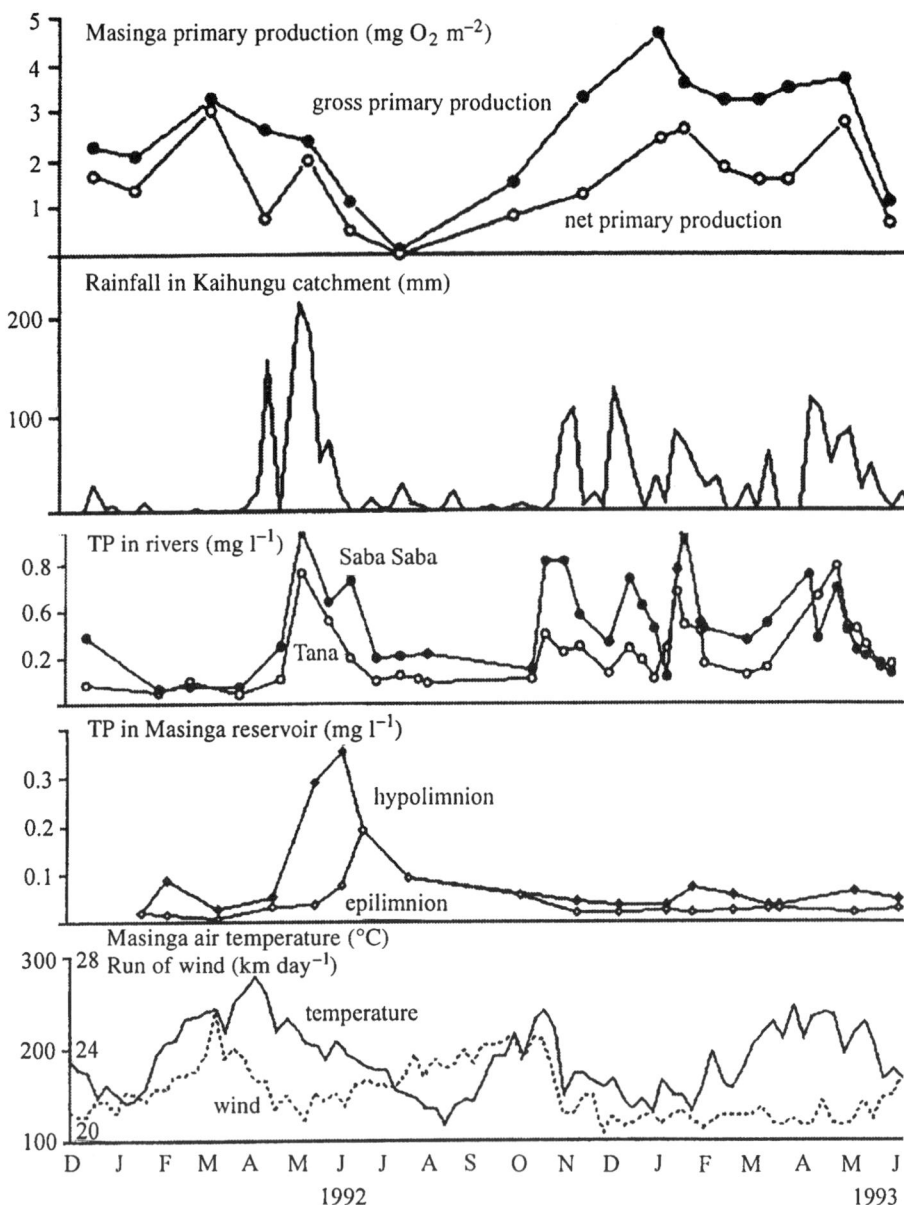

Figure 15.5   Illustration of the controls upon primary productivity in Masinga Reservoir

pools of different nutrients, with two main processes probably responsible: nitrogen fixation by cyanobacteria and phosphorus adsorption by the bottom sediments.

In the downstream reservoirs, hydrological conditions are much more constant due to the buffering effect of Masinga: drawdown is reduced and a richer littoral zone is established. The morphometries and ages are very different, however, and this leads

N/P

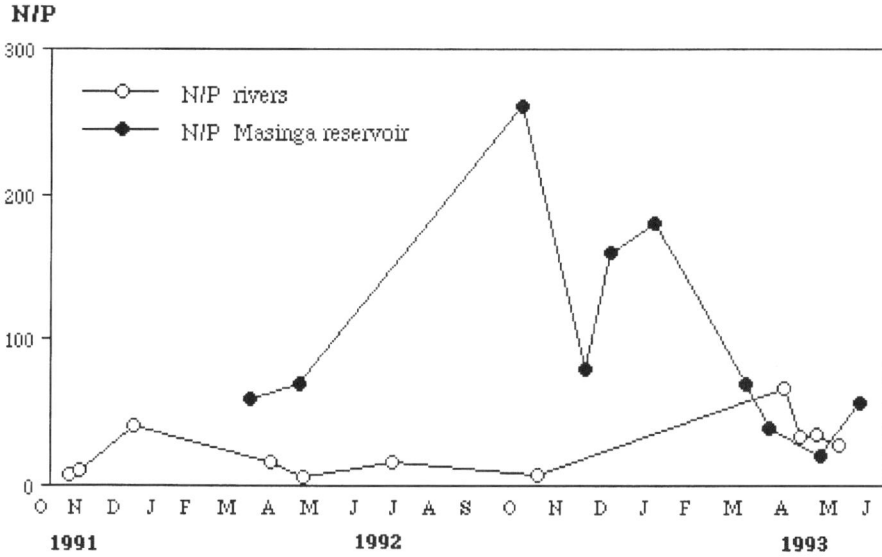

Figure 15.6   Ratios of N/P in river samples and the epilimnion of Masinga Reservoir

to differences in biological character (Table 15.1). This may be seen in the dominant species of phytoplanktonic algae, which are related to water column stability (Figure 15.7): the reservoirs which stratify are dominated by cyanobacterial species whilst those which are more thoroughly mixed are dominated by heavier diatom species which cannot tolerate stable, stratified, conditions.

The relative nutrient status of the reservoir cascade is affected primarily by age and by retention time which controls sedimentation: the accumulation of sediments in the basin tends to strip the reservoir of its incoming phosphorus, as a relationship between reservoir age and trophic status shows (Figure 15.8). An additional factor is reservoir morphometry which determines the surface of sediments which, when anoxic, are in contact with the water column, influencing nutrient cycling. Masinga has relatively little compared to Kamburu which experiences severe anoxia and phosphorus release, and supports a phytoplankton biomass double that of Masinga. Thus Kamburu appears to have a lower N/P ratio than its age would predict. Although other factors interrelate with nutrient ratios, the significant relationship ($R^2 = 0.887$) suggests a general trend towards the progressive strengthening of phosphorus limitation.

The lack of a shoreline ecotone and physical characteristics such as short residence time, turbidity, reservoir depth and the relatively steep banks, combine to create an unproductive aquatic habitat. The Masinga fishery is relatively poor. Fish catches are mainly of eels (*Anguilla mossambica*) and catfish (*Clarias gariepinus*) fished with traps and long lines. Few nets are set in Masinga for the capture of *Tilapia* species of high commercial value due to their relative scarcity. There are few open-water plankti-vores, dominated by *Alestes* sp., and several benthic feeders; among these *Cyprinus*

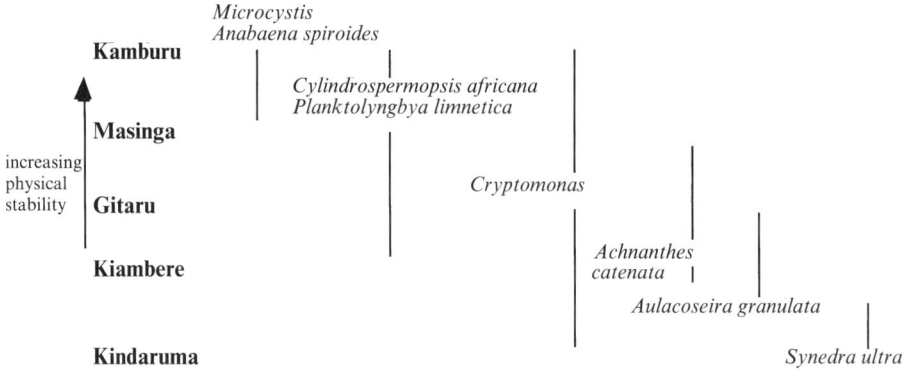

Figure 15.7   Generalised pattern of phytoplankton composition according to water column stability in the reservoirs of the Seven Forks cascade

carpio, *Barbus tanensis* and *Mormyrus kannume* are common in the peri-littoral area while *Labeo cylindricus* and *Clarias gariepinus* occur in profundal waters.

Conditions in the other reservoirs are different. Kamburu, Kindaruma and Gitaru are shallow ecosystems with more developed fringing communities. A traditional *Tilapia* fishery developed on the shores of Kamburu which led to the development of a small settlement (Kisumu Ndogo) of Luo fishermen who migrated here from Lake Victoria. Part of Kamburu's fertility derives from the nutrient-rich inflows of the Thiba river which drains the fertilised Mwea rice scheme. Kindaruma's morphology changed dramatically as a consequence of siltation; the reservoir became an ideal environment for crocodile and hippopotamus and so is little used by humans.

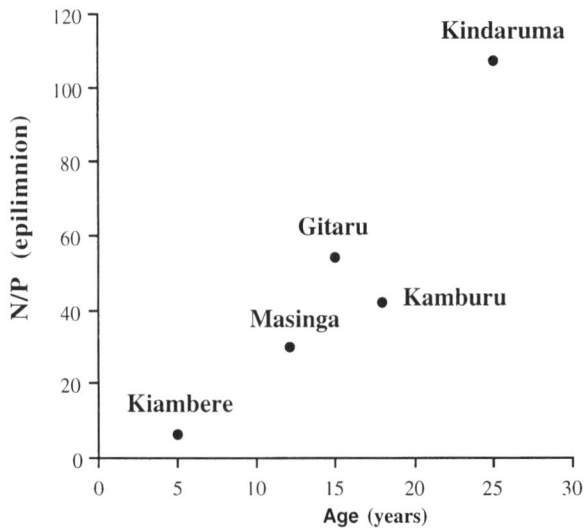

Figure 15.8   Relationship between age and N/P ratio in the reservoirs of the Seven Forks

THE LOWER CATCHMENT

Along the banks of the lower Tana the Pokomo and Malekote tribes carry out a form of traditional agro-silviculture which probably represents the only example of flood-irrigation culture in Kenya. They fish and grow maize, bananas and vegetables in small clearings adjacent to the river with little impact on the forest. Large mango trees found in the forest today were planted by the Pokomo/Malekote over 100 years ago (Ledec, 1987). The dry bush around the floodplain is traditionally used for cattle ranching by the Orma and the Somali who recently started crossing the river and immigrating into the region.

The colonial government initiated development in the lower catchment with the opening of an irrigation scheme of 846 ha at Hola in the 1950s to accommodate prisoners captured during the Mau Mau revolution. A similar irrigated rice production scheme was opened at Mwea, today situated on the northern shore of Masinga Reservoir in the upper catchment. The success of these schemes cannot be assessed on economic grounds as production was based on forced labour and living conditions below the level of subsistence. The location was chosen for optimal conditions of captivity, away from other settlements, and not on the basis of sound land management (Saha, 1982).

Plans for more extensive irrigation started in the 1960s with water resource assessment studies carried out by FAO/UNDP. A National Irrigation Board (NIB) was created in 1966 to co-ordinate development at Hola and Mwea and to plan new irrigation schemes. On the basis of a study conducted by the TRDA cited by Saha (1982), the land–population balance of the lower catchment at the time was considered favourable for an immigration of settlers from other regions. In the rangeland of the lower basin a theoretical subsistence level had been set at 92 ha per person, the corresponding figure being 0.31 ha in the high-potential agriculture soils of the upper catchment (Murang'a district). However, while the high potential areas were overcrowded, with some 75% of the holdings in Murang'a being below the subsistence size, and an average household of five living on a 1.46 ha plot (Jaetzold and Schmidt, 1983), the lower Tana rangeland was considered underpopulated.

At the Hola and Mwea rice plantations, profit was based on the exploitation of the new settlers who had no property rights, were obliged to live under strict discipline and survived at a level of subsistence that was far below Kenya's average per capita income. In terms of capital invested per tonne of rice produced, the scheme was considered unsuccessful (Saha, 1982). Despite the negative financial and social aspects of such irrigation developments, which are typical of large-scale schemes in Africa (Adams, 1995), there were a number of political incentives driving basin development. Among these were:

- Response to the country's dramatic population growth rate in a manner favoured by international donors.
- An assertion of central government control over a region which actually served historically as a northern frontier with Somaliland.
- A desire to ameliorate, at all costs, the unfavourable balance of trade at a time when food imports showed a sharp increase.

A new Bura irrigation scheme started in the 1970s. By 1985, 2,900 ha were farmed. The original project planned for 12,000 ha to be irrigated with groundwater abstraction, diversion of a part of the river with a canal and the building of a small impoundment (Arnold, 1981). The scheme was planned to allow the resettlement of 40,000 people. The production was oriented primarily towards cash crops such as cotton and rice. Only the first phase of the project was realised, with large economic losses. Low wages, poor health conditions and malnutrition progressively convinced many settlers to abandon the scheme before 1982 (Ledec, 1987).

Cultivation in a pastoral semi-desertic area with high evaporation rates is always a problem for agronomists. Soil salinity, sodicity, hydraulic characteristics, nutrient content and other factors relating to soil fertility demand careful consideration and experimentation. Crops grown in a region that offers scarce resources to the animal community come under a higher threat of depredation by wildlife than in a similarly cropped but well-vegetated area. Not last come insect pests and plant diseases. To face this challenge the settlers were invited to make use of an organochlorine teratogenic biocide (endosulphane) that was extensively applied at Bura without implementing the planned impact monitoring programme. Environmental effects of endosulphane could include extensive damage to the environment and threats to the health of human populations. Biocide spraying was often followed by extensive fish kills in the drainage canals while the settlers were left to collect and eat the dead fish (Ledec, 1987).

The building of a 46 km irrigation supply canal negatively affected river flows and disturbed the necessary wild animal and cattle migration between the floodplain and the surrounding savanna. The canal itself proved unneccessary as the second phase of the project (Bura Phase II) was never started (Ledec, 1987).

Beside the direct effects of extensive land clearing (7,300 ha), irrigation and cultivation of the soils in a semi-desert area, important damage to the environment was caused indirectly by the interaction of the settlers and the surrounding environment and by induced changes in behaviour of the local original population.

The Bura irrigation scheme was located just 5 to 10 km from the lower Tana floodplain forest. Extensive wood cutting occurred almost immediately following the immigration of the new settlers as they were not provided with adequate lodging and cooking facilities (Saha, 1982; Hughes, 1985; Ledec 1987). A wood plantation programme had been planned but had not been carried out as it was soon realised that developers had been over-optimistic about the possibility of tree growth in the surrounding land. Wood cutting and land clearing continued and was progressively extended to an area of some 10 km around the settlement causing severe erosion.

The proximity of the forest induced the settlers to take advantage of a wide range of forest resources by collecting wood, fruits and by poaching. The exploitation of the fragile forest environment was exacerbated by the fact that the settlers were coming from the well-watered highland regions of Kenya, had no experience of the area they moved into and had difficulty with adapting to the new living conditions (poor water supply, higher temperatures, malnutrition, diseases).

The creation of a large new settlement, in a region which, traditionally, had low population densities, formed a centre of attraction for the local nomadic populations. The progressive settlement of Orma pastoralists caused severe overgrazing in the area and some damage to the scheme plantations (Ledec, 1987). The settlers, the local populations and the wildlife came progressively into starker competition for scarce resources.

The Bura scheme was a disaster for the local people and the settlers themselves. A proposal for the creation of more forest reserves that would include the Nanigi, Bura and Milalulu forest tracts had been put forward by the World Bank but was never realised as most of the land on the floodplain was claimed by families of the indigenous Pokomo/Malekote.

LONG-TERM ENVIRONMENTAL IMPACTS OF DAM CONSTRUCTION

The strongest impact of development upon the river system can be attributed to the building of Masinga Dam, which was large enough to provide managers with an absolute control over river flows. From then onwards the customary hydrological cycle became lost with the disappearance of high-magnitude floods and the appearance of increased low flows. An artificial rhythm imparted to the water levels by dam operation can be observed within the lower floodplain at Nanigi with a daily fluctuation of up to 1 m. A weekly cycle with lower water levels on Tuesdays as a consequence of the Sunday reduction in electricity consumption and dam closure, can be observed at Kora (Hughes, 1985).

Sediments originated in the fertile volcanic slopes of Mt Kenya and the Nyandarua Ridge are today deposited in Masinga and Kamburu. Reservoir siltation has been underestimated by developers and current estimates of the life of Masinga Reservoir (about 200 years) are significantly lower than originally expected (520 years). Reservoir siltation also has a negative effect on the floodplain as it deprives the lower catchment of the fertilising action of the sediment deposition during inundation.

Changes in the physico-chemistry of the lower Tana river following the building of the dams occur as follows:

- nutrient stripping by the uptake of algae and by sediment deposition in the reservoirs;
- increased conductivity (higher dissolved salt concentrations) due to higher evaporation rates; and
- higher water temperature.

The combination of these characteristics has contributed to the formation, below the reservoirs, of erosive flows with a high scouring and sediment mobilising potential. Such changes are likely to provoke new instability in the river channel and damage to the Hameye swamp environment, a wetland located in the higher part of the floodplain which is believed to play an essential mitigating role in the hydrology of the lower Tana river (Hughes, 1985).

As a consequence of the dams, the river water coming into the lower floodplain today has a higher salt content and is therefore less suitable for irrigation as it may bring faster soil salinisation – one of the main growth limiting factors in the region. The sediments brought onto the floodplain do not originate any more in the fertile highlands but come from the scouring of the river bed and erosion of the adjacent soils. These soils, formed on the Precambrian metamorphic basement which runs along the East African coast, offer poor hydraulic properties and low nutrient content. The change in the material transported is bound to affect the morpho-edaphic evolution of the floodplain soils.

The building of the reservoir caused the submersion of vegetated riparian zones of the mid-upper river valley and the loss of the land/water ecotone there. There is no information about the riverine habitat preceding the building of the reservoir since no pre-impact investigation was carried out. Consequently, it is impossible to comment in this respect about the impacts of development. The oral tradition of the local indigenous people has always associated the Tana river with a wild and dangerous part of the country. No settlements were ever erected in close proximity of the river; prior to the recent development the riparian zone of the Tana river was the unrestricted domain of wildlife.

Today the survival of animals, plants and indigenous people is threatened by river regulation brought by the dams. The closing of the Kiambere Dam coincided with a water shortage which caused severe crop failures among the traditional Malekote agriculturists. There seems to be no future for traditional farming systems; indigenous inhabitants tend to leave their activities and move to the proximity of new centres around the irrigation schemes where they sell fuelwood, charcoal and timber to the settlers (Johansson, 1992).

ECOLOGY AND MANAGEMENT OF THE LOWER TANA FLOODPLAIN

The Tana River National Primate Reserve (TNPR) was established in 1976 to preserve the best remaining riverine forests along the Tana river and the primary populations of the endangered Tana river red colobus (*Colobus badius rufomitratus*) and crested mangabey (*Cercocebus galeritus galeritus*) (Marsh, 1979). Both primates are endemic to the isolated mosaic of forests along the river.

The uniqueness of the faunal and floristic composition of the Tana floodplain forest ecosystem derives from its long period of isolation from a continuous Central African forest belt which separated during a dry period some 4000 BC (Medley and Hughes, 1996). A detailed floristic study revealed that the Tana forest functioned over millennia as a refuge for animals and plants which migrated there from three different directions: the Congolean rainforests, the East African coastal forests and the semi-arid Sahelian region. With time the forest underwent progressive shrinking under the impact of the Sahelian drought and as a result of human impact (wood cutting, burning, grazing, agriculture). Today the forest is reduced to a thin strip of vegetation running along the riverbanks.

Studies conducted by Hughes (1985, 1988, 1990) demonstrated a tight link between the forest survival and the Tana river floods. The riverine forest persists in a state of dynamic equilibrium in which the flood regularly rejuvenates the floodplain soils preventing salt accumulation. Salinisation would otherwise occur as a consequence of the high evaporation rates and the origin of the parent material (Muchena, 1987). The sands brought by the river floods improve drainage and allow the leaching of salts out of the upper soil horizon permitting luxuriant tree growth (Andrews et al., 1975). Field observations demonstrate a significant relationship between the proximity of the river channel and the presence of sandy, basic but non-saline soils (Hughes, 1985). On the other hand, the presence of the forest is equally essential to soil properties. When the forest is not able to sustain itself due to a change in moisture conditions or due to land clearing, evaporation induces salts to come to the surface carried by matrix flow, reducing soil aggregation, nutrient availability and promoting the formation of sterile saline to sodic soils.

The successive formation of new habitats within the floodplain by the scouring and sediment depositing action of the flood, contributed to create a high diversity of vegetation types characterised by different ages of formation, different substrates and different moisture regimes (Hughes, 1985). Habitat diversity is essential to preserve a high species diversity and, for each species, the simultaneous presence of different development stages. Both these factors are responsible for the temporal availability of fruit resources which are considered critical for the life histories of the primate fauna (Kinnaird, 1992).

Floods are vital for recharging the floodplain aquifer and providing medium-term moist conditions necessary for tree regeneration. Calculations demonstrate that some trees such as the Tana red poplar (*Populus ilicifolia tanensis*) regenerate only in high-magnitude floods (approximately 80-year return time). The Tana river poplar constitutes today the base structure of much of the floodplain, so the arrest of its regeneration will not be without important changes (Hughes, 1985).

The importance of the hydraulic regime is not restricted to morpho-edaphic processes but determines life within the floodplain in a more direct manner. Flowering phenology, for example, is generally related to the timing and spell of dry conditions, in such a way that the fruit can then ripen and fall at the onset of the rains to provide optimal conditions for seed germination. In the tropics, where temperature variations over the year are mild, flooding rather than rainfall and temperature patterns sets biological cycles of animals and plants (Payne, 1986). In the Amazon floodplain trees undergo a physiological winter as the floods cause oxygen stress (Worbes, 1985, 1986). In the lower Tana floodplain, growth and maturation respond to a complex combination of hydrologic signals. Local rainfall is scarce and fruiting phenologies are correlated to the timing and spell of the low water levels in the river. A sudden modification of the intimate relationship between the hydrology and the phenological response can have wide-reaching consequences for fruit production and tree regeneration. The disruption of these ancient equilibria will exacerbate differences between periods of resource abundance and scarcity which are relevant for the survival of the floodplain animal community (Kinnaird, 1992).

CURRENT ISSUES

It is well established and widely accepted that competing water uses are restricting options for future development in the Tana basin. The use of hydro-electric power resources coupled with irrigated agriculture downstream is internally contradictory and calls for careful economical decisions. Part of the failure of the Bura scheme lies in water shortage caused by dam operation upstream.

The chances of new irrigation projects in the Tana floodplain area are remote after the World Bank officially removed support from the Bura irrigation scheme and the President of Kenya cited it publicly as one of the worst examples of agricultural failure. One of the environmentally most disruptive options is the proposal for an inter-basin diversion scheme that would bring the water of the Tana to the more developed Athi basin. Such development would have the technical advantage of bringing the water resource closer to a more developed part of the country which would make any future project, including irrigation, more cost-efficient. An inter-basin diversion might also alleviate the water supply deficit of the port city of Mombasa where shallow freshwater aquifers formed in coastal sediments are being exhausted by the expanding tourist industry (see also Chapter 21). Such a development would also mean the destruction of the coastal mangroves along the Tana delta and wide salt-water intrusions into the floodplain.

The main alternatives for the supply of Mombasa are the enlargement of the pipeline bringing high-quality water from the Mzima Springs (Tsavo National Park) and the damming of the Athi river with a suitable programme of soil conservation measures to reduce siltation (Wain, 1983).

The Tana delta region itself has attracted great interest during the past 5 years. Several projects were debated including irrigated rice plantations and crustacean salt-water aquaculture. Developers believe that the delta offers 17,000 ha of land suitable for cash crop irrigation schemes (Rowntree, 1990). A recent proposal for the development of a large shrimp aquaculture was turned down at the last moment by the intervention of the President of Kenya. There is support within the country for the development of the delta as a National Park and Kenya's third Ramsar site, and it could be a sign of change in central government policy after the negative experiences with past irrigation projects in the Tana. It is indeed a good time for a wetland management proposal that would put natural resource conservation hand in hand with support of local human activities.

The future of the Tana floodplain forest and the survival of several endemic species seem bleak. The main reasons seem to stem from the dependency of the biological cycles on the rejuvenating effects of flood, changes in soil formation and the lack of regeneration of *Populus ilicifolia* which constitutes the base structure of much of the floodplain forest.

Current planning for a new reservoir, Grand Falls, at the downstream of the cascade includes a proposal to create enough storage capacity for the generation of an artificial flood which could include the sediment and nutrients needed for soil rejuvenation (Acreman and Hollis, 1996). It seems unrealistic to expect artificial flood releases generated by a new reservoir at Grand Falls to be able to help the situation in the forest, accommodate irrigation schemes, generate power and still be cost-effective. It was the building of Masinga Dam with its sediment trap efficiency estimated over 90% (Maingi, 1991) which has removed sediment and nutrients from the lower floodplain, accumulating more than $40 \times 10^6 \, \text{m}^3$ of sediment within 10 years (TARDA, pers. comm.). The Tana flood peaks have disappeared since its construction (Pacini, 1994). Sediment discharge from the dam is limited to fine clays and colloids released through the overspill at times when periods of turnover within the reservoir coincide with high water levels (Pacini et al., 1993).

THE FUTURE

The variety of environmental problems already experienced means that only the management of the catchment as a single entity has any hope of mitigating the potential disasters that await it.

In the upper catchment, intensive rain-fed agriculture supports a highly populated region. Continued support for soil and water conservation schemes including development of integrated programmes such as agroforestry, and infrastructural support (transport links) will minimise erosive soil loss (Chapters 2 and 3).

In the reservoir cascade, multipurpose wetland management could yield greater income and employment in commercial fishing, fish and crocodile farming and tourism. To support this, further limnological investigations are necessary to better understand mechanisms which determine habitat quality in the reservoirs.

Evidence from a series of samples of limnetic zooplankton, analysed for species composition and size structure, together with three investigations of the reservoir fisheries development by gill-netting and subsampling of fisheries catches, indicate that the reservoirs are at present under-exploited. Zooplankton species composition is dominated in all reservoirs by cyclopoid copepods but includes both calanoids and cladocerans. The relative scarcity of smaller-bodied forms indicate a low fish predation pressure upon the zooplankton.

There are at present several problems in the management of the reservoir fisheries which are being addressed by the responsible agencies. Chief among these is that most of the landed catch goes unrecorded due to time differences between the landing (earliest and most successful boats beach around 03:00 AM in order for the catch to reach Nairobi in fresh condition) and the recording (dawn). Hence the data collected from the fishery cannot realistically be used in any management strategy.

Multipurpose reservoir uses would require that Masinga Reservoir water levels be reduced to allow for more natural flow patterns within the river and for a reduction of water level change within the reservoir itself. A survey of the consequences of water level fluctuations in a large reservoir of central Africa suggest a variation of 3.5 m as the maximum amplitude which can allow macrophyte development (Bernacsek, 1984). The reservoirs constitute large water masses in a region which was otherwise semi-arid. Species of birds have altered their migration routes to benefit from the Seven Forks food and shelter resource. Populations of resident birds seem to increase every year. A study of migration patterns and bird ecology within the Seven Forks could pin-point important habitats for sustainable tourism, the potential for which in Kenya is still great (Visser, 1992).

Sustainable management of the lower catchment requires careful management for soil and water conservation and protection of the Hameye swamp. Novel solutions to the long-term deterioration of the lower Tana floodplain forest are required, such as consideration of the building of a small reservoir in a suitable location within the floodplain to accommodate traditional indigenous agricultural practices in the area and sustain forest regeneration. A necessary option would be the extension of the protected area to include a larger part of the floodplain forest for an equilibrated management of wildlife and human activities. Local people should be directly involved. The form of protection should not exclude them and their activities from the land but preserve their traditional way of life in the respect of their traditions, provide them with better living conditions, shelter them from foreign newcomers. Resident populations are the ones that more than any other can appreciate the value of their environment and the importance of preserving it for the future generations. However it is not surprising that these populations are hostile to the creation of a national Forest Reserve which *de facto* excludes them from the forests' natural resources.

The TNPR itself needs greater protection, transport and staff to ensure higher security within the region, to stop poaching and illegal depredation of the woody resource, and to develop the area for tourism and wildlife. Its status should be raised, perhaps by linkage to that of National Park, with a buffer zone in which careful management agriculture and cattle ranching was exercised with particular control on grazing and wood cutting.

Currently the decision-making in catchment management is both centralised and relatively closed. The management agency, TARDA, would benefit from closer collaboration with other Ministries concerned with wildlife management and sustainable tourism and with Kenyan university departments, so that it could develop internal knowledge and experience about land and water management making it less dependent on external donor agencies and consultants for strategic thinking.

CONCLUSIONS

Phosphorus runoff from the upper catchment is strongly linked to sediment; soluble phosphorus in this system is negligible. As a consequence, the pattern of primary productivity in the upper reservoir, Masinga, is controlled by the seasonality of rainfall and sediment input superimposed upon the temperature- and wind-controlled stratification cycle. The reservoir thus experiences alternate periods of phosphorus- and light-limited productivity. A more careful management of water level fluctuations could benefit the development of a littoral area within the reservoir, leading to growth of fish habitats as well as a more natural pattern of discharges downstream.

Food-web links through zooplankton to fish populations and the fishery suggest an under-harvest at present, although the limited observations and recording of catches at Kamburu demonstrate that the reservoirs potentially have a high yield of commercial fish such as carp, tilapias and eels. The value of this resource has never been compared with the loss of hydropower necessary to optimise a fishery. Inadequate monitoring, both of the fishery and of the hydrochemistry of the reservoirs, inhibit further recommendations for exploitation.

This is particularly important in the context of the cascade, where downstream reservoir limnology is controlled by physical morphology (determining thermal stability) and age (determining phosphorus accumulation in sediments). Hence the importance of hydraulic flushing upon nutrient cycling and algal biomass is underestimated and the role of stratification and deep water curents (e.g. providing habitat for *Claria* and *Anguilla*) is unquantified.

The reservoir cascade has impacts upon downstream ecosystems of international importance, which although not yet apparent are potentially severe. Mitigation possibilities exist if the planning of the final two dams in the cascade takes account of the environmental impact of discharge changes and provides the flexibility for "environmental flows".

ACKNOWLEDGEMENTS

This project was funded by the European Community's STD2 programme, contract TS2-A-0256-UK. We thank colleagues at the University of Nairobi for logistical support throughout the fieldwork.

REFERENCES

Acreman, M.C. and Hollis, G.E. 1996. '*Water Management and Wetlands in Sub-Saharan Africa*'. IUCN, Gland, Switzerland.

Adams, W.M. 1995. '*Wasting the Rain*'. Earthscan, London.

Andrews, P., Groves, C.P. and Horne, J.F.M. 1975. 'Ecology of the lower Tana river floodplain (Kenya). *Journal of the East African Natural History Society and National Museums*, **15**, 1–31.

Arnold, G. 1981. *Modern Kenya*. Longman, London.

Bernacsek, G.M. 1984. '*Guidelines for Dam Design and Operation to Optimise Fish Production in Impounded River Basins*'. CIFA Technical Paper 11, FAO, Rome.

Grove, A.T. 1978. 'Geographical introduction to the Sahel'. *Geographical Journal*, **144**, 407–415.

Hughes, F.M.R. 1985. 'The Tana River Floodplain Forest, Kenya: Ecology and impact of development'. Unpublished PhD Thesis, University of Cambridge.

Hughes, F.M.R. 1988. 'The ecology of African floodplain forests in semi-arid zones: a review. *Journal of Biogeography*, **15**, 127–140.

Hughes, F.M.R. 1990. 'The influence of flooding regimes on forest distribution and composition in the Tana river floodplain, Kenya'. *Journal of Applied Ecology*, **27**, 475–491.

Jaetzold, R. and Schmidt, H. 1983. '*Farm Management Handbook of Kenya: Natural Conditions and Farm Management Information*'. Ministry of Agriculture, Nairobi.

Johansson, S. 1992. 'Irrigation and Development in the Tana River Basin'. In *African River Basins and Development Crises* (ed. M.B.K. Darkoh), pp. 97–112. Uppsala University, Sweden.

John, D.M. 1986. 'The inland waters of tropical West Africa. *Archiv für Hydrobiologie (Ergebnisse)*, **23**, 1–244.

Kinnaird, M.F. 1992. 'Phenology of flowering and fruiting of an East African riverine forest ecosystem. *Biotropica*, **24**, 187–194.

Ledec, G. 1987. 'Effects of Kenya's Bura irrigation settlement project on biological diversity and other conservation concerns'. *Conservation Biology*, **1**, 247–258.

Maingi, J.K. 1991. 'Sedimentation in Masinga Reservoir'. Unpublished MSc Thesis, University of Nairobi.

Marsh, C.W. 1979. Ecology and social organisation of the Tana River red colobus, *Colobus badius rufomitratus*. Unpublished PhD Thesis, University of Bristol.

Medley, K.E. and Hughes, F.M.R. 1996. 'Riverine forests'. In *East African Ecosystems and their Conservation* (eds T.R. McClanahan and T.R. Young), pp. 361–384. Oxford University Press, New York.

Muchena, F.N. 1987. 'Soils and irrigation in three areas in the lower Tana region, Kenya'. Unpublished PhD Thesis, Agricultural University of Wageningen, The Netherlands.

Muthuri, F.M. 1985. 'The ecology of *Cyperus papyrus* in Kenya'. Unpublished MSc Thesis, University of Nairobi.

Njuguna, G.S. 1992. 'Tana River Delta wetlands'. In *Wetlands of Kenya: Proceedings of the KWWG Seminar on Wetlands of Kenya* (eds S.A. Crafter, S.G. Njuguna and G.W. Howard). National Museums of Kenya, Nairobi, Kenya, 3–5 July 1991, IUCN Wetlands Programme, IUCN, Gland, Switzerland.

Odingo, R.S. 1979. '*An African Dam, Ecological Survey of the Kamburu/Gitaru Hydroelectric Dam Area*', Ecological Bulletins No. 29, Swedish National Research Council, Stockholm.

Pacini, N. 1994. 'Coupling of land and water: phosphorus fluxes in the upper Tana river catchment, Kenya'. Unpublished PhD Thesis, University of Leicester.

Pacini, N., Harper, D.M. and Mavuti, K.M. 1993. 'A sediment-dominated tropical impoundment: Masinga Dam, Kenya'. *Verhandlungen Internationale Vereinigung für Theoretische und Angewante Limnologie*, **25**, 1275–1279.

Payne, I. 1986. '*The Ecology of Tropical Lakes and Rivers*'. Longman, London.

Rowntree, K. 1990. 'Political and administrative constraints on integrated river basin development: an evaluation of the Tana and Athi Rivers Development Authority, Kenya'. *Applied Geography*, **10**, 21–41.

Saha, S.K. 1982. 'Irrigation planning in the Tana basin of Kenya'. *Water Supply and Management*, **6**, 261–279.

Thompson, J.R. 1996. 'Africa's floodplains: a hydrological overview'. In *Water Management and Wetlands in Sub-Saharan Africa* (eds M.C. Acreman and G.E. Hollis), pp. 5–20. IUCN Wetland Programme, Gland, Switzerland

Visser, N.W. 1992. 'Wetlands and tourism'. In *Wetlands of Kenya: Proceedings of the KWWG Seminar on Wetlands of Kenya* (eds S.A. Crafter, S.G. Njuguna and G.W. Howard). National Museums of Kenya, Nairobi, Kenya, 3–5 July 1991, The IUCN Wetlands Programme, IUCN, Gland, Switzerland.

Wain, A.S. 1983. 'Athi river sediment yields and significance for water resource development'. In *Soil and Water conservation in Kenya* (eds D.B. Thomas and W.M. Senga). Institute for Development Studies, University of Nairobi, Occasional Paper No. 42.

Ward, Ashcroft & Parkman (East Africa) Ltd. 1976. 'Upper reservoir pre-construction environmental study'. Cited in Hughes F.M.R. 1985.'The Tana River Floodplain Forest, Kenya: Ecology and impact of development'. Unpublished PhD Thesis, University of Cambridge.

White, W.R. 1990. 'Reservoir sedimentation and flushing'. In *Hydrology of Mountainous Regions. II Artificial Reservoirs; Water and Slopes*. Proceedings of a symposium held in Lausanne, Switzerland, IAHS Publication No. 194, 129–139.

Worbes, M. 1985. 'Structural and other adaptations to longterm flooding by trees in central Amazonia'. *Amazoniana*, **9**, 459–484.

Worbes, M. 1986. 'Lebensbedinungen und Holzwachstum in zentralamazonischen Überschwemmungswäldern'. *Scripta Geobotanica*, **17**, 112–134.

CHAPTER 16

Information for the Sustainable Management of Shallow Lakes: Lake Naivasha, Kenya

GEOFF JOHNSON

and

DAVID HARPER
Department of Biology, University of Leicester, UK

KENNETH MAVUTI
Department of Zoology, University of Nairobi, Kenya

INTRODUCTION

Shallow tropical lakes are inherently unstable, relying upon unpredictable amounts of seasonal rains, and those in Africa are probably the best known. Some, such as Lake Chad, have well-described phases of high and low water levels whilst others, such as Lake Chilwa, have known dry phases. Despite this unpredictability, shallow tropical lakes and their associated wetlands are valuable natural resources (Podolsky and Conkling, 1991). In an otherwise arid environment, their water supply is essential for human life and livelihood. Farming is close to the lake edges in order to use irrigation water and so may directly threaten the health of lake/wetland through elevated concentrations of pesticide and fertiliser residues in runoff.

Lake Naivasha, in the Eastern Rift of Kenya (Figure 16.1), has been the subject of scientific investigation at various times since the late 1920s (Melak, 1996, summarises the history of this and other more saline lakes in the chain). Up to the early 1970s its ecology remained relatively stable with the exception of its fish community, which had been interfered with for "sport" since 1926. Plant communities and distribution, and lake/wetland areas changed in response to water level changes, but without major upsets in species abundance patterns. A small commercial fishery operated and the lake was primarily used for recreation and moderate-intensity farming.

There have always been small-scale conflicts between commercial interests and those of environmental conservation; one recorded during this time was between the

---

The Sustainable Management of Tropical Catchments. Edited by David Harper and Tony Brown.

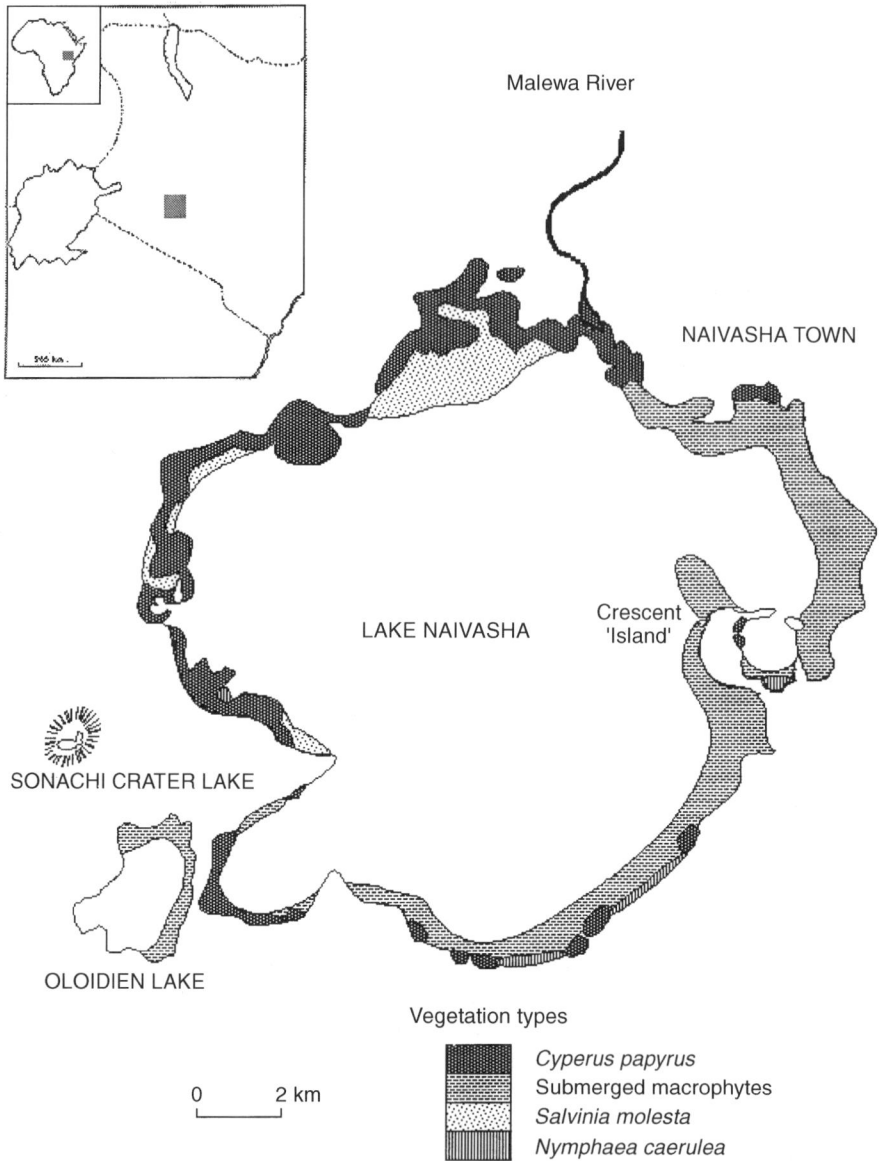

Figure 16.1   Location map of Lake Naivasha, showing aquatic vegetation surveyed by boat in
1987 (Harper, 1992).

hippopotamus (*Hippopotamus amphibius*) population and farming, with the proposal
from some quarters that the population of *c.* 400 animals (sometimes inflated to
1500!) should be shot out. Another was between hunters of duck (principally *Anas
undulata*, but up to 20 species of African and palearctic Anatidae) who believed that
the coot (*Fulica cristata*) was successfully competing with duck for food, principally
underwater plants, and should be reduced by shooting.

The major changes in the lake, however, occurred from the mid-1970s. Most damaging have been those which were the result of alien species appearing by accident or deliberate introduction. Four species, between them, have destroyed (probably for ever) the former lake community and replaced it with a quite different one (Harper et al., 1990). Of least importance is the coypu (*Myocastor coypus*), a South American fur-bearing mammal brought to Kenya in the 1950s. A population from escaped animals had established at Naivasha by the mid-1970s and the animal was blamed for the disappearance of the famous blue water lily (*Nymphaea caerulea*) which dominated the shallow edges of the lake. This is unlikely, as the coypu population crashed within a decade and now only occasional individuals are seen. Coypu were, however, closely followed by the deliberate introduction of the Louisiana crayfish (*Procambarus clarkii*), which was promoted as a species which would enhance and diversify the commercial fishery. This has been an almost unmitigated disaster, as crayfish are voracious feeders on submerged vegetation and in the two decades since the species introduction no growth of *N. caerulea* other than scattered individual seedlings has been recorded and there has been a clear, inverse cyclical fluctuation in crayfish numbers and in area covered by submerged vegetation (Harper et al., 1995). Crayfish also climb up commericial gill nets and damage the catch, whilst the market – in tourist hotels and Europe – is quite fickle.

Two floating plant species, renowned throughout the Old World tropics for their negative environmental impact, have played a secondary role to crayfish in the ecological deterioration. The first, *Salvinia molesta*, appeared in the 1970s and was able to take advantage of the late 1970s/early 1980s water level rise which coincided with the disappearance of submerged and floating vegetation caused by the initial population burst of crayfish. Large floating mats of *S. molesta* disrupted fishing and helped to prevent recovery of water lilies during periods of crayfish population crashes until the mid-1990s, when they disappeared through a combination of water level decline causing stranding and rapid dessication, and the introduction of the herbivorous weevil, (*Cyrtobagus salvinii*), in 1993. By then, however, the second species, water hyacinth (*Eichhornia crassipes*), had arrived, also by accident (though freely available as an ornamental pond plant in garden centres). This species, although not capable of forming extensive mats (the altitude of Naivasha, 1850 m, results in a temperature regime which slows both growth rate and maximum biomass of *E. crassipes*) has totally dominated the littoral shoreline. Unlike *S. molesta*, it is widely consumed by species from coot to hippopotamus, but its ability to survive on wet mud and even amongst littoral grassland in a prostrate form means that it is very tolerant of water level changes. It, too, is probably responsible for the failure of native vegetation to recover when crayfish populations crash (lake mud contains an adequate bank of seeds from all species of aquatic vegetation formerly present in the lake, evident from their successful germination in absence of crayfish and alien floating plants in garden ponds and irrigation lagoons along the lakeshore).

By the late 1980s then, the submerged vegetation ecology of the lake was totally changed from that described as a typical tropical freshwater/wetland succession by Gaudet (1977). The fringing wetland swamp vegetation had also changed dramatically, for different reasons (Harper, 1992). Here, water level recession between a high of 1982 and a low of early 1987 had coincided with a rapid increase in the

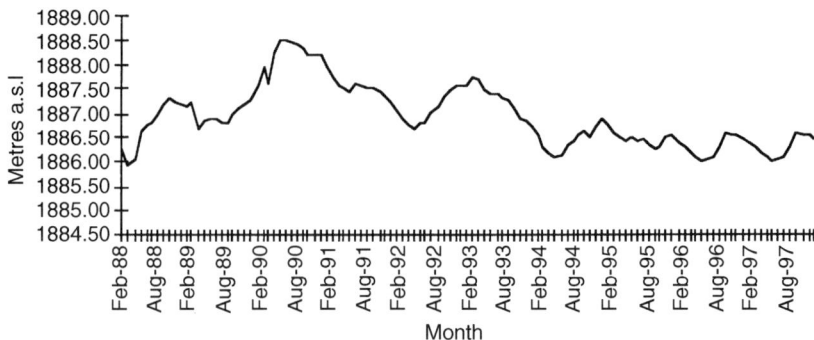

Figure 16.2   Recent water level changes at Lake Naivasha

development of irrigated agriculture and horticulture, and large areas of papyrus (*Cyperus papyrus*) had been cleared and developed. At the same time, burning, either deliberately to improve cattle grazing or accidentally as a consequence of fishermen's fires, had destroyed substantial stands. In early 1988 substantial rains, which brought to an end the mid-1980s drought experienced by most of eastern Africa, resulted in a 1 m rise in water level, equivalent to between 500 and 800 m horizontally in the shallow western and northern shores of the lake. These areas almost immediately germinated new swamp (initially of mixed species but within 2 years papyrus monoculture) (Harper, 1992; Harper et al., 1995), but over the next 9 years the pattern of papyrus destruction was repeated as water level receded once more (Figure 16.2).

Over the past 5 years a local initiative from the association of lakeside owners (then the Lake Naivasha Riparian Owners Association, LNROA; now the Lake Naivasha Riparian Association, LNRA) to conserve and sustainably manage the lake has led, with international assistance, to the Government's 1995 declaration of the lake as Kenya's second Ramsar site (a wetland of international importance under the 1971 Ramsar Convention), after Lake Nakuru. A management plan, proposed by the LNROA after a three-phase review of the pressures on the lake (Goldson, 1993), has been accepted by the District and Provincial administration and Kenya Wildlife Services (the central government's Ramsar agent) and is in the early stages of implementation. Acceptance of the need for sustainable management has been driven by the widespread realisation that a continued increase in irrigated area under horticultural crops, coinciding with the development of the neighbouring Olkaria geothermal power station drawing water for drilling and cooling from the lake, against a background of 5 years' falling water levels 1992–97, could, if unmanaged, totally destroy both the ecology and the available water from the lake. Pressures in the catchment from increasing density of human settlement leading to larger volumes removed from the rivers for domestic and subsistence agricultural use, combined with the beginnings of trans-catchment water diversion to augment domestic supplies of Gilgil and Nakuru combine to make the possibility of the lake's disappearence in the near future a reality.

THE INFORMATION NEEDED FOR SUSTAINABLE MANAGEMENT AT LAKE NAIVASHA

The Ramsar management plan for the lake identifies a number of priority areas for action (LNROA, 1996), the most important of which is an accurate water balance both for the natural inputs and the human offtakes. Such a balance has never been accurately achieved, although estimates have been made since the 1930s (Sikes, 1935). The reason is the porous nature of the volcanic and sedimentary deposits in the catchment, giving difficulty of measuring subsurface inflows from the catchment and and subsurface outflow, since the lake has never had a surface outflow yet has always remained fresh (Gaudet and Melack, 1981). A study funded by the Netherlands government is under way to try to resolve this problem (see also Chapter 20). Most of the other management issues require accurate, relevant and frequently updated information about vegetation change.

All the threats to the lake, from the early problems of coot and duck to the present-day problems of alien species and increased settlement and development, involved vegetation changes. Initially, such change in the lake was measured by aerial and boat surveys, and manual estimation of area from the resultant mapping (Watson and Parker, 1969; Harper et al., 1990). The speed of change, however, necessitates more automatic information on vegetation distribution. In this respect Naivasha is no different from wetlands elsewhere (Norton and Snecker, 1990; Dobson and Bright, 1991) and aerial photography has been successfully used to measure changes in other wetland ecosystems (e.g. Lyon and Greene (1992) for Lake Erie; Bakker et al. (1994) in the "Polder Westbroek" (The Netherlands)). However, wetland present challenges to effective monitoring and quantification. For example, wetland types are diverse, ranging from small tributary streams, shrub/scrub and marsh communities, to open water (Jensen and Mackey, 1993). In addition, the type and spatial distribution of wetlands can change dramatically especially where there are seasonal variations (Mackey, 1990). For these reasons, satellite remote sensing is now more often used to obtain information on the spatial distribution and biophysical conditions of catchments and wetlands (Jensen et al., 1991; Roughgarden et al., 1991).

The Multispectral Scanner (MSS) flown on the Landsat series of satellites was the first high spatial resolution sensor to be employed for any form of terrestrial mapping. This instrument produced 185 km^2 images with individual pixels corresponding to $79 \times 79 \text{ m}$ squares on the ground. It collected data from three regions in the visible and one region in the near-infrared part of the spectrum. Glimmer et al. (1980) used MSS data in conjunction with higher resolution airborne data to enumerate prairie wetlands in the USA. A simple two classification system of land and water was derived using the near-infrared wavelengths. Butera (1983) demonstrated the diversity of applications for MSS data in quantifying different types of wetlands. Examples used were the delineation of *Phragmites australis* in the Mississippi delta; communities, principally mangroves, in Florida; and forested and non-forested wetland cover types in the south-east tidal/riverine system of the Savannah River.

Since 1984, the Landsat Thematic Mapper (TM) has replaced MSS. This instrument has a maximum resolving power of $30 \, \text{m}^2$. Spectral sensitivity is greater in three visible and one near-infrared band with increased functionality due to a further two short-wave and one mid-infrared wavelength. Ackleson and Klemas (1987) used both MSS and TM data in assessing areal extent of *Zostera marina* beds in lower Chesapeake Bay.

REMOTE SENSING METHODOLOGY

In order to provide a baseline for change at Lake Naivasha, July 1987 and March 1989 Landsat TM images (Path 169, Row 060) were acquired from the National Remote Sensing Centre (NRSC) at Farnborough, UK. The 1987 image was geometrically corrected to the Kenyan national grid and the 1989 image registered to the 1987. Atmospheric correction was carried out using a Dark Area Subtraction technique, as described by Chavez (1988). Imagery was processed using four techniques: (a) photo-interpretation of hard copy false colour composites, (b) Single Pass and (c) Iterative Optimisation unsupervised classifications, and (d) a Maximum Likelihood supervised classification (Richards, 1993). Training data site selection was augmented by the collection of multispectral data collected with a Milton Multiband Radiometer over target cover types in March and April 1994 and by examination of photographs taken during flights over the study area around the same time as image acquisition. Both Single Pass and Iterative Optimisation unsupervised algorithms produced poor classifications and will not be discussed further.

The decision process of which classes to include in the photo-interpretation and Maximum Likelihood supervised classifications was based on a series of considerations. The first was that the resulting area cover maps should have as many different classes as possible. Secondly, for the classifications to be of use in further studies, it is important that vegetation types with specific growth and reproduction characteristics be distinguishable. There are however species of the same genus, such as carnations, which although they appear distinct in the image, have similar biology and therefore do not need to be separated in the classification. Class choice was ultimately governed by the ability to identify, by eye, discrete groups in the imagery. Classes were identified due to their relative colours, location relative to the lake margin and, in the case of agricultural cover types, regular field geometry.

MultiSpec is a multispectral and hyperspectral image analysis program designed for PC and Macintosh computers, rather than more powerful workstations. It has been developed by David Landgrebe and Larry Biehl in the School of Electrical and Computer Engineering and Laboratory for Applications of Remote Sensing at Purdue University, West Lafayette, IN, USA. Further details can be obtained from: http://dynamo.ecn.purdue.edu/~ biehl/MultiSpec/.

The software uses a number of the more common classification procedures. Of these, the Maximum Likelihood Classification is the most common supervised classification procedure used. The algorithm is based on Bayes' classification which can be described as follows:

$$\omega_i, i = 1, \ldots, M \tag{1}$$

where M is the total number of classes. In trying to determine the class or category to which a pixel at a location x belongs it is the conditional probabilities

$$p(\omega_i|x), i = 1, \ldots, M \tag{2}$$

that are of interest. The position vector x is a column vector of brightness values for the pixel. It describes the pixel as a point in multispectral space with co-ordinates defined by the brightness of individual image bands. The probability $p(\omega_i|x)$ gives the likelihood that the correct class is ω_i for a pixel at position x. Classification is performed according to

$$x \in \omega_i \text{ if } p(\omega_i|x) > p(\omega_i|x) \text{ for all } j \neq i \tag{3}$$

For further information see Richards (1993).

Whilst this procedure was found to produce the more accurate land cover maps for the individual scenes, another approach would have been to try to identify differences between the different date images irrespective of the cover type. This could have been done using Principal Components Analysis (PCA). For this technique, pairs of multi-date images bands (i.e. band 4 from 1987 and band 4 from 1989) are combined, and the differences measured as variability between pixel vectors. For example an area that is water in 1987, but papyrus in 1989, will have a lower band 7 reading for 1987 than 1989, because water absorbs the near-infrared more than vegetation. Plots of the resultant PCA images would highlight changes in cover types. This would work very well for determining the change in water coverage for example, but would require further analysis for determining differences in agricultural practices or other land-based variations (Richards, 1993).

PHOTO-INTERPRETATION OF IMAGES

Figures 16.3 and 16.4 compare the result of photo-interpretation of the two images. Clearly visible are the homogeneity of the *Salvinia molesta* beds, in the north-west region of the lake, and papyrus, forming a fringe round the shore and rafts in the western half of the lake. The three lake basins: (a) the inside of Crescent Island in the east of the lake, (b) the main lake itself and (c) Oloidien (the smaller, higher conductivity basin in the south-west of the image), are also visible, as is Sonachi, the small, isolated saline crater lake in the south-west of the image. Interesting areas include the large circular irrigation unit and the regularly shaped horticultural developments to the north and south of the lake.

Visual identification of hard copy output resulted in the discrimination of 15 specific cover types, one class for cloud and one encompassing those areas which could not be accurately assigned (Table 16.1). Total classifications account for 670.9 and 671.4 km^2 for the 1987 and 1989 images respectively. The discrepancy in total area covered can be attributed to the different areas covered by boundaries. This in turn may be a function of greater heterogeneity within the 1987 image, inconsistencies during the tracing or scanning of the image, or a combination of all three.

Examination of the area change figures indicate large variations between the

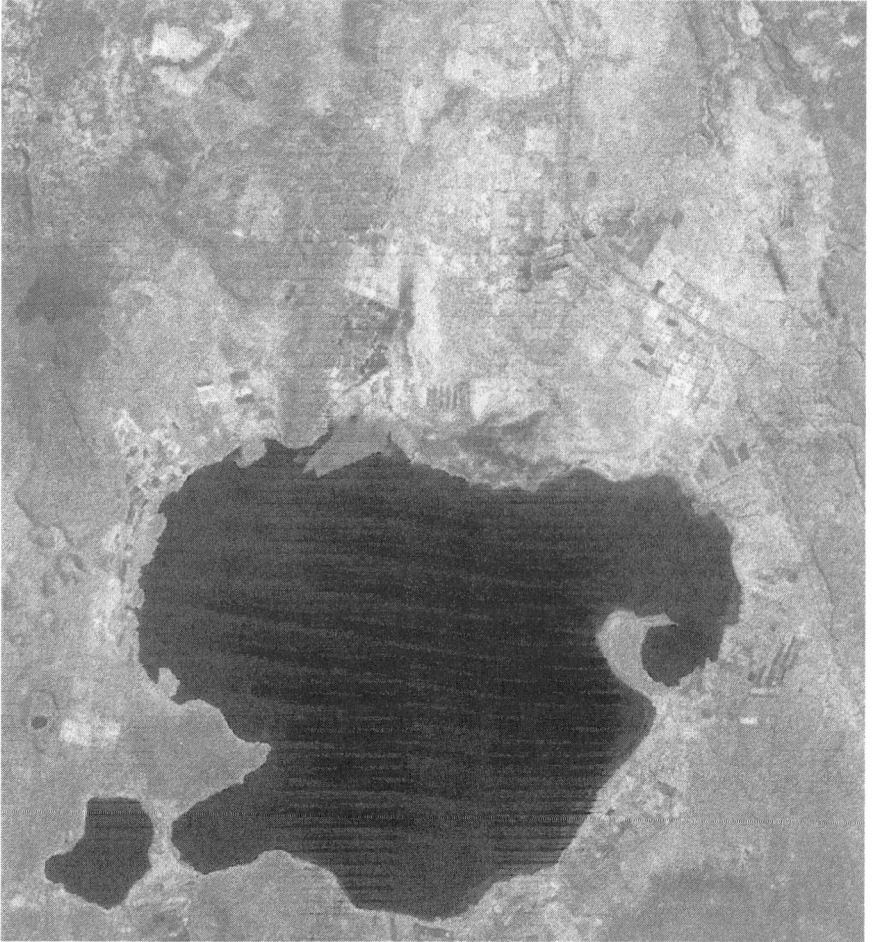

Figure 16.3   July 1987 Landsat TM Band 4 (760–900 nm) image of the Lake Naivasha basin (image size 25 × 27 km)

images for particular cover types. The value for *Salvinia* $(+8.6\,\text{km}^2)$ confirms the increase following the 1988 water level rise. There was also a change in location between the years. In 1987 *Salvinia only* formed a large aggregation in the south-west region of the North Swamp. In 1989 the more mobile mats extended further south and westwards on the western shore. The 7.5 km^2 increase in papyrus between the years is due to the widening of the fringe, rather than an increase in any one specific location, and is also the result of the 1988 rains (Harper, 1992).

Submerged vegetation was restricted to areas to the north and west of Crescent Island and was only visible on the 1989 interpretation. However, from 1984 to 1987 *Potamogeton octandrus* spread throughout the eastern half of the lake, starting from scattered plants in sparse beds confined to the shallow mud shores on the outside of Crescent Island to extensive fringes in shallow water in 1987 (Harper, 1992).

Figure 16.4   March 1989 Landsat TM Band 4 (760–900 nm) image of the Lake Naivasha basin (image size 25×27 km)

However, they were not easily delineated on the image and have consistently been misclassified as papyrus or *Salvinia molesta* due to the fact that they trap surface-floating mats or small islands.

Identification of the water types of the lake reveal relatively small changes. There is a decrease in the area classified as "shallow water", and a subsequent increase in the value for "main lake water". The differences may be due to an increase in water level between image acquisition dates. However, examination of aerial photographs of the eastern half of the lake for 1987 reveals extensive amounts of particulate matter making the water turbid. This turbidity may have interfered with an accurate assessment of the specific water classes. The increased papyrus cover for 1989, in the north of the lake where the Malewa river enters, resulted in less particulate matter entering, thus decreasing the turbidity of the water.

Table 16.1   Results of photo-interpretation of the 1987 and 1989 images

Information class	1987 area		1989 area		% change 1987–98
	km^2	%	km^2	%	
Deep water	1.95	0.29	1.95	0.29	0.85
Shallow water	51.82	7.71	47.45	7.06	−7.93
Main lake water	116.60	17.35	119.76	17.82	3.35
Oloidien	6.85	1.02	6.99	1.04	3.46
Sonachi water	0.20	0.03	0.13	0.02	−30.00
Volcanic scrub	153.90	22.90	166.6	24.79	8.90
Euphorbia woodland	11.09	1.65	8.06	1.20	−27.16
Intensive agriculture	30.24	4.50	29.57	4.40	−1.73
Horticulture	15.05	2.24	20.83	3.10	38.93
Other agriculture	15.26	2.27	9.95	1.48	−34.59
Salvinia	2.35	0.35	10.95	1.63	371.22
Papyrus	6.99	1.04	14.58	2.17	110.14
Unclassified grassland	202.15	30.08	176.68	26.29	−12.07
Grazed wetland	4.23	0.63	4.57	0.68	−8.80
Fallow/grazed agricultural land	10.82	1.61	10.55	1.57	−1.71
Acacia woodland	42.68	6.35	43.01	6.40	1.42
Potamogeton beds	0.00	0.00	0.40	0.06	100.00

The apparent 0.07 km^2 decrease in the size of Sonachi Crater lake over the study period is probably due to the relatively small total area and the resultant inaccuracies in calculation. The area of *Acacia* woodland stayed relatively constant between the interpretations (0.33 km^2 increase). Areas of *Euphorbia* woodland and unclassified grassland seemed to shrink in area; however, these can be accounted for by the increase in land judged to be "volcanic scrub".

Visual interpretation is best employed when there is some feature other than spectral response which is the key to good classification. Agricultural fields have a regular shape which helps in this. Values for the total amounts of agriculture classified in each interpretation are constant (60.6 for 1987 and 59.9 km^2 for 1989). Discrepancies between differences in "horticulture', "intensive agriculture" and "other agriculture" may be due to either changes of growing regimes between the years, discrepancies arising from imagery being recorded at different times of the year (March 1987 and July 1989), but more likely from misinterpretation of the spectral response differences between the agricultural types. Interpretation appears to be most variable in the area northern area of the basin. Here identification is hampered by two features. The first is the use of non-regular field shapes, the second is the high density of *Acacia* cover which has a spectral signature very similar to the agricultural stands.

THE MAXIMUM LIKELIHOOD SUPERVISED CLASSIFICATION

Table 16.2 shows the area cover values derived from the Maximum Likelihood supervised classification of the images. Sixteen clusters are identified corresponding directly to those identified using the photointerpretation techniques. Areas classified

Table 16.2   Results of the Maximum Likelihood classification of the 1987 and 1989 images

Information class	1987 Area (km^2)	1989 Area (km^2)	Change (km^2)	%
Deep water	8.98	13.64	4.66	51.89
Shallow water	33.58	14.52	19.06	−56.76
Main lake water	111.20	106.65	4.55	−4.09
Oloidien	7.12	5.67	1.45	−20.37
Sonachi water	1.30	0.12	1.18	−90.77
Volcanic scrub	90.70	95.33	4.63	5.10
Euphorbia woodland	16.60	18.27	1.67	10.06
Intensive agriculture	60.66	26.22	34.44	56.78
Horticulture	38.09	43.52	5.43	14.26
Other agriculture	43.47	61.71	18.24	41.96
Salvinia	6.32	10.56	4.24	67.09
Papyrus	13.02	11.63	1.39	−10.68
Unclassified grassland	165.28	191.34	26.06	15.77
Fallow/grazed agricultural land	15.23	4.67	10.56	−69.34
Acacia woodland	58.21	54.04	4.17	−7.16
Potamogeton beds	−	14.15	−	−

as deep water within the image cover an area of 8.98 km^2 in comparison to an area of 1.95 km^2 identified using photo-interpretation. There are two reasons for this increase. Areas of papyrus and *Acacia* woodland to the south and west of Crescent Island, along the shore, have been misclassified and, secondly, there is banding within the image producing lines of dark (low DN value) pixels across the data which have been misclassified as deep water.

This banding accounts for the reduction in area of main lake water within the classification. There is a small degree of error within the "shallow water" and "Oloidien" fractions, but they tend to be between these types, and are probably associated with local differences in phytoplankton composition. There is some level of misinterpretation between Sonachi and the shallow water classes, but this is more likely to be due to regions of relatively low macrophyte cover being identified as water with high phytoplankton biomass.

The areas of papyrus appear to have been well delineated in the north and west regions of the lake; however, there is confusion between papyrus and *Acacia* wood-land in areas to the south and east. The total area recognised as papyrus is 13.02 km^2, double that of the photo-interpretation. It is likely that this increase is to some extent due to misclassification, but perhaps more interestingly as a result of the automated classification regime identifying very small regions of papyrus unrecognisable by conventional techniques.

The area of "*Salvinia*" is calculated as 6.3 km^2 in comparison to 2.4 km^2 from the photo-interpretation. There do not appear to be any regions of misclassification. It would seem therefore that the value derived from the automated classifier is more accurate due to small areas being identified. The advantage is apparent in the recognition of *Salvinia* on the fringes of the papyrus areas on the western shore and in the southern part of the eastern shore.

The classification of *Acacia* appears accurate other than the misrepresentation with papyrus as previously mentioned. The area classified is 58.2 km^2 some 16 km^2 more than from photo-interpretation. The majority of the narrow bands surrounding the lake are identified, as are the large expanses of the trees to the north of the North Swamp area. The same levels of accuracy appear to be prevalent for regions of *Euphorbia* woodland. The area of volcanic scrub classified is 90.7 km^2 which is much less than the 153.9 km^2 identified by photo-interpretation.

The total area attributed to agriculture is 142.2 km^2 compared to 60.6 km^2 from the photo-interpretation. Examination of the thematic map again suggests accurate identification for these classes. The increases in area cover for many of the classes already identified has to be accounted for in one of the cover types. It is not surprising therefore that the most heterogeneous land type categorised via photo-interpretation, unclassified grassland, has a smaller area cover using the automated method; an area of 165.28 km^2 in comparison to 202.15 km^2.

The classification resulting from Maximum Likelihood analysis of the 1989 image produces results similar to those for 1987. There are the same areas of misclassification for papyrus and *Acacia* woodland, with the area of *Potamogeton* beds providing another class for potential error. Again there are significant discrepancies between areas identified as unclassified grassland and volcanic scrub as discrete regions of known cover types are delineated.

DISCUSSION

It is difficult to relate this work to similar studies. In the past, collection of ground truth data has occurred contemporaneously with satellite data recording or there has been some other stable inventory available for assessing the efficiency of methods used. This work was initiated because of the dynamic nature of the lake ecosystem and the difficulty associated with providing an accurate inventory by conventional techniques. There are thus no reference data available to quantitatively assess the procedures used here. The aim of this study was to explore techniques for classifying a greater number of wetland and agricultural classes than in previous work. The Maximum Likelihood classification produced a 15-class thematic map of accurate inventory information and two further classes of cloud and unclassified grassland. These results are better than published classifications. Glimmer et al. (1980), Jensen et al. (1986) and Browder et al. (1989) all used a two-class system, where regions of water were separated from other cover types. Ackleson and Klemas (1987) attempted greater differentiation with the identification of percentage cover values of *Zostera marina*. Butera (1983) had three study regions, but at best could only delineate a maximum of six cover types. Similar studies through the 1980s and early 1990s (Jensen et al., 1984, 1987, 1991; Christensen et al., 1985, 1988) yielded accurate thematic maps with less than 10 cover types.

Most of the literature deals with work carried out in the temperate regions, typically the US, with seasonality and subsequent phenological differences used to enhance classifications. As there are not such specific growing seasons in tropical Kenya this method is less applicable, although it may be possible to use the effect of the rainy season to improve results. A second consideration is that of scale. Most studies concentrated on regions much smaller than the $670 \, \text{km}^2$ area of this site. In contrast to these potentially inhibiting factors, there are also features of the Lake Naivasha basin which augment a supervised classification inventory. The first is that although the study site is relatively large, much of its area ($\sim 220 \, \text{km}^2$) is water. The second is that the cover types of most interest form huge homogeneous zones allowing relatively large areas to be selected for training data.

The visual interpretation techniques produced reasonable classifications for the groups selected. Problems in identifying the smaller fields resulted in better identification of the aquatic groups in relation to agricultural cover types for both images. The Maximum Likelihood supervised algorithm resulted in the more accurate classifications. The 1987 image produced a good thematic map with accurate delineation of papyrus and *Salvinia molesta*. The classification of the 1989 image produced misrepresentation of the *Potamogeton* beds as papyrus and *Salvinia molesta*. This misclassification means that change detection between years is not accurate for specific aquatic groups but changes in total aquatic macrophyte cover and surface area of water are detectable. There were no apparent problems with the classification of agricultural or woodland types, however, poor ground truth data for these classes limited accuracy assessment.

The Maximum Likelihood algorithm produced satisfactory classifications for the imagery studied. The success of this approach is dependent on several factors. The lake and margin are relatively homogeneous regions where selection of training data is easy. In contrast the agricultural regions are more dynamic and heterogeneous making training data selection difficult. There is no method of testing the accuracy of the classifications further because no records exist for the nature of the agriculture in these years. The next step is to repeat the exercise on a contemporaneous image with concurrent, extensive ground survey, to test the accuracy of the automated classification for the future. Once this is done, subject to the cost of acquisition of up-to-date images, a management plan can be informed and assisted by accurate information about vegetation changes in this highly dynamic system.

The effectiveness of using this technique to inventory other wetlands and lake basins, many in Africa more remote than Naivasha, is feasible but its success would depend upon the following:

- the heterogeneity of the study site;
- collection of good, simultaneous ground truth data;
- determination of a realistic range of cover type classes; and
- selection of accurate training data.

The value of this approach in Naivasha is that it provides a relatively rapid, ecosystem-wide classification of water, wetland vegetation and land uses. Its two main drawbacks are the lack of definition of submerged plants, which were known from contemporary studies to be present at the time of the 1987 image in particular, and the expense of a time-series of satellite images, particularly up-to-date ones. Nevertheless, it is likely that any future international aid for the management of the Ramsar site, defined as the area of land and lake inside the Moi North and Moi South Lake Roads, would be of sufficient scale to permit annual, contemporary images. These could be analysed using a system such as that described here on locally-based laptop computers.

ACKNOWLEDGEMENTS

The authors are grateful to the members of the Lake Naivasha Riparian Owners Association (now the Lake Naivasha Riparian Association) who permitted access to their property for ground measurements of radiation and to Angus Simpson for his piloting skills in flying over natural vegetation types whilst accurate signatures were obtained. The authors are also grateful for access to the unpublished management plans and reports of the LNROA, but opinions expressed in this chapter are the authors' own and in no way reflect either LNRA, or Kenya Government policy.

REFERENCES

Ackleson, S.G. and Klemas, V. 1987. 'Remote sensing of submerged aquatic vegetation in lower Chesapeake Bay'. *Remote Sensing of Environment,* **22**, 235–248.

Bakker, S.A., Vandenberg, N.J. and Speleers, B.P. 1994. 'Vegetation transitions of floating wetlands in a complex of turbaries between 1937 and 1989 as determined from aerial photographs with GIS'. *Vegetation,* **114**, 161–167.

Browder, J.A., May, L.N., Rosenthal, A., Gosselink, J.G. and Baumann, R.H. 1989. 'Modelling future trends in wetland loss and brown shrimp production in Louisiana using Thematic Mapper imagery'. *Remote Sensing of Environment,* **28**, 45–59.

Butera, M.K. 1983. 'Remote sensing of wetlands'. *IEEE Transactions on Geoscience and Remote Sensing,* **21**, 383–392.

Chavez, P.S. 1988. 'An improved dark-object subtraction technique for atmospheric scattering correction of multispectral data'. *Remote Sensing of Environment,* **24**, 459–479.

Christensen, E.J., Jensen, J.R. and Sharitz, R.R. 1985. 'Remote sensing of wetlands at the Savannah River Plant'. *Proceedings of the Fifth Department of Energy Environmental Protection Information Meeting, 6–8 November 1984*, pp. 607–619. Department of Energy, Alberquerque, New Mexico.

Christensen, E.J., Jensen, J.R., Ramsey, E.W. and Mackey, H.E. 1988. 'Aircraft MSS data registration and vegetation classification for wetland change detection'. *International Journal of Remote Sensing,* **9**, 23–38.

Dobson, J.E. and Bright, E.A. 1991. 'Coast watch – detecting change in coastal wetlands'. *Geographical Information Systems,* **1**, 36–40.

Gaudet, J.J. 1977. 'Natural drawdown on Lake Naivasha, Kenya, and the formation of papyrus swamps'. *Aquatic Botany,* **3**, 1–47.

Gaudet, J.J. and Falconer, A. 1982. *'Remote Sensing for Tropical Freshwater Bodies: The Problem of Floating Islands on Lake Naivasha'.* Report from the Regional Remote Sensing Facility, Box 18332, Nairobi.

Gaudet, J.J. and Melack, J.M. 1981. 'Major ion chemistry in a tropical African lake basin'. *Freshwater Biology,* **11**, 309–333.

Glimmer, D.S., Work, E.A., Colwell, J.E. and Rebel, D.L. 1980. 'Enumeration of Prairie wetlands with Landsat and aircraft data'. *Photogrammetric Engineering and Remote Sensing,* **46**, 631–634.

Goldson, J. 1993. 'A three phase environmental impact study of recent developments around Lake Naivasha'. Unpublished report to the Lake Naivasha Riparian Association, Naivasha, Kenya.

Harper, D.M., 1992. 'The ecological relationships of aquatic plants at Lake Naivasha, Kenya'. *Hydrobiologia*, **232**, 65–71.

Harper, D.M., Mavuti, K.M. and Muchiri, S.M. 1990. 'Ecology and management of Lake Naivasha, Kenya, in relation to climatic change, alien species' introductions and agricultural development'. *Environmental Conservation*, **17**, 328–336.

Harper, D.M., Adams, C. and Mavuti, K.M. 1995. 'The aquatic plant communities of the Lake Naivasha wetland, Kenya: pattern, dynamics and conservation'. *Wetlands Ecology and Management*, **3**, 111–123.

Jensen, J.R. and Mackey, H.E. 1993. 'Measurement of seasonal and yearly cattail and waterlily changes using multidate SPOT panchromatic data'. *Photogrammetric Engineering and Remote Sensing*, **59**, 519–525.

Jensen, J.R., Christensen, E.J. and Sharitz, R.R. 1984. 'Non tidal wetland mapping in South Carolina using airborne multispectral scanner data'. *Remote Sensing of Environment*, **16**, 1–12.

Jensen, J.R., Hodgson, M.E., Christensen, E.J., Mackey, H.E., Tirmey, L.R. and Sharitz, R.R. 1986. 'Remote sensing inland wetlands: a multispectral approach'. *Photogrammetric Engineering and Remote Sensing*, **52**, 87–100.

Jensen, J.R., Ramsey, E.W., Mackey, H.E., Christensen, E.J. and Sharitz, R.R. 1987. 'Inland wetland change detection using aircraft MSS data'. *Photogrammetric Engineering and Remote Sensing*, **53**, 521–529.

Jensen, J.R., Narumalani, S., Weatherbee, O. and Mackey, H.E. 1991. 'Remote sensing offers an alternative for mapping wetlands'. *Geographic Information Systems*, **1**, 46–53.

LNROA, 1996. *The Lake Naivasha Management Plan*. Lake Naivasha Riparian Owners' Association.

Lyon, J.G. and Greene, R.G. 1992. 'Use of aerial photographs to measure the historical areal extent of Lake Erie coastal wetlands'. *Photogrammetric Engineering and Remote Sensing*, **58**, 1355–1360.

Mackey, H.E. 1990. 'Monitoring seasonal and annual wetland changes in a freshwater marsh with SPOT HRV data'. *American Society for Photogrammetry and Remote Sensing*, **4**, 283–292.

Melack, J.M. 1996. 'Saline and freshwater lakes'. In *East African Ecosystems and Their Conservation* (eds T.R. McClanahan and T.P. Young), pp. 171–190. Oxford University Press, New York.

Njuguna, S.J. 1991. 'Water hyacinth: the world's worst aquatic weed infests Lakes Naivasha and Victoria'. *Swara*, **14**, 8–10.

Norton, D.J. and Snecker, E.T. 1990. 'The ecological geography of EMAP'. *Geographical Information Systems*, **1**, 33–43.

Podolsky, R. and Conkling, P. 1991. 'Satellite search aids wetlands visualization'. *GIS World*, **4**, 80–85.

Richards, J.A. 1993. '*Remote Sensing Digital Image Analysis: An Introduction*' (2nd edition). Springer-Verlag, New York.

Roughgarden, J., Rurming, S.W. and Matson, P.A. 1991. 'What does remote sensing do for ecology?' *Ecology*, **72**, 1918–1922.

Sikes, H.L. 1935. 'Notes on the hydrology of Lake Naivasha'. *Journal of the East African Natural History Society*, **13**, 74–89.

Watson, R.M. and Parker, I.S.C. 1969. 'The ecology of Lake Naivasha'. Unpublished report to the East African Wildlife Society, Wildlife Services Ltd, Nairobi.

SECTION C

Modelling: An Essential Tool for Sustainable Management

Introduction

TONY BROWN

In this section the chapters illustrate that modelling is used for a wide variety of applications in tropical environments. The approach taken is essentially similar in each chapter, being largely conceptual and physically based mathematical modelling. Chapter 19 discusses catchment dynamics first simplified through the development of a perceptual model. This can identify some initial problems where there is a lack of pertinent knowledge in the tropics, an example being the importance of translatory or piston flow in deeply weathered soils which may be laterised. Indeed, it would seem likely that the anisotropy of subsurface flow is common in tropical soils due to impeding layers at depth, such as plinthite or clay accumulations. Not all the processes identified at the perceptual stage can be formalised, so the conceptual modelling will represent the closest fit compromise between reality and existing or new modelling methods. The chapter goes on to test a distributed, cell-based model based on TOPMODEL, in the sub-tropical region of the Nepal Middle Hills. Two important conclusions from the work are that the effect of terracing on soil moisture deficits needs to be simulated and secondly that, in spatially variable catchments with highly localised rainfall intensities, a similarity of response can arise from different parameter sets – the classic problem of equifinality. The only way to resolve the problem of equifinality is to identify the key catchment processes and this will generate a demand for specific field data which can be used to tailor the original model to the catchment under study.

One area of variability in tropical catchments is infiltration, and Chapter 17 explores the application of fuzzy numbers to the incorporation of uncertainty into infiltration modelling. The chapter shows that the method presents advantages in time and cost over traditional stochastic modelling of uncertainty and there is the potential to model ideas of the shape of uncertainty distributions (if we know them) through the variety of shapes of fuzzy number matrices. In the distributed model presented in Chapter 18, a single-event model is applied through a cellular structure to an entire basin. Based on rainfall excess and standard hydrological theory (Richards and Yalin equations), the two variables solved by calibration, delay time and maximum surface storage, are clearly in reality related to micro-topography and land use patterns.

Although operational at a different scale, the land reference units used in the hydrological budget modelling of lakes in Chapter 20 is potentially applicable at a finer spatial resolution. The approach used here has similarities to Chapter 3, but it still remains for GIS/remote sensing approaches such as these two to be integrated with cell-based distributed models of the kind used in Chapter 19.

At an even larger spatial scale, Chapter 21 uses a steady-state two-dimensional hydrogeological model to determine groundwater discharge patterns along three tropical coasts with mangrove forests. Groundwater is a particular problem in many tropical catchments where deep but discontinuous percolation can occur. Because mangroves are sensitive to changes in the brackish-water micro-environments, it is necessary to adequately model both the coastal–inland hydrogeological conditions and recharge patterns.

As a prelude to modelling and construction of a conceptual model some major differences between tropical and temperate catchments are worth noting. These include:

1. The character of tropical soils and tropical soil variation, especially with depth and including the occurrence of parameters outside the normal (i.e. well-known) range.
2. The inevitable coincidence of seedling establishment and high intensity rainfall during the monsoon cycle and the high local variability in rainfall intensities making regional averaging unreliable.
3. The ubiquitous occurrence of terracing; this completely alters local slopes and makes map-derived slopes/DTMs of questionable validity. This affects many parameters including interception, evapotranspiration, surface storage, infiltration capacity and runoff.
4. The small size of land use units (plots, fields, forest patches). This implies that spatial pattern of land relation to slope distribution is itself an important variable.
5. The high productivity potential of many tropical soils and the desirability of this potential being maximised (not a contempory constraint for much of the temperate zone).

These factors suggest that variable cell size and a variable connectivity with a network of preferential pathways which are not purely slope-driven would be desirable. Indeed the approach suggested by the morphology and variability of small, steep, tropical catchments is essentially that which has been applied to urban catchments, with which there are functional similarities. However, this presents an obvious problem as a model becomes more tailored to the configuration of a catchment so it loses its general applicability and increases the costs of a modelling approach. One way round this may be to use process-based models which bolt on to individual spatial catchment models. A guide for this approach is urban catchments in which the spatial configuration is made up of a number of cell types (e.g. road, garden, roof, etc.) and a flow network (the drainage system).

The chapters in this section probably represent an early stage in the modelling of tropical catchments, but it is only through trial and error that further process work can be efficiently directed and conceptual advances be made in our understanding of the dynamics of these catchments.

CHAPTER 17

The Importance of Soil Infiltration Dynamics and Data Uncertainty: Field Studies on Soils in Zimbabwe

LORENZO BORSELLI, STEFANO CARNICELLI, GIOVANNI AURINDO FERRARI and UGO GALLIGANI

Università di Firenze, Dipartimento di Scienza del Suolo e Nutrizione della Pianta, Firenze, Italy

INTRODUCTION

Soil and water conservation are a central problem of land management in tropical countries. The combination of tropical climates, characterised by irregular high-intensity storms, and of high pressure on land, due to socio-economic reasons, is such that the majority of agricultural land in tropical areas should be considered as subject to high erosion risk, and efficient water conservation for agriculture is essential. At catchment level, the negative effects of soil erosion are magnified: in addition to soil degradation are the problems of siltation and pollution of water bodies lying downstream of areas where soils are seriously eroded.

Water runoff on the soil surface is one of the processes that determines soil erosion and sediment transport. In tropical environments, runoff amount is especially critical; high amounts of runoff create high risks of rill and gully erosion, and the limiting factor of splash erosion is often the runoff transport capacity, rather than the, usually higher, rainfall detachment ability. Runoff rate from soils is furthermore a fundamental quantity of catchment hydrology, necessary to improve forecasts of water flow.

The infiltration capacity of soils determines the amount of runoff, through the partition of rainfall water between infiltration and runoff itself. Although in the past models for predicting soil erosion have been developed and used without explicitly taking account of runoff (Wischmeier and Smith, 1978), such models are today acknowledged as not really suitable for assessing effects of management techniques

The Sustainable Management of Tropical Catchments. Edited by David Harper and Tony Brown.
© 1998 John Wiley & Sons Ltd.

and the potential for innovations in soil conservation (Stone et al., 1996). In fact, many classical or newer methods of reducing soil erosion are actually aimed at reducing runoff by increasing infiltration, as reviewed by Levy (1996) and Unger (1996).

Consequently, soil infiltration is today considered to be a critical parameter for the building and running of hydrological and soil erosion models designed to support catchment management in tropical countries. Such models then usually include infiltration as a submodel, but their building and use is hampered by theoretical and practical problems in the measurement of soil infiltrability. Methods aimed at predicting infiltration from soil properties, such as saturated hydraulic conductivity or infiltration by double-ring infiltrometer, present several shortcomings for predicting infiltration under rainfall, as reviewed, among others, by Bowyer-Bower (1993). More sophisticated recent approaches, frequently used in soil erosion models, estimate infiltration as a complex function of some basic soil properties, such as soil texture, porosity or matric potentials (Nearing et al., 1989; Woolhiser et al., 1990; Morgan et al., 1992). Infiltration, anyway, is a dynamic process, both within a single rainstorm and on a seasonal basis, and such approaches only partially take into account dynamic aspects (Stone et al., 1996), as they assume constant soil properties with time. Neglecting intrinsic soil dynamics is a specially relevant shortcoming when dealing with soils that, under natural rainfall, are subjected to surface sealing or crusting, causing rapid changes in soil physical properties.

An alternative approach is to model single rainstorm dynamics by some established infiltration law, such as those of Horton (1940) or Philip (1957). Use of this type of law requires that infiltration be measured by applying simulated rainfall; this is a complex approach and sets some practical limitations to the amount of data that can be produced. The need then arises for techniques suitable for dealing with uncertainty in datasets of limited size. Ideally such techniques, especially in the model-building phase, should allow easy handling and understanding of the links between data and prediction uncertainties.

Uncertainty in models is usually treated by statistical or stochastic techniques, but such techniques are not free of problems and limitations. Limited amounts of data do not usually allow a reliable assessment of the distribution laws followed by the random variables and, furthermore, stochastic modelling is very demanding on computing power and time. Consequently, a less rigorous heuristic kind of approach appears attractive. One such approach is the use of possibility distributions represented by fuzzy numbers. A fuzzy number, in this context, is used as a representation of an uncertainty distribution, defined in the field of real numbers; fuzzy arithmetic (Kaufmann and Gupta, 1991) allows use of such distributions to perform calculations in a much faster and simpler way than possible with probabilistic techniques, and to obtain results directly expressing the uncertainty originated at the parameter level.

In Zimbabwe, significant agricultural areas are covered by soils strongly susceptible to sealing and crusting (Nyamapfene and Hungwe, 1986; Borselli et al., 1996a, 1996b). The behaviour of these soils creates serious problems, in terms of both erosion and siltation in downstream water bodies. A series of rain simulation experiments were conducted on a representative soil of this type, to explore the implications, in terms of soil and surface hydrology, of single and multiple rainstorm dynamics of infiltration. The use of fuzzy numbers, as a way to explicit uncertainty in both experimental results and potential predictions obtainable by them, were explored.

BASIC THEORY

The representation by fuzzy numbers stems from the theory of fuzzy mathematical techniques. The core of the fuzzy mathematical techniques is the theory of fuzzy logic, introduced by Zadeh (1965) in his seminal papers.

In classical logic and sets theory a set has a crisp boundary and a generic object or element may only be a member or not a member of the set. This behaviour is described by a function that may take only two values: 1, when an element belongs to a set, and 0 when it does not. This type of function is called a membership function.

$$\text{Crisp set} \quad A \subseteq X$$

$$\mu_{A(x)} = \begin{cases} 1 & x \in A \\ 0 & x \notin A \end{cases} \tag{1}$$

Conversely, fuzzy logic is a multi-value logic, and the membership function may take continuous values in the interval [0.0, 1.0]. As a consequence of this concept the considered set has no crisp boundary. It is a fuzzy set, where the degree of belonging of an element to the set varies continuously. When the grade of membership reaches 1.0, the element completely belongs to the set. Lower grades only assure a partial belonging to the set, down to the value of 0.0 where there is no belonging.

$$\text{Fuzzy set} \quad F \subseteq X$$

$$\mu_{F(x)}: X \rightarrow [0, 1] \tag{2}$$

Fuzzy numbers are a special type of fuzzy set, defined in the space of the real numbers, R. A formal definition of a fuzzy number (abbreviated as FN) may be given in different ways (Dubois and Prade, 1979; Mizumoto and Tanaka, 1979; Kaufmann and Gupta, 1991). Kaufmann and Gupta (1991) introduce the definition of fuzzy number by the concept of interval of confidence, familiar for anyone having to deal with imprecise data. In the space of real numbers, R, an interval of confidence is expressed as a closed interval, A.

$$A = [a_1, a_2] \quad \text{for} \quad a_1 < a_2 \tag{3}$$

Considering a parameter a variable between 0 and 1 or $\alpha \in [0, 1]$, the interval of confidence, A, may be generalised as:

$$A(a) = [a_{1(a)}, a_{2(a)}] \tag{4}$$

This is a monotonic decreasing function of α such that for each $\alpha' > \alpha$ there will be an $A'(\alpha)' \subseteq A(\alpha)$.

The resulting mapping functions $a_1(\alpha)$ and $a_2(\alpha)$ represent the fuzzy number, and α becomes equivalent to the value of the membership function of the fuzzy number A.

The only prerequisite for the membership function to be consistent with the previous definition is the monotonicity and continuity (decreasing or increasing) of both the left and right side of the fuzzy number. At the same time, these properties define a basic property for an FN: the convexity. Following Mizumoto and Tanaka (1979) an FN A in the real line R is said to be convex if for any real numbers $x, y, z \in R$ with $x \leq y \leq z$:

$$m_A(y) \geq \min[m_A(x), m_A(z)] \tag{5}$$

where the symbol min[,] stands for "minimum between".

Another important property is the normality. An FN A is called normal if the maximum possible membership grade is 1. The fuzzy numbers (Figure 17.1) may be defined as normal convex FNs. The analytical definitions of the of normal convex FN give an idea of the possibility of obtaining a great variety of shapes with an FN can be built. The possibility of FNs that are symmetric, asymmetric or with a flat region helps to model some common ideas of the reality of uncertain distributions.

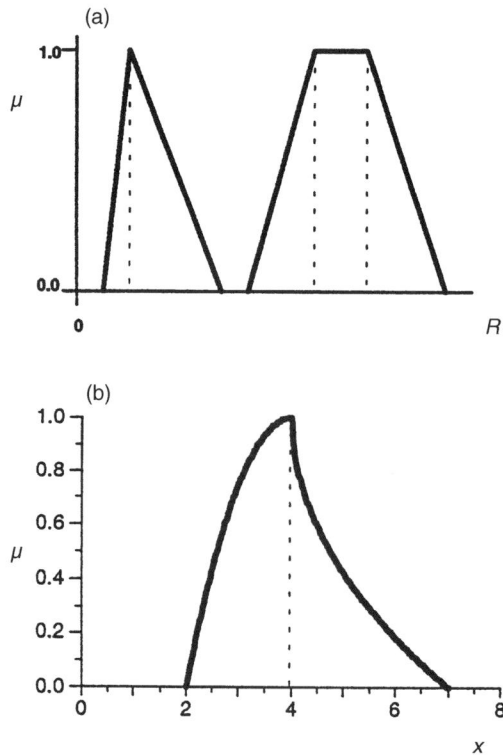

Figure 17.1   Some examples of normal convex fuzzy numbers: (a) the two most classical forms of FNs, triangular and trapezoidal; (b) a generalised, asymmetrical fuzzy number

Zadeh (1978) posed the basis for a formal and practical equivalence between the membership function and the possibility distribution. In a representation of data uncertainty based on fuzzy numbers, a given value of membership function (membership grade) gives the possibility that the corresponding interval of confidence be representative of the interval of confidence (or unique values) at membership grade 1. From a point of view of experimental data, a higher or lower membership grade gives the higher or lower possibility that the associated interval of confidence be representative of the measure at level 1, that is the highest possible value for the measure.

One of the most important properties of FNs is the possibility to manipulate them with arithmetic, algebraic and general mathematical functions as ordinary classical numerical values or variables. This possibility was formulated for the first time by Zadeh (1975) with the theorems of the so-called "extension principle". The basic arithmetic operations between FNs may be done easily using the concept of interval of confidence as used by Kaufmann and Gupta (1991) for the previous definition of the FN itself.

The choice of appropriate membership functions and fuzzy number types to model real data remains a crucial problem in the application of fuzzy mathematical techniques (Lai and Hwang, 1992). Various types of membership function are available in the literature (Klir and Folger, 1988; Zimmermann, 1990; Lai and Hwang, 1992); nevertheless, a general way does not exist. The approach remains greatly dependent upon the kind of applications and in some cases an optimum formulation does not exist.

METHODS

A typical crusting soil from Zimbabwe (Nyamapfene and Hungwe, 1986) was used. The soil was located in the north-east of the country, and was classified as a Rhodic Kandiustalf (Soil Survey Staff, 1990); its main characteristics are summarised in Table 17.1. Management experience and field observations had shown the soil to be structurally unstable and highly susceptible to crusting.

The rainfall simulator was a simplified, single full-cone nozzle version of that described by Panini et al. (1993), giving a rainfall intensity of about $55\,mm\,h^{-1}$ and a rainfall KE of $13.3\,J\,m^{-2}\,min^{-1}$. High-quality borehole water, with EC $0.25\,dS\,m^{-1}$ and SAR < 1, was used. The plots used were of 1×1 m size, with a 50 cm wide buffer zone. Natural slope, about 2%, was exploited.

Table 17.1 Main characteristics of sample soil – the Ap horizon

Soil classification	Rhodic Kandiustalf
Texture	Clay 30.6%; Silt 45.4%; Sand 24%
pH ($CaCl_2$)	5.0
ESP	0.7
ECEC ($cmolc^+\,kg^{-1}$)	7.1
OM($g\,kg^{-1}$)	0.15
Clay mineralogy	illite and kaolinite dominant

A preliminary series of experiments had shown a potential effect of phosphogypsum, spread at the rate of $5\,\mathrm{mg\,ha^{-1}}$, in modifying the shape of infiltration curves. Consequently, the main experiment was conducted in duplicate, on both untreated and phosphogypsum-treated plots. Four plots were prepared by hand-hoeing to 15 cm depth, to simulate a medium tilth seed-bed; phosphogypsum was spread on two of them. The plots were then subjected to four successive simulated rainstorms, all of 40 minutes duration, with a minimum between-storm interval of 48 h, enough to achieve drying given the constant, hot and windy, weather conditions. During each experiment, runoff sampling was effected at intervals and for lengths of time adjusted to get the most accurate possible tracing of the runoff hydrograph. In the preliminary experiment, four different plots had been subjected to simulated rainstorms carried on until attainment of a stabilised runoff.

Rainfall intensity and duration were selected, resting on the study published by the Zimbabwe Department of Meteorological Services (ZDMS) (1975), to represent heavy storms with a probability of occurrence of once in a year. Instantaneous infiltration was estimated by the balance of rain minus runoff ($I = P - Q$). This neglects surface storage, but it was observed that almost all ponding water readily infiltrates when rain was stopped.

The infiltration curves were successfully fitted to a modified version of the Horton-type equation proposed by Morin and Benyamini (1977):

$$i = i_f + (i_0 - i_f)e^{-R/K} \qquad (6)$$

where i = instantaneous infiltration rate in $\mathrm{mm\,h^{-1}}$; i_f = final infiltration rate in $\mathrm{mm\,h^{-1}}$; i_0 = initial infiltration rate in $\mathrm{mm\,h^{-1}}$; R = cumulative rainfall in mm, equivalent to pt in Morin and Benyamini (1977); while K, $\mathrm{mm^{-1}}$, equivalent to $1/\gamma$, is a coefficient intrinsic to soil conditions that, for a given rain intensity, is inversely proportional to the rate of infiltration decay.

This is just a formal modification, adopted for ease of use with the available software. Attempts to fit experimental data to different forms of the Philip law resulted in poor fittings, and it was concluded that this law does not satisfactorily describe the data.

INFILTRATION PARAMETERS

The experimentally obtained infiltration parameters are summarised in Tables 17.2 and 17.3. While initial infiltration rate appears to be highly variable, final infiltration rate looks much more robust, with the exception of a single, anomalous, value. Neither appreciable trends with successive rainstorms nor significant difference between treatments appear; non-parametric statistical tests (Kolmogorov–Smirnov) show a high probability that all i_f data belong to the same population. The infiltration decay coefficient behaves in a different way, showing both a well-defined decreasing trend with successive rainstorms and a general trend for values to be higher in gypsum-treated plots; it is worth remembering here that a higher K indicates a slower decay of infiltration rate, and then a lower total runoff.

The decreasing trend of K in successive rainstorms was well-enough defined that it

Table 17.2   Rainfall and infiltration parameters, untreated plots

Untreated plots	Total rainfall, R (mm)	Previous cumulative rainfall, R_p (mm)	Initial infiltration rate, i_0 (mm h^{-1})	Final infiltration rate, i_f (mm h^{-1})	Infiltration decay coefficient, K (mm)
Plot 1	47.7	0.0	241.1	3.38	8.95
Plot 1	33.7	47.7	121.8	13.22	4.16
Plot 1	33.1	81.1	321.0	14.05	1.50
Plot 1	49.3	114.1	2030.8	7.95	0.86
Plot 3	39.6	0.0	990.0	5.50	11.33
Plot 3	34.8	39.6	425.3	15.48	2.74
Plot 3	38.7	74.4	281.0	44.04	6.64
Plot 3	45.1	113.1	280.0	16.65	2.44

Table 17.3   Rainfall and infiltration parameters, treated plots

Treated plots	Total rainfall, R (mm)	Previous cumulative rainfall, R_p (mm)	Initial infiltration rate, i_0 (mm h^{-1})	Final infiltration rate, i_f (mm h^{-1})	Infiltration decay coefficient, K (mm)
Plot 2	39.9	0.0	180.1	6.00	25.04
Plot 2	37.9	39.9	158.1	19.40	4.74
Plot 2	34.4	77.7	142.8	16.21	3.43
Plot 2	42.0	112.2	213.9	13.53	2.88
Plot 4	49.3	0.0	155.7	2.00	15.55
Plot 4	34.1	49.3	143.5	16.33	3.71
Plot 4	42.0	83.4	437.9	21.18	2.02
Plot 4	43.0	125.4	156.3	20.79	3.55

was possible to represent it mathematically, by fitting the cumulative rainfall against K in relation to an exponential decay-like function, of the shape:

$$K = a + (b - a)\mathrm{e}^{-R_p/c} \tag{7}$$

where a, b, c = fitting parameters; and R_p = cumulative rainfall in previous storms.

This representation is shown in Figure 17.2, which also demonstrates the slower evolution of K in the treated plots.

These results prompt some considerations about the nature of the variables defining the infiltration curve; the i_0 and i_f values basically behave like random variables, i.e. they may be taken as representative of the combination of soil and rainfall used for the experiments, and their variability treated as due to experimental errors and spatial variability of the soil. On the contrary, the K parameter cannot be treated in the same way, as it shows a defined dependence on the cumulative amount of rainfall applied.

Regarding the relative weight of K and i_f in determining the hydrological balance

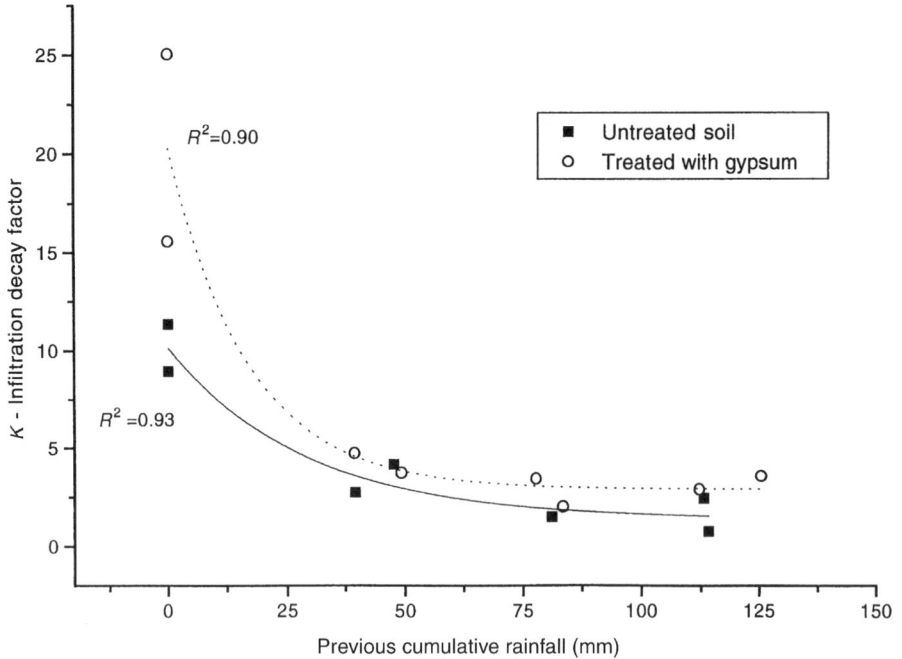

Figure 17.2 The function expressing the dependence of K by the cumulative amount of rainfall

of the soil, it is apparent that they will vary according to the nature of the rainfall events; in general, the weight of K will be greater the shorter the rainstorm is, but, furthermore, the dynamic nature of K will cause this parameter to take different weights according to the total amount of rainfall to which the soil has been exposed.

CONSTRUCTION OF A NUMERIC SIMULATION

In order both to show quantitatively the above considerations and to explore the potential of fuzzy logic and arithmetic in dealing with uncertainty in infiltration parameters, a numeric simulation procedure has been built, able to show the outcome of different successions of rainstorms on the soil used as a sample.

It is to be stressed that this simulation is not a true infiltration model, able to predict the outcome of real events, but a procedure designed only to quantify the potential influence of the dynamic aspects of infiltration in multiple storms of different nature; in this context, the simulated rainstorm sequences have been kept within the parameters of the experimental rainstorms, i.e. the same intensity and the possibility for the soil to dry between each rainstorm.

The building of the numeric simulation involved two steps:

1. Obtaining appropriate fuzzy numbers for the values of i_0 and i_f from the experimentally derived values.

2. Writing of a series of simple PASCAL routines, using equation (6) with K defined by equation (7) and i_0 and i_f in fuzzy form, to calculate potential infiltration rate and actual runoff rate for different series of $60 \, \text{mm} \, \text{h}^{-1}$ rainstorms of different duration.

The data obtained could then be easily converted into runoff data and actual infiltration in either instantaneous or cumulative form.

OBTAINING FUZZY NUMBERS FOR INFILTRATION PARAMETERS

To obtain fuzzy numbers for i_0 and i_f from the experimentally derived data the following approach was used:

1. Range, median and quartiles of experimental data were obtained and techniques of exploratory data analysis (Tukey, 1977) were employed for discrimination of outliers and possible range corrections.
2. An empirical cumulative density frequency (CDF) of the observed, and if necessary corrected, data was built.
3. The empirical CDF was fitted by a two-parameter incomplete *beta* function using a non-linear fitting method with some constraints.
4. The first derivative of the fitted function was performed numerically and then normalised between 0 and 1.

The function so obtained is itself a numerical representation of a fuzzy number. The fuzzy numbers representing the field of variation for the infiltration parameters are shown in Figure 17.3. The fuzzy i_0 is more asymmetrical in its distribution, with values about $280–300 \, \text{mm} \, \text{h}^{-1}$ for membership grades about 1, the most possible conditions during the simulation. The fuzzy i_f shows a quasi-symmetrical distribution with values around $13–15 \, \text{mm} \, \text{h}^{-1}$ for membership grades about 1.

RUNNING THE SIMULATIONS FOR INFILTRATION IN MULTIPLE STORMS

For running the simulations, some input variables are defined in terms of constant fuzzy numbers (i_0, i_f), while others are defined in functional continuous form (K), or fixed as constants (rainfall intensity and duration). During the simulation, fuzzy algebraic manipulations are performed involving fuzzy and ordinary variables; the basic fuzzy state variables are potential infiltration rate and runoff rate; from this last, other results are derived in fuzzy form as: actual infiltration rate and cumulative amounts of runoff and infiltration (see Table 17.4 for details of the procedure).

During the simulation the state of the system is updated every 1 mm of rainfall (the adopted discrete step). At every step both cumulative rainfall fallen in the previous storms and cumulative rainfall in the present storm are checked, so that values for K and potential fuzzy infiltration may be calculated.

Afterwards, the amount of rainfall resulting in runoff or infiltration is computed. To achieve this it is first necessary to evaluate the degree of satisfaction of

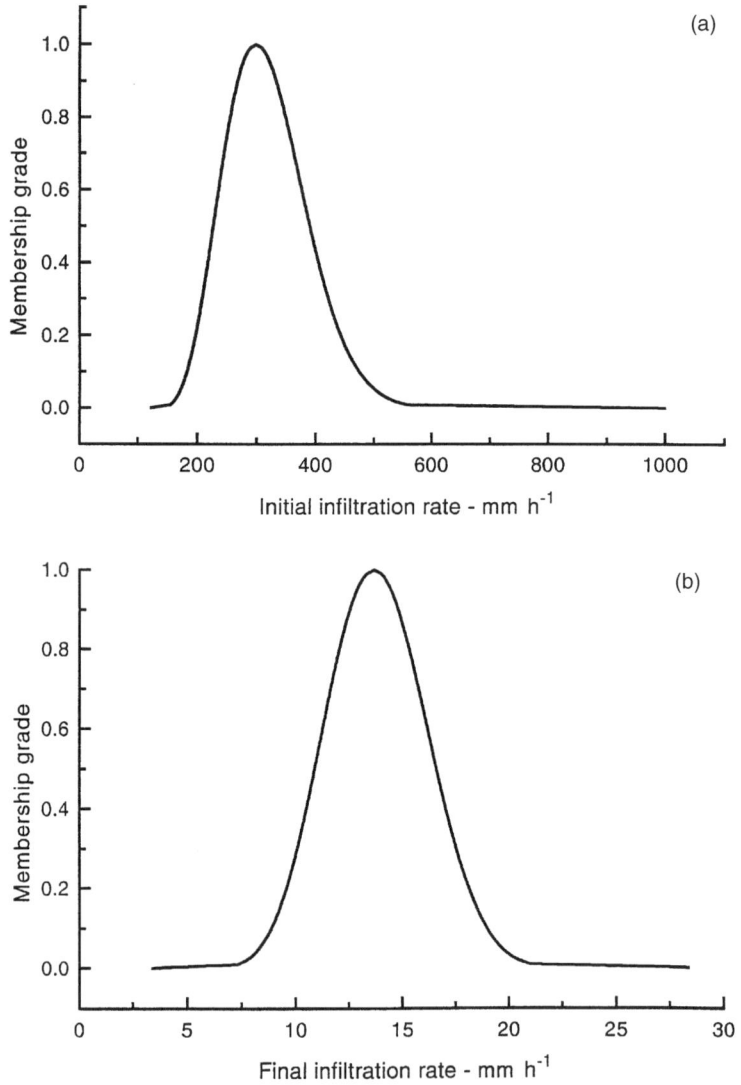

Figure 17.3 Fuzzy numbers expressing (a) initial infiltration i_0 and (b) final infiltration i_f

the condition that rainfall intensity be greater or equal to fuzzy potential infiltration rate. The result of this evaluation may be expressed in terms of degree of possibility, comprised between 1 and 0. For degree 1 the condition results are completely true, for 0 completely false and for intermediate values a condition of partially true and false.

This relation may be formalised in the following relational equation:

$$\mu(R): int \geq FInf \rightarrow [0, 1] \qquad (8)$$

Table 17.4   Pseudocode for the computer simulation

// *Pseudocode describing the simulation process for a multistorm*
// *dynamic infiltration process in a crusting soil from Zimbabwe.*
// *A fuzzy logic and arithmetic approach is adopted.*
// *by L.B. 1994*
Hydrcrust Simulation
   // *Called functions during the simulation*
function get_k (input: Cumulate rainfall – in previous storm (CPR)): get_k-1.375+(10.128–1.375)*exp(–CPR/28.8)
   function get_fuzzy_inf(input: IO, If, K, Cumulate rainfall (CR)): get_fuzzy_inf = if + (IO–if)*exp(–CR/K)
   // *start of the simulation*
   begin
   input_phase of main simulation parameters:
      input fuzzy variables:
         Initial infiltration rate
         Final infiltration rate
      input simulation constants:
         Rain intensity
         Unitary rainfall amount
         Total rainfall
         Simulation step (e.g. 1 mm of rainfall)
   Simulate infiltration ((bi)main procedure):
   begin
      initialise some variables at zero:
         fuzzy cumulate runoff
         Instantaneous previous fuzzy runoff rate (IPFRR)
         intialise at zero the number of previous storms
   for each simulation's step until total rainfall is reached do:
      begin
        Update the cumulative rainfall in the previous storms (CPR)
        Compute cumulative rainfall in the present storm (CR)
        Update the K value (*call function get_k*)
        Compute fuzzy instantaneous infiltration rate (*call function get_fuzzy_inf*)
        Eval. degree of possibility that rainfall intensity >= fuzzy instantaneous infiltration rate
            if degree=0 (infiltration entirely greater than intensity) then:
            begin
              instantaneous fuzzy runoff rate (IFRR) = 0
              instantaneous fuzzy runoff amount (IFRA) = 0
            end
            else if degree = 1 (infiltration completely less than intensity) then:
            begin
               instantaneous fuzzy runoff rate (IFRR) = Intensity – instantaneous fuzzy infiltration rate
              instantaneous fuzzy runoff amount (IFRA) = ((IFRR)+(IPFRR))/(2 intensity)
            end
            else (intermediate cases) do:
            begin
               instantaneous fuzzy runoff rate (IFRR) = Intensity – instantaneous fuzzy infiltration rate
              find corrected instantaneous fuzzy runoff rate (CIFRR) as positive part of (IFRR) × degree
              instantaneous fuzzy runoff amount (IFRA) = ((CIFRR+(IPFRR))/(2 intensity)
            end
        instantaneous previous fuzzy runoff rate (IPFRR) = instantaneous fuzzy runoff rate(IFRR)
        fuzzy cumulate runoff = fuzzy cumulate runoff + instantaneous fuzzy runoff amount (IFRA)
        fuzzy cumulate infiltration = cumulate rainfall – fuzzy cumulate runoff
      end
   end
   exports results of simulation
   end // *end of the simulation*

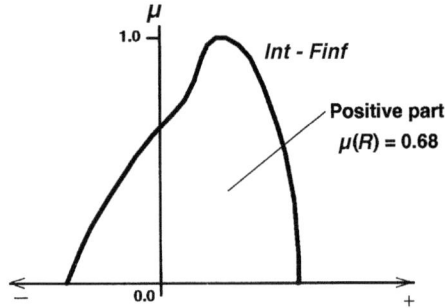

Figure 17.4   Evaluation of the degree of possibility that, at a moment during numerical simulations, the condition that rainfall intensity ≥ infiltration rate is satisfied

The value of $\mu(R)$ is obtained from the arithmetic difference between the two variables:

$$Fdiff = int - FInf \qquad (9)$$

and computing the positive fraction of the total area of the resulting fuzzy number *Fdiff*:

$$\mu(R) = \frac{AFdiff^{+}}{AFdiff} \qquad (10)$$

   An example of resulting fuzzy numbers is given in Figure 17.4.
   For $\mu(R) = 0$ (the resulting *Fdiff* lying entirely in the negative domain) potential infiltration rate is always greater than rainfall intensity and the fuzzy runoff rate is zero. Conversely, for $\mu(R) = 1$ the opposite situation arises and the fuzzy runoff rate is calculated as difference between intensity and fuzzy infiltration rate. For intermediate cases with $0 < \mu(R) < 1$ a corrected fuzzy runoff rate is derived by multiplying the positive part of the fuzzy runoff rate by $\mu(R)$. Finally, the fuzzy cumulative runoff and infiltration for the multiple storms are updated.
   In Table 17.4 a form with the pseudocode for the simulation used is presented.

RESULTS OF THE NUMERIC SIMULATIONS

Among the different simulations tried, two have been selected as the most representative for the purpose of the exercise. The first case, simulation 6060, includes three rainstorms of 60 mm while the second, simulation 1560, is based on a series of 12 rainstorms of 15 mm; total rainfall applied is then the same in both cases, 180 mm. Among the various possible representations of the results, we have selected to plot the cumulative infiltration and the instantaneous runoff rate. The first kind of data is interesting for agricultural purposes, relating to the efficiency of soil water storage, while the second is of obvious significance in hydrology.

The most evident effect of infiltration dynamics on the outcome of different kinds of rainfall event appears in the cumulative infiltration data; as represented in Figure 17.5. The same total amount of rainfall, falling at the same intensity, results in two quite different values of total infiltrated water: this simply as a consequence of considering the dynamic aspect of infiltration and varying rainstorm duration. It is interesting to note that, according to ZDMS (1975), the short-rainstorm sequence is much more probable in this environment.

The uncertainty of this result clearly accumulates with time, resulting in a spread of possible values that widens with increasing total rainfall; total spread is independent from central value, as it is only influenced by the spread of the FNs expressing i_0 and i_f; the actual small differences between the two runs are caused by the different weights taken by the more uncertain i_0, but this will be dealt with later on.

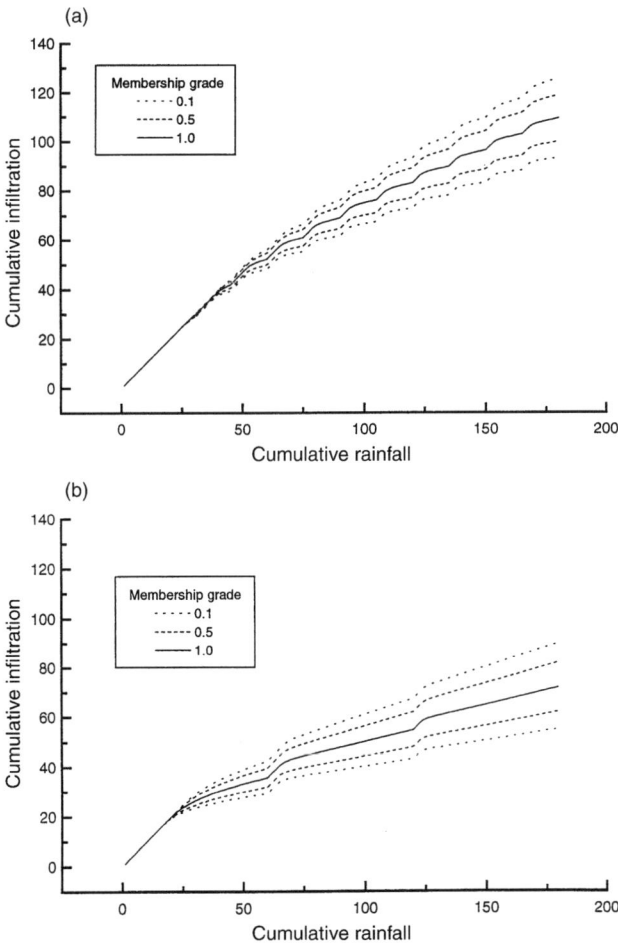

Figure 17.5   Results from the numerical simulations plotted as cumulative infiltration: (a) run 6060, three rainstorms of 60 mm each; (b) run 1560, 12 rainstorms of 15 mm each

Figure 17.6 shows the results of the two simulations in terms of instantaneous runoff rate. In run 6060, most of the runoff regime is actually controlled by i_f only (flat portions of the curves) and, in fact, no major error would be entailed in use of these data alone. A quite different situation is shown for run 1560, that clearly visualises how, in a climatic regime dominated by short rainstorms, the dynamic phase of infiltration ought not be neglected in hydrological models. Use of final infiltration only would result, in this kind of situation, in a gross overestimation of runoff for more than 50% of total rainstorm time.

On the other hand, use of a single, constant value for K would produce different errors according to the conditions in which this value is determined: should it be measured in conditions likely to give a high K, as, for instance, on a freshly tilled soil,

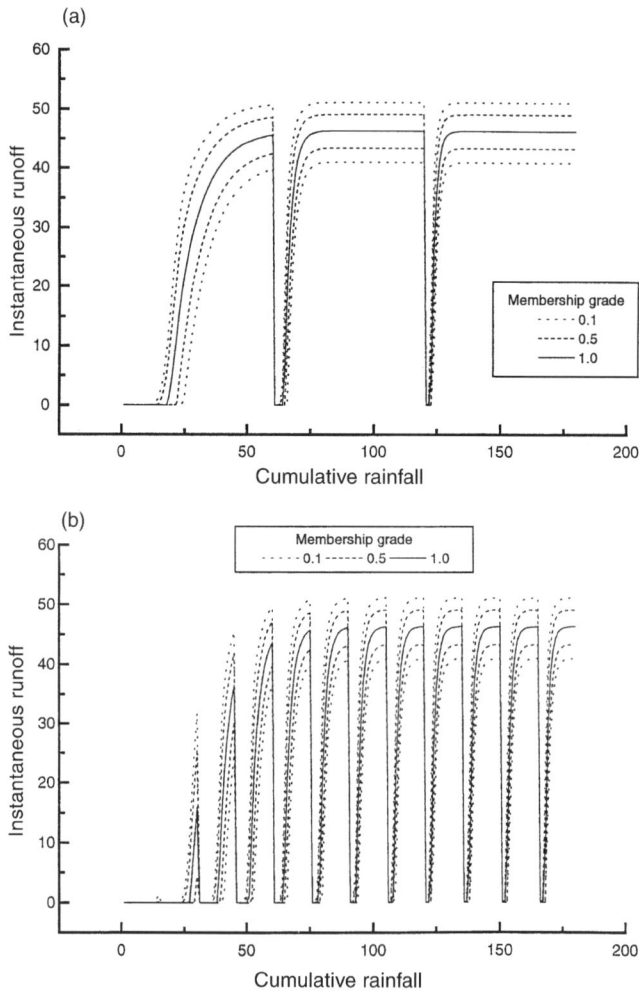

Figure 17.6   Results from the numerical simulations plotted as instantaneous runoff rate: (a) run 6060, three rainstorms of 60 mm each; (b) run 1560, 12 rainstorms of 15 mm each

significant underestimation of runoff could easily occur. A value measured in more average conditions would lead to a more evenly distributed error, but would still entail overestimation of runoff in the first rainstorms of a season.

Uncertainty in calculation of instantaneous runoff rate follows a cyclical pattern: of the two fuzzy variables used, i_0 is more uncertain than i_f: consequently, uncertainty varies according to the relative weight of the two parameters. Given the nature of equation (6), the weight of i_0 is maximum at the beginning of any event and then decreases exponentially down to zero at the moment in which steady infiltration is attained.

In the calculation of potential infiltration rate, uncertainty should then decrease exponentially with time down to a minimum equivalent to the uncertainty of i_f. In the actual calculations, the corrections made to avoid a negative runoff rate curtail uncertainty of the results, that then is initially very low, rising rapidly up to the point where all possible runoff values are positive with no need of correction and from then decreasing exponentially.

Analysis of the i_0 data obtained in our experiments suggests that their high uncertainty is produced by a limited number of anomalously high values; such values could well result from excessive sensitivity of the curve fitting performed to small variations in the experimental points defining the first portion of infiltration curves. Strong uncertainty of initial infiltration is then probably linked to the mathematical artefact nature of this parameter, that has actually little physical significance and robustness, but is nevertheless necessary to calculate instantaneous infiltration with Horton-type approaches.

CONCLUSIONS

The results obtained by both experiments and numerical simulations point to the significance of the dynamic phase of infiltration; in environments characterised by short and intense rainstorms, accounting for infiltration dynamics promises a significant progress in the ability to predict hydrological soil behaviour and, consequently, soil erosion and catchment hydrology. As prevalence of short, strong rainstorms is a widespread occurrence in tropical regions, such progress would be of particular value in these areas.

In the evaluation of management strategies for soil and water conservation, the ability to correctly estimate water infiltration and runoff is decisive; an appropriate weight attributed to infiltration dynamics would substantially improve this ability in tropical environments.

A more dynamic view of infiltration would lead to more favourable prospects for techniques able to improve soil response in non-steady phases. As it happens for the example tried out in our work, with the use of phosphogypsum, such techniques could show themselves to be cheaper, simpler and more feasible than techniques addressed to final infiltration rates, that appear to be harder to modify. Cheapness, simplicity and feasibility are all decisive requisites for soil conservation strategies to be implemented in tropical regions.

The approach to infiltration dynamics based on empirical infiltration laws is simple, attractive and undoubtedly useful for raising the question; however, the treatment of data uncertainty experimented with in our work has elucidated how applicability of Horton-type laws is reduced by insufficient robustness and physical significance of some parameters. This is probably a result of infiltration laws having not been originally developed for this purpose.

The need for improvement on this point has been evidenced, and the analysis of parameter uncertainty points to existing shortcomings and possible solutions. The simplest possibility would be improvements in experimental data collections and fitting techniques, to reduce sensitivity of results; a more general solution would probably be a redefinition of infiltration laws, in terms able to give adequate physical meaning to all parameters.

Analysis of uncertainty by fuzzy numerical techniques has met some of the expectations; it has been shown to be easy to use and has allowed rapid and efficient inspection of sources of uncertainty. Though it is undoubtedly premature to comment on its potential as a general modelling technique, it has already demonstrated a strong potential in the model building phase, for the evaluation of the robustness and sensitivity of input parameters, and clearly warrants further exploration of possible applications.

REFERENCES

Borselli, L., Carnicelli, S., Ferrari, G.A., Pagliai, M. and Lucamante, G. 1996a. 'Effect of gypsum on hydrological, mechanical and porosity properties of a kaolinitic crusting soil'. *Soil Technology*, **9**, 39–54.

Borselli, L., Biancalani, R., Carnicelli, S., Giordani, C. and Ferrari, G.A. 1996b. 'Effect of gypsum on seedling emergence in a kaolinitic crusting soil'. *Soil Technology*, **9**, 71–81.

Bowyer-Bower, T.A.S. 1993. 'Effects of rainfall intensity and antecedent moisture on the steady-state infiltration rate in a semi-arid region'. *Soil Use and Management*, **9**, 69–76.

Dubois, D. and Prade, H. 1979. 'Fuzzy real algebra: some applications'. *Fuzzy Sets and Systems*, **2**, 327–348.

Horton, R.E. 1940. 'An approach toward a physical interpretation of infiltration capacity'. *Soil Science Society of America Proceedings*, **5**, 399–417.

Kaufmann, A. and Gupta, M.M. 1991. '*Introduction to Fuzzy Arithmetics – Theory and Applications*'. Van Nostrand Reinhold, New York.

Klir, G.J. and Folger, T.A. 1988. '*Fuzzy Sets, Uncertainty, and Information*'. Prentice-Hall, New York.

Lai, Y.J. and Hwang, C.L. 1992. '*Fuzzy Mathematical Programming – Methods and Applications*'. Lecture Notes in Economics and Mathematical Systems, No. 394, Springer-Verlag, Berlin.

Levy, G.J. 1996. 'Soil stabilizers'. In *Soil Erosion, Conservation and Rehabilitation* (ed. M. Agassi), pp. 267–299. Marcel Dekker Inc, New York.

Mizumoto, M. and Tanaka, K. 1979. 'Some properties of fuzzy numbers'. In *Advances in Fuzzy Set Theory and Application* (eds N.N. Gupta, R.K. Ragade and R.R. Yager), 153–164. North Holland, New York.

Morgan, R.P.C., Quinton, J.N. and Rickson, A.J. 1992. '*EUROSEM Documentation Manual, Version 1*'. Silsoe College, Silsoe, UK.

Morin, J. and Benyamini, Y. 1977. 'Rainfall infiltration into bare soils'. *Water Resources Research*, **13**, 813–817.

Nearing, M.A., Foster, G.R., Lane, L.J. and Finker, S.C. 1989. 'A process-based soil erosion model for USDA–Water Erosion Prediction Project Technology'. *Transactions of the ASAE*, **32**, 1587–1593.

Nyamapfcnc, K.W. and Hungwe, A.P. 1986. 'Nature, distribution and management of crusting soils in Zimbabwe'. In *Assessment of Soil Surface Sealing and Crusting* (eds F. Callebaut, D. Gabriels and M. De Boodt), pp. 320–329. Proceedings of the International Symposium held in Ghent, Belgium, 1985.

Panini, T., Sanchis, M.P. and Torri, D. 1993. 'A portable rainfall simulator for rough and smooth morphologies'. *Quaderni di Scienza del Suolo*, **5**, 47–58.

Philip, J.R. 1957. 'The theory of infiltration 4, sorptivity and algebraic infiltration equation'. *Soil Science*, **84**, 257–264.

Soil Survey Staff 1990. *Keys to Soil Taxonomy*. SMSS Technical Monograph No. 19, Virginia Polytechnic Institute and State University.

Stone, J., Renard, K.J. and Lane, L.J. 1996. 'Runoff estimation on agricultural fields'. In *Soil Erosion, Conservation and Rehabilitation* (ed. M. Agassi), pp. 203–238. Marcel Dekker Inc, New York.

Tukey, J.W. 1977. '*Exploratory Data Analysis*'. Addison-Wesley, Reading, MA.

Unger, P.W. 1996. 'Common soil and water conservation practices'. In *Soil Erosion, Conservation and Rehabilitation* (ed. M. Agassi), pp. 239–266. Marcel Dekker Inc, New York.

Wischmeier, W.H. and Smith, D.D. 1978. 'Predicting rainfall erosion losses – a guide to conservation planning'. *USDA Handbook* No. 537.

Woolhiser, D.A., Smith, R.E. and Goodrich, D.C. 1990. '*KINEROS, A Kinematic Runoff and Erosion Model, Documentation and User Manual*'. USDA–ARS, Springfield, MA.

Zadeh, L.A. 1965. 'Fuzzy sets'. *Information and Control*, **8**, 338–353.

Zadeh, L.A. 1975. 'The concept of a linguistic variable and its application to approximate reasoning'. *Information Sciences*, **8**, 199–249; 301–357; **9**, 43–80.

Zadeh, L.A. 1978. 'Fuzzy sets as a basis for a theory of possibility'. *Fuzzy Sets and Systems*, **1**, 3–28.

Zimbabwe Department of Meteorological Services (ZDMS), 1975. 'Rainfall intensity in Zimbabwe'. *Rainfall Handbook Supplement No. 7*. ZDMS, Harare.

Zimmermann, H.J. 1990. '*Fuzzy Set Theory and its Applications*'. Kluwer Academic Publishers, Boston.

CHAPTER 18

Distributed Numerical Modelling of Surface Runoff and Soil Erosion in Arid Catchments

KAPIL DEV SHARMA
Central Arid Zone Research Institute, Jodhpur, India

JAP HUYGEN and MASSIMO MENENTI
The Winand Staring Centre for Integrated Land, Soil and Water Research, Wageningen, The Netherlands

ALBERTO VICH and PEDRO FERNANDEZ
Centro Regional Andino – INCYTH, Mendoza, Argentina

INTRODUCTION

In the arid and semi-arid lands of the world, watersheds are used for grazing, wood collection and localised intensive agriculture. By overexploitation of available resources and expansion of certain land use practices into unsuitable environments, due to increasing population, these activities have severely degraded vegetation cover, soils and slopes in many countries through wind and water erosion.

In many regions land degradation can still be reversed, however, and programmes have been initiated by national and local governments to develop practical guidelines to recover these lands in a self-sustainable way through reforestation and by construction of small works in a pattern to modify the overall hydrological behaviour of the watershed.

Quantitative integrated computer simulation of surface runoff and soil erosion, where output from a hydrologic model is to be used to predict sediment transport, has recently become feasible and has rapidly developed into a powerful tool in designing watershed amelioration projects as it allows for the evaluation of alternative strategies for improved land management. De Roo (1993) has reviewed current modelling aproaches. A model is a simplified representation of a complex system. If the behaviour of the system is represented by one or a set of equations one speaks of a mathematical model. Mathematical models are well suited for computer implementation and in this chapter the term "model" implicitly presumes a mathematical model.

The Sustainable Management of Tropical Catchments. Edited by David Harper and Tony Brown.
© 1998 John Wiley & Sons Ltd.

Models can be stochastic or deterministic. Variables included in stochastic models are regarded as random variables having distributions in probability. If all variables are free from random variation, the model is called deterministic. Hydrological models can be lumped or distributed. Lumped models are simple input–output models which calculate "effective values" for the entire area and whose parameters can rarely be interpreted in terms of physical processes. Distributed models are able to represent the spatial variability of parameters and/or processes within catchments.

Four major areas which offer the greatest potential for the application of distributed models can be identified (Beven, 1985). These are:

- forecasting the effects of land use change;
- forecasting the effects of spatially variable inputs and outputs
- forecasting the movement of sediments; and
- forecasting the hydrological response of ungauged catchments where no data are available for calibration of a lumped model.

Furthermore, models can be conceptual or empirical. Conceptual models (also named "physically based models") usually consist of linked process equations derived from physics principles (conservation of mass, momentum and energy) at a point, laboratory or plot scale. Empirical models, such as regression models, are by strict definition based on observation and experiment, not on theory.

The application of distributed-parameter, physically based models to real catchments presupposes the following:

- that physical processes can be represented in a deterministic way and that the overall catchment response can be represented by the combined action of the constituent process algorithms;
- that the spatial variability of a catchment can be represented by distributed values of the model parameters; and
- that model parameters can be meaningfully measured or estimated at the element scale of the model (Parsons and Abrahams, 1992).

Distributed-parameter models that give estimates of flow over a catchment are the basis of many sediment transport models. The flow velocity and depth predictions are used in equations to estimate sediment transport capacity and sediment detachment that are then used to estimate the amount of soil removed from an area.

Such a linked distributed-parameter, physically based hydrologic–soil erosion model has recently been developed at the Winand Staring Centre and successfully validated using data collected in a pilot catchment in the Piedmont of Mendoza, Argentina.

COMPONENTS OF THE SIMULATION ENVIRONMENT

A Geographical Information System (GIS) was used for processing spatial input data to produce input tables for the numerical simulation models and for displaying the results of the simulations.

Surface runoff was simulated with a transient one-dimensional finite-difference model for the unsaturated zone. The model is based on the combination of one-dimensional calculation of soil water flow, by numerical integration of the Richards' equation, with simulation of rainfall-excess concentration using a digital terrain model. The watershed was divided into a number of cells and the one-dimensional calculation done for each cell, while the mechanism of concentration was described using a simple time-lag parameter, i.e. the residence time of rainfall-excess in each cell. Soil removal was calculated cell-wise on the basis of rainfall-excess, as calculated by the hydrologic model, and slope by means of a physically based soil erosion model.

DESCRIPTION OF THE STUDY AREA

The study basin, Divisadero Largo ($5.47\,km^2$), is located within the piedmont and precordilleran areas of the Andes mountain range in the west of Mendoza, Argentina. The altitude ranges from 950 m in the east to 1,450 m in the west. The basement of the area is formed by Triassic and Tertiary sediments, which are covered discordantly by alluvial fan deposits from Quaternary age. The region is intersected by steeply eroded gullies and rock outcrops. The soils are shallow, undeveloped and consist of medium to fine sand. The vegetation comprises low shrubby pastures ranging from 5 to 45% cover depending upon the steepness of the slope.

The area lies in a sub-tropical arid climate and is characterised by convective summer thunderstorms. The annual precipitation is 200 mm, 77% of which is received within the summer months of October to March. The average annual temperature is 13°C. The hydrological network comprises 32 automatic rain gauges and two stream gauging stations covering an area of $600\,km^2$ within the region and the data have been recorded through a telemetry network since 1983. The present study is based on the data recorded between 1983 and 1992.

MODELLING THE RAINFALL–RUNOFF PROCESS

In the arid upland basins virtually all the runoff occurs in the form of overland flow which is generated when rainfall intensity exceeds the surface infiltration rate. The paucity of rainfall and limited rainfall amounts, coupled with high infiltration losses in the dry channels, prevent the development of saturated or nearly saturated conditions in arid watersheds. Hillslope runoff is thus regarded as the main contributor to storm channel runoff, representing an essential stage in the initiation and development of any channel flow. Hillslope areas responding most rapidly to rainfall are represented by bedrock outcrops with a low infiltration capacity. At the same time, unconsolidated colluvial or alluvial sediments absorb all rainwater for most rainstorms, thus providing little or no runoff (Yair, 1983; Yair and Lavee, 1985). In the channel headwater area extensive bedrock outcrops, almost devoid of any effective soil cover and vegetation, respond quickly to rainfall and frequently generate high runoff rates per unit area as flash floods.

Owing to the climatic characteristics of the Mendoza region, the stream channels with drainage areas up to a few tens of square kilometres are usually dry valleys. Runoff on such watersheds generally follows periods of thunderstorm rainfall during the summer months. At other times the stream channels are normally dry. The substantial infiltration losses and the steep slope of the channels tend to produce sharply peaked runoff hydrographs. The resulting hydrograph consists of a fairly narrow triangular peak followed by a relatively longer recession of low flow. The time to peak (time from beginning of runoff to the hydrograph peak) is usually shorter than the recession time. There is little baseflow to be separated from the hydrograph. This characteristic shape of the runoff hydrograph suggests that the discrete event models can be appropriate for simulating the rainfall–runoff process within the region. Understanding and properly modelling the upland flow characteristics is essential in the calculation of flow velocity, routing of runoff hydrographs and in developing the process-based soil erosion models.

In the present study, for the purpose of modelling, the rainfall–runoff process is divided into the excess rainfall process and the runoff concentration process. The former focuses on estimating the infiltration process and the latter on calculating the basin outlet hydrograph from excess rainfall using a routing technique.

POINT SIMULATION OF EXCESS–RAINFALL WITH SWAMIN

A one-dimensional dynamic simulation model, describing the distribution of precipitation and runon among runoff, surface storage, soil profile storage and deep percolation, has been derived from the well-known water balance model SWATRE (Feddes et al., 1978; Belmans et al., 1983).

SWATRE has been modified in several ways:

- The time scale of the rainfall–runoff process (a few hours only) permits neglect of processes like interception of rainfall by the canopy and evapotranspiration.
- Maximum simulation duration has been limited to 24 hours making it essentially a single-event model.
- Input of rainfall data may have a time resolution of one minute.

These modifications have given rise to a new name for the program: SWAMIN.

In SWAMIN, vertical water flow in unsaturated or partly saturated soils is described with Richards' equation (Richards, 1931):

$$C(h)\frac{\delta h}{\delta t} = \delta \frac{\left[K(h)\left(\frac{\delta h}{\delta z}+1\right)\right]}{\delta z} \tag{1}$$

where h = soil water pressure head (cm), t = time (days), C = differential moisture capacity, $\delta q/\delta h$, with q being the volumetric soil water content, z = vertical co-ordinate, with origin at the soil surface, directed positive upwards (cm), and K = hydraulic conductivity (cm day^{-1}).

Because the hydraulic conductivity K and the differential moisture capacity C are non-linear functions of the dependent variable h, serious problems arise in trying to solve equation (1). Mathematical solutions are known for special cases only and are usually restricted to very simple boundary conditions. To handle more general flow situations a numerical solution is needed. In SWAMIN a finite-difference scheme as proposed by Haverkamp et al. (1977) is employed. The finite-difference scheme is implicit and applies an explicit linearisation.

RUNOFF ROUTING WITH SWAMREG

To handle three-dimensional water flow, a shell has been built around SWAMIN, that takes care of rainfall-excess concentration and that routes the combined runoff flow through the catchment. The shell and SWAMIN together are named SWAM-REG.

OPERATING PRINCIPLE OF SWAMREG

The mechanism of concentration is described by using a simple time-lag parameter: runoff from a pixel is supposed to be released after a time period called the translation time, in a rate equal to runoff depth divided by translation time. A simulation with SWAMREG starts with a run for a period of one translation time: for each pixel SWAMIN is executed for a time period corresponding with the number of minutes in the translation time. The list with pixels is processed sequentially and rainfall excess from each pixel is written to a temporary output file.

The simulation is repeated for a period of two translation times: possible first-translation-time runoff from neighbouring pixels is added to the rain falling during the second translation period and cumulated runoff at the end of the second translation time is again written to a temporary output file. This procedure is iterated for an increasing number of translation times until the whole simulation period is covered. Runoff from the outlet pixel during the final simulation run is considered as the outflow hydrograph of the catchment.

MODELLING SOIL EROSION

Soil erosion by water is one of the major soil degradation processes in the world. Arid regions particularly have the potential to generate and transport large quantities of sediment. The low-altitude, hot arid regions in the world are characterised by a single rainy season during the summer, followed by 9-10 months of dry season. In general, the rainfall is localised and is received in short, intense bursts. Since there is a sparse vegetative cover to shield the surface, these storms act on the loose surface material generated during the dry period due to intense weathering and biotic interference, and cause very high soil erosion rates. Therefore, the soil erosion within these arid regions is governed by the transport capacity of the runoff rather than the availability of the erodible material within the basin since this is always available in abundance.

To predict the spatial and temporal distribution of actual soil loss within a drainage basin, a process-based soil erosion model is preferable, because it can be extrapolated to a broad range of conditions which may not be practical or economical to test in the field.

Many process-based models of sediment transport by water in upland basins dynamically route sediment by solving the continuity equation for sediment transport (Bennett, 1974). The solution of this equation is generally accomplished using numerical methods which are not only unstable, but also uncertain due to friction losses. In the present study a closed form solution to the governing differential equation under steady-state condition is attempted.

Sediment movement downslope obeys the principle of continuity of mass expressed by a steady-state sediment continuity equation as (Nearing et al., 1989):

$$\frac{\delta Q_s}{\delta x} = D_f + D_i \tag{2}$$

where Q_s = sediment transport rate per unit of width ($kg\,s^{-1}\,m^{-1}$), x = downslope distance (m), D_f = net flow detachment rate ($kg\,s^{-1}\,m^{-1}$), and D_i = net rainfall detachment rate ($kg\,s^{-1}\,m^{-1}$).

In the present analysis D_i is neglected since the transport capacity of rainsplash is very low. Lu et al. (1989) also found that the effect of raindrop impact on sediment transport relationship was insignificant. The value of D_f in the arid regions, where the initial potential sediment load is in excess of the transport capacity (Foster et al., 1980; Sharma et al., 1992), can be estimated by a first-order reaction model of the type:

$$D_f = G(T_c - Q_s) \tag{3}$$

where G = a first-order reaction coefficient (m^{-1}), and T_c = flow transport capacity ($kg\,s^{-1}\,m^{-1}$).

By combining equations (2) and (3) the equation for the simplified soil erosion model can be derived as:

$$\frac{\delta Q_s}{\delta x} + GQ_s - GT_c = 0 \tag{4}$$

Equation (4) is a linear non-homogeneous first-order differential equation whose closed form solution is:

$$\ln(T_c - Q_s) = -Gx + \ln C \tag{5}$$

where C = constant of integration ($kg\,s^{-1}\,m^{-1}$) being equal to $T_c - Q_s$ at $x = 0$, thereby explaining the discrepancy between the sediment transport capacity and the actual soil loss at the point of initiation of runoff, i.e. the ridge line of the basin.

Model parameters G and C are associated with the soil erosion process and must be determined experimentally. According to equation (5) the sediment transport is proportional to the difference between the transport capacity and the actual transport. If flow transport capacity T_c is known, soil loss rate Q_s can be calculated.

Transport capacity of flow determines the upper limit of soil erosion by water. Several generalised formulae have been developed for computing sediment transport capacity. Many of the equations were developed for stream flows, and were later applied to shallow overland and channel flows. However, Alonso et al. (1981), who evaluated nine sediment transport formulae, concluded that the Yalin equation (Yalin, 1963) provided reliable estimates of transport capacity for shallow overland and channel flows. Foster and Meyer (1972) also concluded that the Yalin equation was most appropriate for shallow flows associated with upland erosion. The Yalin equation is defined as:

$$\frac{T_c}{SGd} \sqrt{\rho_w \tau_s} = 0.635\delta \left(1 - \frac{1}{\beta} \ln(1 + \beta) \right) \tag{6}$$

with:

$$\beta = 2.45 SG^{-0.4} \sqrt{Y_{cr}} \, \delta \tag{7}$$

$$\delta = \frac{Y}{Y_{cr}} - 1 \quad \text{(when } Y < Y_{cr}, \delta = 0) \tag{8}$$

$$Y = \frac{\left(\frac{\tau_s}{\rho_w} \right)}{(SG - 1)gd} \tag{9}$$

$$\tau_s = \gamma h S \tag{10}$$

and where SG = soil particle specific gravity, obtained from detailed soil analyses, ρ_w = mass density of water (kg m^{-3}), d = mean particle diameter (m), obtained from soil analyses, Y = dimensionless shear stress, Y_{cr} = dimensionless critical shear stress from Shield's diagram (ASCE, 1977), g = acceleration of gravity (m s^{-2}), τ_s = flow shear stress (kg s^{-2} m^{-1}), γ = specific weight of water (kg s^{-2} m^{-2}), h = flow depth (m), obtained from the runoff simulation, S = slope, and β and δ are parameters as defined by equations (7) and (8).

The Yalin equation computes sediment transport capacity as a function of flow hydraulics and sediment diameter and density. The equation applies to any point on the landscape provided that the estimates of hydraulic and sediment properties are available.

The procedure adopted to model soil erosion is shown in Figure 18.1.

NON-SPATIAL INPUT DATA

EROSION MODEL PARAMETERS OBTAINED FROM FIELD EXPERIMENTS

In the study basin small plot studies were conducted between 1982 and 1992 to assess the soil loss under different conditions of plant cover using a rainfall simulator (Vich, pers. comm.). Between 1990 and 1992 experiments were carried out in larger plots

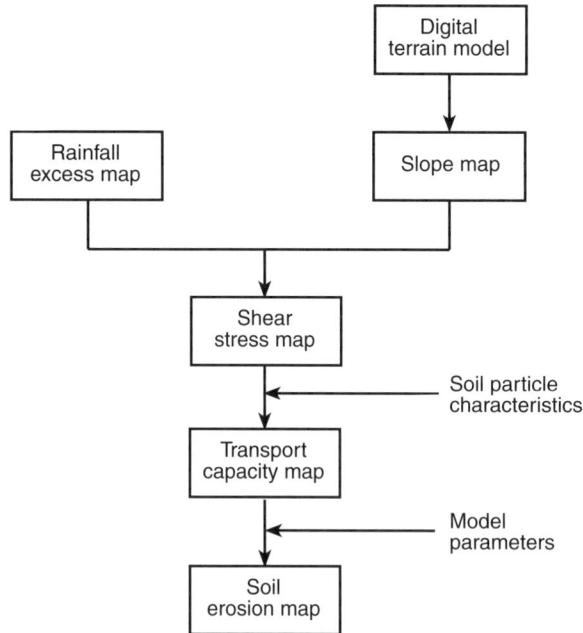

Figure 18.1   Representation of the procedure for modelling soil erosion

(0.5 ha) where the total sediments removed were measured. Fitting a least squares technique to these data, the parameter G was found to be $0.036\,\mathrm{m}^{-1}$. Singh and Regl (1983) suggested a reasonable value of G as $0.030\,\mathrm{m}^{-1}$. The parameter C was found to be equal to $0.73\,\mathrm{kg\,s}^{-1}\,\mathrm{m}^{-1}$ which was well within the range of 0.34 to $2.42\,\mathrm{kg\,s}^{-1}\,\mathrm{m}^{-1}$ as reported by Sharma et al. (1993).

SOIL PARTICLE CHARACTERISTICS

Ligtenberg et al. (1992) conducted detailed soil analyses in the study area. Soil particle characteristics such as representative particle diameter (0.22 mm) and specific gravity (2.65), required by the erosion model, were taken from their study.

SOIL HYDRAULIC PROPERTIES

Soil water retention and hydraulic conductivity characteristics for the relevant soils found in the study area and required by the hydrologic model were also obtained from Ligtenberg et al. (1992).

SPATIAL INPUT DATA

For the distributed simulation approach a regular grid was laid over the area with cells of $120 \times 120\,\mathrm{m}$. Although most input data were available with a higher resolution, this rather large cell size was chosen because of limited computer processing power facilities. Relevant attributes for each cell were determined by processing a number of basic maps with the help of GIS procedures.

SATELLITE DATA

A Landsat TM image (path 232, row 083, date 22 June 1986) was acquired to serve as a source for data concerning vegetation cover.

DIGITAL ELEVATION MODEL

From the topographic map of the Divisadero Largo basin (1:5,000 scale) 5 m contour lines were digitised and interpolated to a continuous elevation map. The digital terrain map thus obtained was processed further for the production of a slope map and a map with local drainage direction. The latter was derived from the slope map taking into account a pit removal algorithm.

SURFACE STORAGE MAP

Landsat TM imagery was analysed to obtain a Green Index (GI) map, a ratio image calculated using the formula: $GI = (TM\ 4/TM\ 3)$. After analysing the histogram of GI values, four classes were identified, which were converted to Fractional Vegetation Cover (FVC) using data supplied by Roby et al. (1990). The FVC map and the slope map were combined resulting in the surface storage map.

SOIL MAP

Aerial photographs were used in combination with field observations to produce a map of hydrological soil groups, where soils within a group are similar with respect to soil physical characteristics as required by the hydrological simulation model.

RAINFALL DISTRIBUTION MAP

Hoefsloot et al. (1992) concluded that with a relatively high density of rain gauges (one station in about $20\ km^2$ area) the determination of rainfall distribution using Thiessen polygons was reasonable and reliable in the Andean precordillera. Accordingly the Thiessen polygons were drawn, knowing the co-ordinates of the rain-gauge stations within the study basin, and digitised.

DERIVED MAP WITH HYDROLOGICAL RESPONSE UNITS

An overlay of the soil map, the Thiessen polygon map and the surface storage map yielded the map with so-called Hydrological Response Units (HRU), representing unique combinations of soil type, rainfall regime and surface storage capacity. In this way 26 HRUs were delineated.

SPATIAL INPUT DATA FOR SWAMREG

From the maps (both basic and derived), tables were constructed by the GIS, which served as input for the three-dimensional model SWAMREG as follows:

- a list with grid-cell (pixel) co-ordinates plus, for each pixel, a pointer to the corresponding hydrological response unit (HRU);
- a list with pixel co-ordinates plus the amount and co-ordinates of the surrounding pixels draining to that pixel; and
- a table (for each unique rainfall event) describing the HRUs in terms of soil physical parameters, rainfall data, surface storage capacity and initial soil moisture conditions.

RESULTS AND VALIDATION

RESULTS OF SURFACE RUNOFF MODELLING

The SWAMREG model was applied to simulate three runoff hydrographs, covering a broad range from low to high runoff volumes. Combinations of the values for delay time and maximum surface storage were selected randomly and simulation runs were made so that the predicted runoff volume was within 10% of the observed runoff volume. A comparison of the observed and simulated hydrographs, ranging from the worst calibration to the best calibration, is shown in Figure 18.2. For further evaluation the goodness of fit of a predicted hydrograph to an observed hydrograph was determined by means of the following criteria:

- Relative squared error (*RSE*), defined as the ratio of the sum of squared differences between simulated and measured discharges at a number of times and the sum of the squared observed discharges. The RSE represents the overall shape of the hydrograph and if RSE = 0 the predicted hydrograph will coincide with the observed hydrograph.
- Relative error in estimated peak (*Ep*), defined as the ratio of the absolute value of the difference between simulated and observed peak discharge and the observed peak discharge.
- Relative error in estimated time base of the hydrograph (*Et*), defined as the ratio of the absolute value of the difference between simulated and measured time base of the hydrograph and the measured time base.
- Relative error in estimated runoff volume (*Er*), defined as the ratio of the absolute value of the difference between simulated and observed runoff volume and the observed runoff volume.

Table 18.1 shows values of these four criteria for all the calibration events. It appears that once the predicted runoff volume is forced to match the observed runoff volume there is also good resemblance between other properties of the simulated and observed hydrographs, notably overall shape, peak flow and time base.

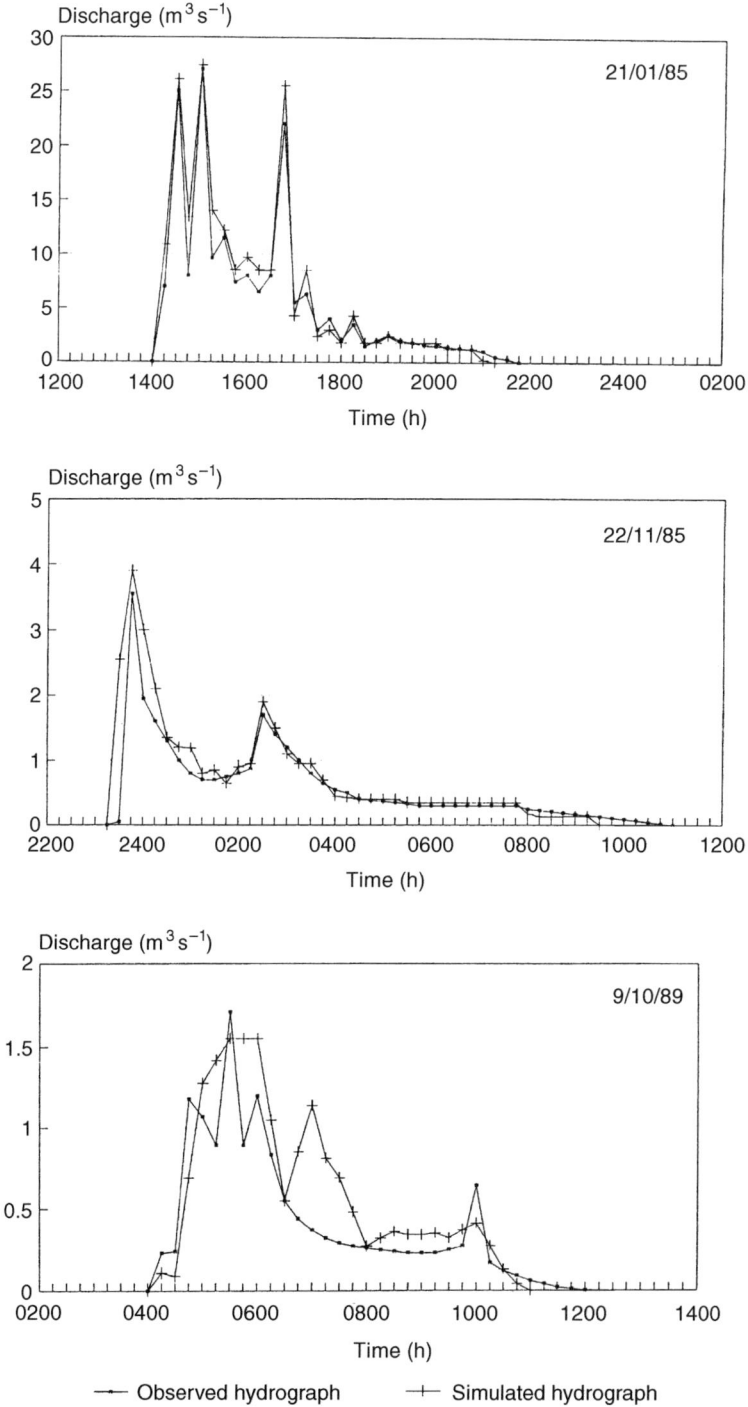

Figure 18.2   Observed and simulated outflow hydrographs for three rainfall events

Table 18.1   Comparison of predicted and observed hydrographs for three calibration events

Date	Δt	SS_{max}	RSE	Ep	Et	Er
20 Jan. 85	40	0	0.13	0.094	0.125	0.061
22 Nov. 85	90	0	0.04	0.015	0.085	0.025
09 Oct. 89	30	0	0.13	0.085	0.130	0.054

Note that all three flow events give the best fit at $SS_{max} = 0$. Also for the best fit of the predicted hydrograph delay time Δt varies from 30 to 90 minutes, thereby giving an average velocity of overland flow between 0.067 and 0.022 m s^{-1} respectively. These values are well within the range of observed overland flow velocities in the field (Emmett, 1978).

RESULTS OF SOIL EROSION MODELLING

The soil erosion model was validated on seven independent rainfall events for which the soil loss data were recorded. A comparison of measured and predicted soil loss shows a good agreement (Figure 18.3). With a coefficient of determination of 0.996 ($P > 0.01$) the predicted and measured soil loss can be regressed by the equation:

$$Y = 1.208X - 0.022 \tag{11}$$

where Y = predicted soil loss (kg m^{-2}), and X = measured soil loss (kg m^{-2}).

The relative error in predicted soil loss, defined as the ratio of the absolute value of the difference between predicted and observed soil loss and observed soil loss was found to be 6.8% on average. The maximum was 16.5% and the minimum 2.7%.

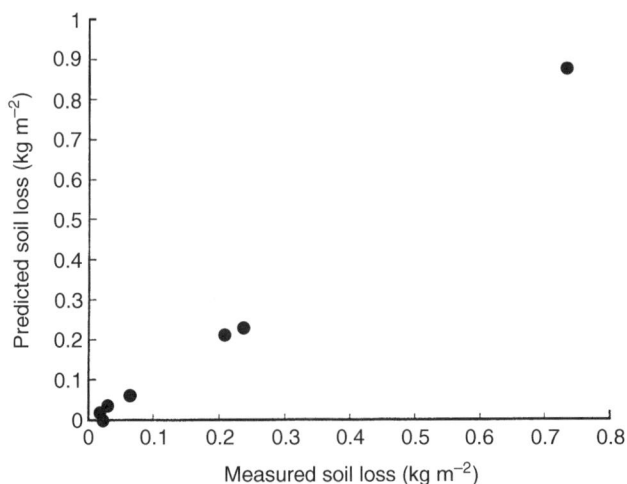

Figure 18.3   Comparison of predicted and measured soil loss, as observed during seven rainfall events

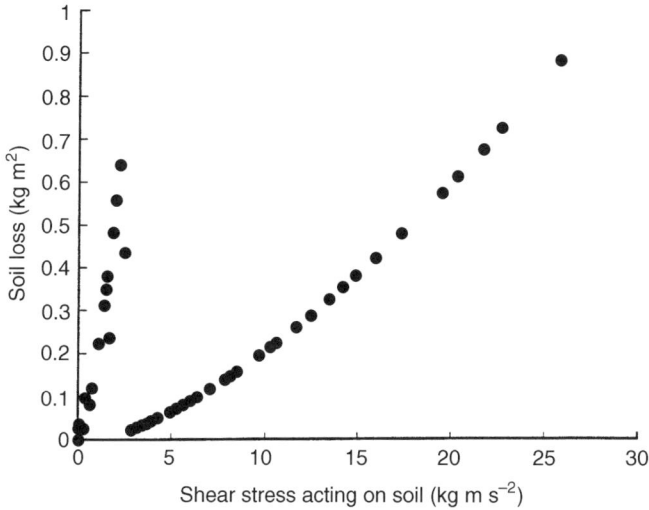

Figure 18.4   Soil loss as a function of soil shear stress for two different kinds of soils occurring within the study basin

An exponential relationship was observed between the actual soil loss and the shear stress acting on soil in the Divisadero Largo basin (Figure 18.4). Such a relationship has been reported by other authors as well (Foster, 1982). However, the interesting part of Figure 18.4 is that this relationship indicates the presence of two kinds of surfaces occurring within the drainage basin: one, a highly erodible surface such as fine to medium sand – depicted by the left-hand curve – and the other, a surface resistant to erosion such as bare rock or a soil surface protected by vegetation cover – depicted by the right-hand curve. As an example: at $1.0\,kg\,m^{-1}\,s^{-2}$ shear stress the former loses $0.22\,kg\,m^{-2}$ soil whereas the latter loses only $0.01\,kg\,m^{-2}$ soil.

Using the physically based soil erosion model in conjunction with a GIS it is possible to predict the spatial variability of soil loss within the drainage basin. In this way maps have been produced showing the spatial distribution of flow shear stress, sediment transport capacity and actual soil loss for a particular storm. Such maps are useful in the identification of vulnerable areas within the drainage basin for the siting of soil conservation measures.

DISCUSSION

The accuracy of the final estimates of soil loss are directly related to the quality of the (overland) flow estimates of the hydrological model. These flow estimates are subject to errors as a result of uncertain antecedent conditions and poor resolution in the data, as well as limitations imposed by the model structure and algorithms. Erosion modelling adds its own uncertainties due to spatial and temporal variability of input variables, which are not sufficiently taken into account, and simplifications in the modelling approach.

UNCERTAIN ANTECEDENT CONDITIONS

An important source of errors in hydrological and soil erosion modelling is the soil moisture content just before the start of the rainfall event. The soil moisture content varies both in space and time. At catchment scale it is not feasible to measure soil moisture content continuously. Therefore sometimes a submodel is used to estimate initial moisture content through time-series analysis of past rainfall events. In the present study, however, initial conditions are just the product of intelligent guesswork.

RESOLUTION IN DATA

Model results are influenced by the grid size. Input data spatial variability is averaged out more rigorously in a large grid than in a small grid, so for improved precision a small cell size is desirable. In this study the relatively large grid size of 120 m was chosen because of limited computing power.

Also the temporal variability in input data is an important issue which however is not touched by the present study. For instance, formation and breaking down of a surface crust may influence infiltration considerably and soil erodibility may vary dynamically due to biotic effects such as those caused by plant roots, soil microfaunal activity and human effects such as fertilisation.

MODEL STRUCTURE AND ALGORITHMS

The theoretical structure of the models employed is limited. The hydrological model lacks several important sub-processes such as subsurface flow, baseflow and saturation overland flow, and does not explicitly differentiate between sheet, rill, ephemeral gully and channel flow and between laminar and turbulent flow. Other processes are simplified, for instance the routing of rainfall excess using a fixed residence time for each grid cell, instead of a distributed residence time being determined by slope and roughness.

CONCLUSIONS

A soil erosion model based on the law of conservation of mass, a first-order reaction rate model and Yalin's sediment transport capacity equation predicts the spatio-temporal distribution of actual soil erosion with an acceptable degree of accuracy. Since this model utilises the rainfall excess, which is spatially varied, the model can calculate erosion/deposition for the case of non-uniform hydrology in the drainage basin. The results are useful in the identification of vulnerable areas within the drainage basin for an economic and effective management of the soil resources.

REFERENCES

Alonso, C.V., Neibling, W.H. and Foster, G.R. 1981. 'Estimating sediment transport capacity in watershed modelling'. *Transactions of the ASAE*, **24**, 1211–1220, 1226.
ASCE, 1977. '*Sedimentation Engineering*'. American Society of Civil Engineers, New York.

Belmans, C., Wesseling, J.G. and Feddes, R.A. 1983. 'Simulation of the water balance of a cropped soil: SWATRE'. *Journal of Hydrology*, **63**, 271–286.

Bennett, J.P. 1974. 'Concepts of mathematical modelling of sediment yield'. *Water Resources Research*, **10**, 485–492.

Beven, K.J. 1985. 'Distributed models'. In *Hydrological Forecasting* (eds M.G. Anderson and T.P. Burt), pp. 405–435. Wiley, Chichester.

De Roo, A.P.J. 1993. 'Modelling surface runoff and soil erosion in catchments using Geographical Information Systems'. PhD Thesis, University of Utrecht, The Netherlands.

Emmett, W.W. 1978. 'Overland flow'. In *Hillslope Hydrology* (ed. by M.J. Kirkby), pp. 145–176. Wiley, Chichester.

Feddes, R.A., Kowalik, P.J. and Zaradny, H. 1978. '*Simulation of field water use and crop yield*'. Simulation Monographs, PUDOC, Wageningen.

Foster G.R. 1982. 'Modelling the erosion process'. In *Hydrologic Modelling of Small Watersheds* (ed. by C.T. Haan, H.P. Johnson and D.L. Brakensiek), pp. 297–360. ASAE Monograph No. 5, American Society of Agricultural Engineers, St Joseph.

Foster, G.R. and Meyer, L.D. 1972. 'Transport of soil particles by shallow flow'. *Transactions of the ASEA*, **15**, 99–102.

Foster, G.R., Lane, L.J., Nowlin, J.D., Laflen, J.M. and Young, R.A. 1980. 'A model to estimate sediment yield from field sized areas'. In *CREAMS – a Field Scale Model for Chemicals, Runoff and Erosion from Agricultural Management Systems* (ed. W. Knisel), pp. 34–64. USDA Conservation Research Report No. 26, Beltsville.

Haverkamp, R., Vauclin, M., Touma, J., Wierenga P.J. and Vauchaud, G. 1977. 'A comparison of numerical simulation models for one-dimensional infiltration'. *Soil Science Society of America Proceedings*, **41**, 285–294.

Hoefsloot, F., de Wit, M., Menenti, M. and Fernandez, P.C. 1992. '*Modelling the Rainfall–Runoff Processes of a Catchment in the Andean Precordillera, Mendoza, Argentina*'. Technical Report, WSC, Wageningen.

Ligtenberg, A., Rijswijk, J.V., Menenti, M. and Fernandez, P.C. 1992. '*Runoff Research – Divisadero Largo*'. Technical Report, WSC, Wageningen.

Lu, J.Y., Cassol, E.A. and Moldenhauer, W.C. 1989. 'Sediment transport relationships for sand and silt loam soils'. *Transactions of the ASAE*, **32**, 1923–1931.

Nearing, M.A., Foster, G.R., Lane L.J. and Finkner, S.C. 1989. 'A process based soil erosion model for USDA – water erosion prediction project technology'. *Transactions of the ASAE*, **32**, 1587–1593.

Parsons, A.J. and Abrahams, A.D. (eds) 1992. '*Overland Flow, Hydraulics and Mechanics*'. University College London Press, London.

Richards, L.A. 1931. 'Capillary condition of liquids through porous mediums'. *Physics*, **1**, 318–333.

Roby, H.O., Maza, J.A., Zuluaga, J.M. and Fornero, L.A. 1990. 'Determinacion de parametros para modelos hidrologios con sensores remotos'. In *Remote Sensing in Evaluation and Management of Irrigation* (ed. M. Menenti), pp. 289–300. Fundacion Branco de Credito, Mendoza.

Sharma, K.D., Dhir, R.P. and Murthy, J.S.R. 1992. 'Modelling sediment transport in arid upland basins in India'. In *Erosion, Debris Flows and Environment in Mountain Regions. IAHS Publ.* **209**, 169–176.

Sharma, K.D., Dhir, R.P. and Murthy, J.S.R. 1993. 'Modelling soil erosion in arid zone drainage basins'. In *Sediment Problems: Strategies for Monitoring, Predictions and Control. IAHS Publ.* **217**, 269–276.

Singh, V.P. and Regl, R.R. 1983. 'Analytical solutions of kinematic equations for erosion on a plane. I: rainfall of infinite duration'. *Advances in Water Resources*, **6**, 2–10.

Yair, A. 1983. 'Hillslope hydrology, water harvesting and areal distribution of some ancient agricultural systems, northern Negev, Israel'. *Journal of Arid Environments*, **6**, 283–301.

Yair, A. and Lavee, H. 1985. 'Runoff generation in arid and semi-arid zones'. In *Hydrological Forecasting* (ed. M.G. Anderson and T.P. Burt), pp. 183–220. Wiley, Chichester.

Yalin, M.S. 1963. 'An expression for bed load transportation'. *Journal of the Hydraulics Division ASCE*, **89**, 221–250.

CHAPTER 19

Hydrological Modelling in Humid Tropical Catchments

JAMES BRASINGTON
Department of Geography, University of Hull, UK

AHMED EL-HAMES

and

KEITH RICHARDS
Department of Geography, University of Cambridge, UK

INTRODUCTION

Whilst research into hydrological processes has continued apace in temperate and semi-arid regions, the humid tropics have received relatively little attention in the last 30 years. However, growing concern over the environmental sustainability of tropical ecosystems, and an increasing awareness of the global climatic importance of these regions, have stimulated a surge of interest in the environmental problems and hydrological processes of the tropics (Bruijnzeel, 1986, 1990; Bonell and Balek, 1993). It is widely accepted that, in many tropical regions, the core of the environmental problems are essentially socio-economic (Bonell and Balek, 1993), caused by the increased pressure placed on natural resources from rapidly expanding populations. Hydrological research has therefore tended to focus upon the hydrological consequences of land use management, and in particular upon the following issues:

- Land surface–atmosphere interactions and the effects of land use change upon global climates, termed macrohydrology, after Klemes (1993) (e.g. Henderson-Sellers and Gornitz, 1984).
- Changes in catchment water yield, storm runoff, and low-flow regimes, following land use change (e.g. Bruijnzeel, 1989).
- The impacts of changes in hydrological regimes upon soil and nutrient losses and to changes in aquatic biology (e.g. Lal, 1983, 1987; Bruijnzeel, 1990; Rundle et al., 1993; Jenkins et al., 1995).

The Sustainable Management of Tropical Catchments. Edited by David Harper and Tony Brown.
© 1998 John Wiley & Sons Ltd.

Despite the increasing interest in the humid tropics, the distribution and scope of empirical studies remains very limited and the lack of long-term detailed environmental records serves to compound the uncertainty surrounding our knowledge of process mechanisms in the region. Recognition of this uncertainty and its subsequent evaluation and analysis has recently developed as a key issue in the study of tropical regions and has been used to provide constructive critiques (Blaikie, 1985; Thompson et al., 1986; Ives and Messerli, 1989) of the scenarios of spiralling environmental decline which gained popular support in the 1970s and 1980s (e.g. Eckholm, 1976; Myers, 1986).

Our scant understanding of environmental processes within the humid tropics has commonly forced the technology transfer of methodologies and models developed in the higher latitudes to application in tropics (Bonell and Balek, 1993). This is a hazardous process and must be made with the utmost care and attention to what little knowledge of tropical hydrology we possess. A key aspect of process studies in hydrology is the development of simulation models. Studies focusing upon hydrological model development in tropics regions have three main advantages. These are:

- Firstly, a well-tested model may provide a useful basis for record extension and the detection of non-stationarity in process response (Blackie and Eeles, 1985).
- Secondly, construction of formal mathematical models necessitates precise specification of ideas and so presents a useful opportunity to investigate our perceptions of the system under study (Howes and Anderson, 1988). Further information about the hydrological system under study may also be gleaned from the development of "physics-based" models, which represent a simplification of the modeller's understanding of key hydrological processes. Detailed model evaluation and the use of sensitivity analysis can then be used to investigate the accuracy of the model. This is particularly useful in areas where long-term data records are unavailable.
- Thirdly, and perhaps most contentiously, physically based models may be used in assisting the planning process, whereby a number of hypothetical scenarios may be constructed and the effects of various changes to catchment conditions simulated.

This chapter describes the development and application of a distributed rainfall–runoff model to a small (4.2 km^2) catchment, the Bore Khola, in the sub-tropical regions of the Nepal Middle Hills. The model has been designed primarily to address the second issue above, although similar models have been adapted to drive secondary models related to planetary boundary layer development (Quinn et al., 1995a) and nutrient and soil erosion processes at the catchment scale (Moore and Burch, 1986; Robson et al., 1992; Vertessy and Wilson, 1992).

THE STAGES OF MODEL DEVELOPMENT

A primary and paramount stage in model development is the setting of modelling goals. Each model application should be tailored towards the desired level of information required. For example, if only daily, lumped catchment, water balance estimates are required, a simple statistical model may suffice. However, where detailed, temporally and spatially variable processes within the system are the focus of attention, a more sophisticated approach may be appropriate. The wide spectrum of goals set by modellers has led to a diverse array of modelling strategies for catchment-scale hydrological investigation.

There is an extensive literature covering the various methodological approaches to modelling rainfall–runoff problems. Wheater et al. (1993) provide an excellent recent review of the subject, suggesting a threefold categorisation of modelling strategies. This includes, firstly, the metric approach, in which observed data are used to characterise both model structure and parameters. Second is the conceptual method, in which the model structure is defined *a priori*, based on the modeller's understanding of the pertinent hydrological processes at work in the catchment. Parameters of such models, are, however, still commonly defined by calibration with an observed rainfall–runoff sequence. The last main approach they identify is based on complex, spatially distributed mathematical descriptions of the laws of continuum mechanics that govern the hydrological cycle. Each of these modelling strategies has a place in hydrologic science. Whilst the threefold classification described above could be viewed as a teleological development of the subject, each of the above methodologies has a range of advantages and disadvantages, that have in turn promoted the development of a range of hybrid modelling strategies. Recent developments in hydrological modelling have therefore been aimed towards a pragmatic tradeoff between physically based representation, computational efficiency, data requirements, mathematical and parametric parsimony and, of course, model accuracy and reliability.

Beven (1991) has suggested that a modelling exercise necessarily involves a sequence of simplifications. The first stage in the construction of conceptual or physics-based models is the development of a "perceptual" model of the system. This perceptual model may represent a far more complex description of the system than that which could be efficiently and succinctly expressed in mathematical form. The development of the perceptual model does, however, provide an excellent opportunity for the modeller to extract the most important features of the hydrological system and begin to consider the important interactions between processes. The second stage in model construction is the specification of a conceptual model, which is an attempt to express the relative complexity of the perceptual model in terms amenable to a formal mathematical description. It should be recognised that this stage incorporates a large degree of necessary simplification. Next, the mathematical model is translated into computer code. This stage may involve the use of numerical approximations for the solution of continuous mathematical functions, again adding further potential sources of inaccuracy into the model. Further stages in the evolution of the model comprise formal model validation and evaluation and, if required, a degree of calibration. Following these final stages of analysis, reformulation of the model may be required and the developmental steps may be repeated in an iterative manner until a final model is either accepted, or rejected as an unsuitable simulator of the system.

THE DEVELOPMENT OF A PERCEPTUAL MODEL

Within the small sub-tropical catchments of this study, the processes leading to runoff production, rather than to flood routing, remain the dominant control upon storm water dynamics. This assumption is indirectly justified by the rapid time-to-concentration in the Bore Khola catchment which lags just half an hour behind peak rainfall. In such catchments, the comments of Cordova and Rodriguez-Iturbe (1983, p.172) appear most relevant: ". . . the problem is more what to route rather than how to route . . . what remains as a crucial and unsolved problem is the description of the infiltration processes at a basin scale".

A full discussion of the mechanisms of runoff generation is beyond the scope of this chapter and salient reviews are given in Kirkby (1985), Anderson and Burt (1990), Wood et al., (1990) and Bonell and Balek (1993). It is now recognised that a spectrum of hydrological processes may be involved in catchment runoff response, the two end-members of which are Hortonian infiltration-excess overland flow and purely subsurface stormflow. In between lie the processes of saturation excess overland flow and various subsurface flow mechanisms dominated by either new (event) or old (pre-event) water (Ward, 1984). The above processes contribute runoff to the stream from only part of the catchment, this active area being referred to as the contributing area, which is observed to expand and contract dynamically over both storm and seasonal time scales (e.g. Dunne and Black, 1970). A geoclimatic distinction, between so-called "arid-area hydrology" and "humid-area hydrology" is often made in reference to runoff generation theories (Bonell and Balek, 1993). This dichotomy is drawn from the differential dominance of Hortonian processes in arid regions and subsurface and saturation excess runoff processes in humid regions. This division echoes the historical development of hydrologic theory, dominated by the American context.

Hydrologic theory suggests that hillslope form may exert a dominant control of the spatial distribution of the contributing area, in which convergent topography is likely to lead to zones of high soil moisture at the base of hillslopes, in the riparian area and in topographic hollows (Anderson and Burt, 1978). High antecedent moisture conditions in such locations result in a greater propensity for rapid subsurface and saturation-excess overland flow. In addition to the macro-scale topographic controls on runoff processes, localised variation in soil characteristics and land cover further influence the distribution of runoff production. Thus, in natural heterogeneous catchments, the spatial variation in hydrologic response is determined by a complex joint probability distribution of the specific topography, soils and land cover which characterise each locality of the catchment. The ensemble of this distribution, together with spatial variations in the rate of precipitation, condition the integrated catchment response.

Observations on the Bore Khola through 1992 revealed a pattern of rapid storm response resulting in a flashy monsoon dominated runoff regime. Such rapid time-to-concentration is indicative of infiltration-excess overland flow. Yet crude estimates of the minimum infiltration rate (using double-ring infiltrometers) were found to be between 60 and 500 mm h^{-1} on all but the most degraded land ($< 1\%$ of total land in the Bore). Rainfall rates in excess of these infiltration rates have rarely been observed even in the most severe storms, and then only for relatively brief periods (Boorman and Gardner, 1995). Analysis of 78 storm events on the Bore in 1992 shows that both the runoff coefficient and total event runoff are only weakly predicted by maximum rainfall intensity. Nonetheless, Hortonian overland flow may play an important role in runoff generation during the early monsoon, when the infiltration capacity of the baked soils may be reduced by surface crusting. Event analysis does however reveal a strong seasonal variation in the runoff coefficient, which peaks during the late monsoon, and thus appears indicative of a typical wetting cycle associated with humid area catchments. This suggests that early monsoon rains are used to recharge the large soil moisture deficit, which is generated by approximately 8 months of drought. The increase in the runoff coefficient over time may reflect an enhanced conductivity of the hillslope system as distal saturated zones become interconnected and form contiguous drainage pathways. This serves to enlarge the effective contributing area through the monsoon period and commensurately enhance the rate of runoff production.

THE DEVELOPMENT OF A CONCEPTUAL MODEL

Clearly the complex set of processes described above is not readily amenable to transcription into a formal computer model. Even if a model could be designed and a suitable mathematical theory for the solution of the model equations existed, much of the information required to initialise and parameterise the model remains unobservable at the scale required for the application. An appropriate conceptual model must, therefore, be aimed at a representation of the "grand-design" or macroscopic nature of the system (Beck et al., 1993).

The spatially variable nature of soils and topography has been highlighted as a dominant control on runoff delivery. A distributed approach to rainfall–runoff modelling is therefore perceived as paramount if the model is to be of any use in deconstructing the various mechanisms and roles of different hydrological phenomena. Recent advances in hydrological modelling have been facilitated by the availability of topographic data in the form of Digital Elevation Models (DEMs). Methods developed for the automated analysis of DEMs have enabled the quantification of various topographic attributes of a landscape (Moore et al., 1991). This analysis has resulted in the identification of a number of topographic indices that have been used to characterise the spatial and temporal distribution of soil moisture (Beven and Kirkby, 1979; O'Loughlin, 1986; Beven, 1986). The derivation of these indices and their incorporation into dynamic models (e.g. TOPMODEL, after Beven and Kirkby, 1979) relies upon the use of a series of simplifying assumptions. Central to this analysis is the presumption of steady-state conditions, in which subsurface conditions are taken to be at drainage equilibrium. The validity of this assumption is clearly dependent upon the time step used in the modelling exercise. However, over the short (half-hour or hourly) intervals often used, this presumption may be questionable. Barling et al. (1994) noted that water tables may actually be discontinuous and, therefore, that many points in a catchment may only receive water from a relatively small proportion of their total upslope contributing area. Beven et al. (1994) suggested that this problematic assumption may explain high values of saturated hydraulic conductivity sometimes used in TOPMODEL applications, as these may serve to compensate for overestimates of upslope contributing areas. A further assumption used in index-driven models is that of spatially uniform recharge rates to the water table. As Beven et al. (1994) note, strict adherence to this assumption precludes the use of spatially distributed infiltration-excess calculations based on soil moisture characteristics.

The model developed here is based on an explicit representation of the processes dependent upon soil–topographic combinations, but seeks to obviate a number of the restrictive assumptions used in index-driven models by explicitly routing flow between points in the catchment. This analysis also allows for spatially variable rates of recharge to the water table, and therefore permits the application of a distributed infiltration excess calculation. Similar approaches to fully distributed catchment modelling have recently been proposed by Wigmosta et al. (1993) and Brasington (1997).

The spatial discretisation used for the model application is based on the regular grid of the available DEM. There have been a recent flurry of investigations into the effects of DEM scale upon topographically driven hydrological models (Quinn et al., 1991, 1995b; Zhang and Montgomery, 1993; Bruneau et al., 1995; Brasington and Richards, 1998). These studies reveal that whilst simulated hydrograph response may be calibrated to compensate for variations in DEM scale, particularly coarse DEMs may distort the representation of flow pathways, and therefore the spatial predictions of soil moisture. Most DEM data are available at 50 m or less resolution and it has been suggested that this should be the limit for reasonable interpretations of distributed results. The model simulation results shown here are based on a 20 m DEM, derived from 1:5,000 contour data at 10 m elevation intervals.

Each "cell" or grid within the DEM is used to represent an individual "patch model" of the hydrological system. Attempts have been made, where possible, to keep the model parameters to a minimum, and make use of accepted hydrological theory in the description of model processes. Spatially uniform rainfall rates have been used here, taken as the arithmetic average of two gauges within the catchment. However, the fully distributed structure of the model provides the opportunity for the use of distributed rainfall fields if the data are available (see Brasington, 1997). A brief discussion of the model is given below, and a more comprehensive description is provided in El-Hames et al. (in prep.).

THE STRUCTURE OF A GRID ELEMENT

Each cell in the model represents an idealised vegetated soil profile, comprising an interception store, a topsoil root zone store, and a dynamic transition zone between the root zone store and the water table. Figure 19.1 depicts the various fluxes into and out from each cell.

(i) Interception store

Rainfall enters the idealised soil profile via the interception store, the capacity of which reflects the surface storage potential of the cell. Evaporation from this store

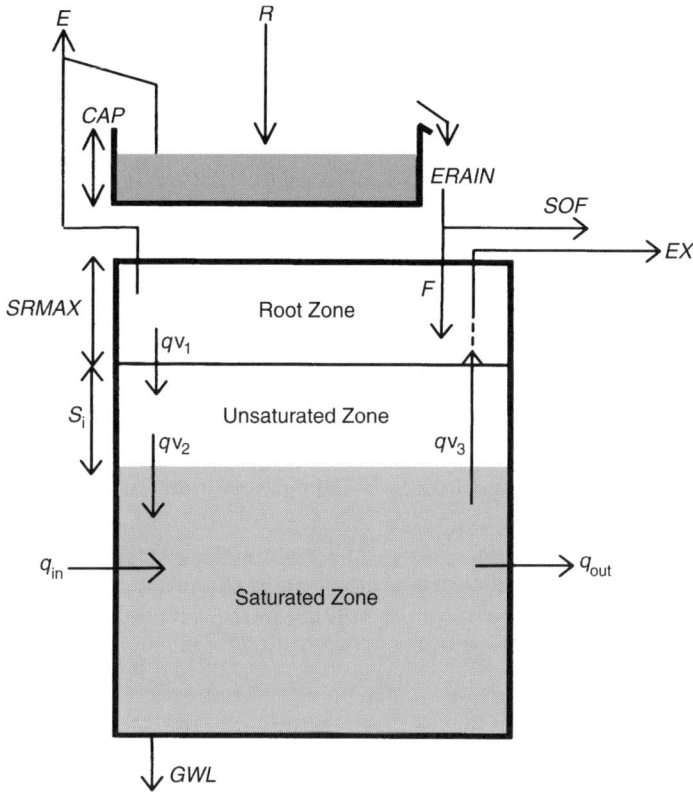

Figure 19.1   Patch model structure showing water fluxes for a grid cell. R is rainfall, E is evaporation, $ERAIN$ is interception excess, F is infiltration, SOF is saturation-excess overland flow, EX is exfiltration runoff. qv_1 is vertical flow from the root zone, qv_2 is vertical unsaturated recharge to the saturated zone, qv_3 is vertical upward water flux representing a rise in the water table, q_{in} is the saturated zone flux into the grid cell and q_{out} is the saturated zone flux out. GWL is groundwater seepage losses, CAP is the capacity of the interception store, $SRMAX$ is the capacity of the root zone store and S is the saturated zone deficit.

continues at the potential rate until the store is exhausted. Any residual evaporative potential may be used to deplete the unsaturated zone. Effective rainfall reaching the soil surface for infiltration is determined from surplus additions to the interception once capacity has been reached.

(ii) Unsaturated zone fluxes

The unsaturated zone comprises a fixed capacity routing store, SRMAX, and a dynamic transition zone between the base of the routing zone and the water table. Recharge to the routing store is determined from the effective rainfall rate, or the infiltration capacity, whichever is lower. Infiltration rates into this store are determined using Philip's equation (1957) applying the time condensation approximation (Milly, 1986), in which the rate is a function of cumulative infiltration volume rather than time. Thus, the infiltration rate, $f(x, y)$ is equal to:

$$f_t(x, y) = cK_{vs}(x, y)\left\{1 + \left[\left(1 + \frac{4cK_{vs}(x, y)I_t(x, y)}{S_t^2(x, y)}\right)^{1/2} - 1\right]\right\}^{-1} \tag{1}$$

where K_{vs} is the saturated vertical hydraulic conductivity, c is equal to 0.5, I_t is the cumulative infiltration depth, S is the sorptivity (after Sivapalan and Wood, 1986), and x, y refer to the grid-cell co-ordinates.

Recharge to the saturated zone is governed by the unsaturated vertical hydraulic conductivity K_v, which is estimated as a function of the soil moisture content and soil hydraulic properties, using the Brooks–Corey relationship, in which:

$$K_v = K_{vs}\left(\frac{\theta}{\phi}\right)^{(2/\lambda)+3} \tag{2}$$

where θ is the available soil moisture, ϕ is the porosity, and λ is the pore distribution index.

Evaporative losses from the unsaturated zone are dependent upon the residual evaporative potential after the interception store is exhausted. These losses are then limited according to the soil exfiltration volume which is determined as a function of cumulative evaporation and sorptivity, after Milly (1986).

(iii) Saturated zone fluxes

Soil moisture accounting in the saturated zone is central to the model. Saturated zone fluxes are calculated individually for each grid cell and local continuity equations apply. Each grid cell can exchange water with its eight neighbours, receiving flows from upslope and discharging downslope. In each cell transient conditions are approximated by a series of local steady-state solutions based on hydraulic gradients estimated from the ground surface slope. Saturated fluxes into and out of each cell are based on the kinematic equation for subsurface flow used by Beven and Kirkby (1979):

$$Q_s = T \tan \beta \exp\left(\frac{-S}{m}\right) \tag{3}$$

where Q_s is the subsurface flux per unit contour length ($m^2 t^{-1}$), T is the profile transmissivity ($m^2 t^{-1}$), $\tan \beta$ is the local gravity induced hydraulic gradient, S is the soil moisture deficit (m) (for which negative values are deficits), and m is a model parameter which defines the exponential relationship of T with depth.

Beven (1984) has demonstrated the exponential relationship between depth and transmissivity above a perched water table to be a reasonable assumption for a wide range of soils. Transmissivity of the profile can then be derived as a function of the hydraulic conductivity of the profile when saturated (K_0), and the recession parameter, m, so that:

$$T = K_0 m \tag{4}$$

It should be noted that here K_0 and K_{vs} are not necessarily equal, and may be differentially parameterised to account for anisotropy.

Flow into a cell occurs from all upslope directions (therefore, negative slopes with respect to the central cell) according to equation (5):

$$Q_{in}(x,y) = \sum_{i=1}^{g} T_i |\tan \beta_i| \exp\left(\frac{S_i}{m_i}\right) \tag{5}$$

where g is the number of upslope cells, and i refers to each of the upslope cells.

Subsurface flow out of the cell to all downslope neighbours is then determined by equation (6):

$$Q_{out}(x,y) = \sum_{i=1}^{\infty-g} T_{x,y} \tan \beta_{x,y} \exp\left(\frac{S_{x,y}}{m_{x,y}}\right) \tag{6}$$

A similar explicit finite-difference scheme for kinematic subsurface flow has been described by Wigmosta et al. (1993). An approximate stability criterion for this explicit solution requires that the predicted wave propagation per unit time remains smaller than the grid cell dimension (see Kirkby, 1997). At each time step the soil moisture deficit is updated, accounting for the summation of lateral flows into and out from each cell, recharge from the unsaturated zone, leakage to deep groundwater systems and surface exfiltration.

RUNOFF GENERATION

Runoff delivery mechanisms are determined explicitly within the model. Three delivery processes are simulated. Firstly, infiltration excess runoff, Q_{ex} which occurs when the effective rainfall rate exceeds the infiltration rate (equation (1)), so that:

$$\begin{aligned} &\text{if} \quad r(x,y) \geq f(x,y) \\ &\text{then} \quad Q_{ex}(x,y) = r(x,y) - f(x,y) \end{aligned} \tag{7}$$

Saturation excess runoff, Q_{se} is produced when the total available storage capacity of the soil profile is exceeded by the vertical recharge depth, so that:

$$\begin{aligned} &\text{if} \quad f(x,y) \geq S(x,y) \\ &\text{then} \quad Q_{se}(x,y) = f(x,y) - S(x,y) \end{aligned} \tag{8}$$

Subsurface contributions to streamflow, Q_s, are explicitly accounted for by the processes exfiltration. This occurs when incoming lateral flows exceed the residual storage deficit of the whole soil profile. The topographic control on flow processes described in the model result in this phenomenon occurring in areas of topographic convergence and at the base of slopes.

The total flow produced from each cell is computed as the summation of the component flows, so that:

$$Q(x, y) = Q_{ex}(x, y) + Q_{se}(x, y) + Q_s(x, y) \qquad (9)$$

FLOW ROUTING

Runoff generated in each cell is routed to catchment outlet using a spatially distributed convolution integral:

$$Q(t) = \int_A \int_0^t Q(x, y, \tau) h(x, y, t - \tau) \, \mathrm{d}\tau \, \mathrm{d}A \qquad (10)$$

where $Q(t)$ is the hydrograph of catchment area A, at time t, and $h(x, y, t)$ is a spatially distributed instantaneous response function. This response function is determined by the nature of the flowpath taken to the outlet. Here the unit hydrograph is identified by dividing the catchment into hillslope and channel areas, based on a digitised river and ravine network. Two parameters, V_h and V_c are used to describe time-averaged flow rates for hillslope and channel areas respectively, and the response function, $h(x, y, t)$ is defined as the Dirac delta function:

$$h(x, y, t) = \delta \left[\frac{l_{h_{x,y}}}{V_h} + \frac{l_{c_{x,y}}}{V_c} \right] \qquad (11)$$

where l_h and l_c are the length of the flowpath over hillslope and channel elements respectively. These are determined directly for each cell of the DEM by summing the distances moving cell-to-cell along a topographically defined path of steepest descent. It should be noted that here, the flow velocities V_h and V_c, and the path lengths l_h and l_c, are assumed to be time-invariant. This simplification clearly fails to account for transient effects of channel expansion and variation in flow velocity with depth (although this will be implicitly compensated by the tendency for exfiltration runoff in near- or in-channel cells). The calibrated flow velocities should therefore be regarded cautiously, and need not necessarily reflect actual observable flow rates, but rather the spatio-temporally averaged wave speeds which will be significantly biased in calibration by the spatial distribution of runoff generation.

Nonetheless, despite the obvious limitations, this two-component distributed unit hydrograph is a conceptual improvement over simple time–area unit hydrographs based on Euclidean flow lengths, which take no account of differential hillslope and channel conductivity. Furthermore, the method is easily parameterised and requires only two calibration parameters and a DEM. More complex approaches along similar lines include the methods of Maidment et al. (1996), which accounts for variable channel flow rates based on reach scale variation in hydraulic roughness, and Garrote and Bras (1995), in which velocities are a function of the discharge at time t_{-1}.

MODEL APPLICATION AND EVALUATION

THE STUDY AREA

Preliminary model testing has been based on an application in a sub-tropical region of the Nepal Middle Hills. The study watershed is the Bore Khola, a small ($c.\,4.2\,\text{km}^2$) sub-catchment of the larger Likhu Khola valley which is a headwater tributary of the Ganges river system. The regional situation of the study catchment, together with a 20 m resolution Digital Elevation Model and a land use map are shown in Figures 19.2 and 19.3. The climate in the region is dominated by the summer monsoon (July–September), in which approximately 95% of the annual 2,000–2,500 mm rainfall occurs. The relief in the catchment is high, with typical slope angles in the region of 25–35%. Maximum elevations in the catchment approach 2000 m, falling rapidly to a base at around 600 m a.s.l. Bedrock geology within the catchment is fairly uniform, dominated by metamorphic crystalline basement rocks of mica-schist. A deep saprolitic weathering profile has developed over much of the catchment, with total depths varying between 1 and 5 m largely influenced by the gradient. Wu and Thornes (1994) have shown that the hydraulic properties of the soil profile are fairly

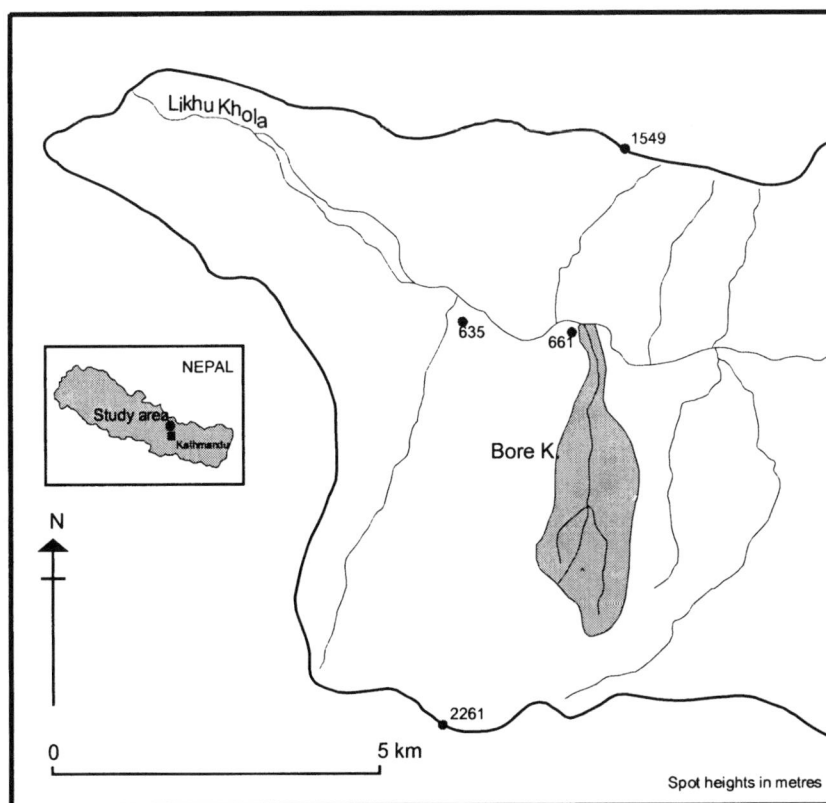

Figure 19.2   The location of Nepal and of the Likhu Khola valley.

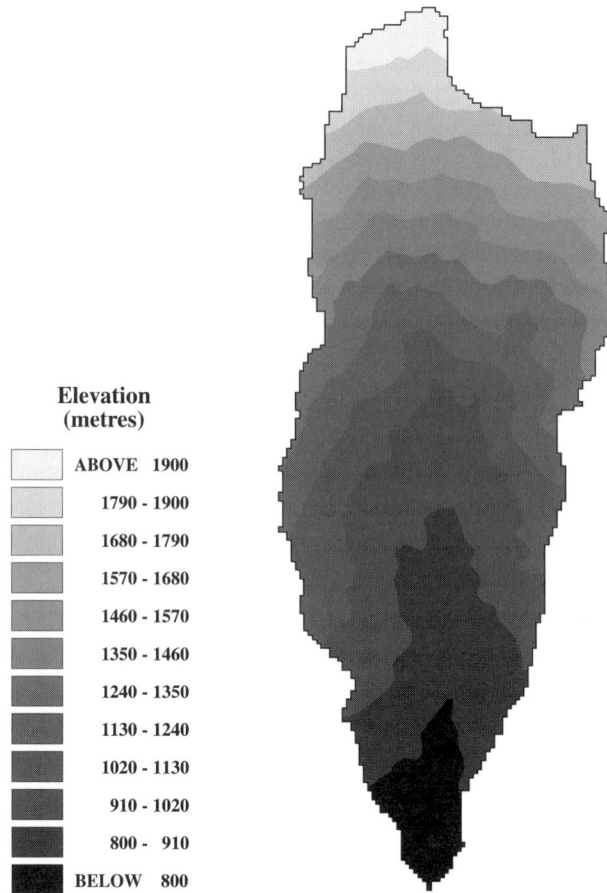

**Elevation
(metres)**

	ABOVE 1900
	1790 - 1900
	1680 - 1790
	1570 - 1680
	1460 - 1570
	1350 - 1460
	1240 - 1350
	1130 - 1240
	1020 - 1130
	910 - 1020
	800 - 910
	BELOW 800

Figure 19.3 (a) Digital Elevation Model of the Bore Khola sub-catchment (elevation scale in metres).

representative of a common saprolite, with high hydraulic conductivity near the surface which decreases rapidly with depth to a kaolinised horizon which acts as an impeding layer. Permeability though this layer is slow relative to the continual rate of recharge during the monsoon period, and this results in the perching of a surface water table, which acts as the hydrologically active zone during the monsoon period. Event analysis shows that runoff coefficients rise throughout the monsoon, reinforcing this conclusion. Slow recharge to deep groundwater through the impeding layer probably provides a major source of recharge for maintenance of the very low perennial flows ($1-5 \, \mathrm{l \, s^{-1}}$).

The catchment headwaters are dominated by second or perhaps third generation tropical hardwood forest (approximately 70% of the catchment), while the lower slopes are cultivated in a series of small (1–2 m in height and width) terraces, upon which a rotation of maize, wheat and rice crops are grown. The terracing presents a number of complications to the description and modelling of the catchment's

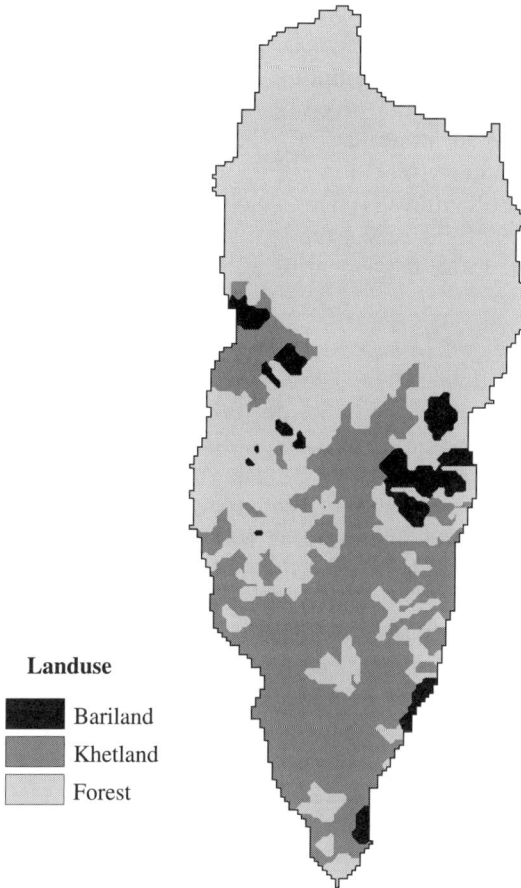

Figure 19.3(*continued*)   (b) The land-use distribution of the Bore Khola

hydrological system. Firstly, terracing effectively lowers slope angles, increasing surface areas, and therefore enhancing infiltration rates. Paddy terraces (locally known as *khet*) are bunded and irrigated during the rice growing season, which occurs twice yearly, between March–June, and, subsequently, July–September. The *khet* terraces therefore serve as semi-permanent runoff source areas which have been observed to overflow in near-channel areas during rainfall events.

DIGITAL TERRAIN MODELLING

A DEM for the Bore Khola was derived at a 20 m resolution by interpolation from digitised contours, mapped at 1:5000 scale with a 10 m interval. Surfaces were fitted using the tension-spline method of Hutchinson (1989). This global fitting procedure ensures the smoothest fit to the sampled contours taking account of specified breaklines such as rivers and ravines. A number of pit and dam features were identified during surface fitting. Most of these were related to near-channel areas and were deemed to be spurious, resulting from errors in the interpolation procedure. Any anomalies were therefore corrected to yield hydrologically continuous drainage surfaces. The choice of a 20 m sampling interval reflects an *a priori* compromise between data efficiency and the precision required to reflect the deeply dissected terrain. Higher resolution interpolation is limited by the fundamental information content of the input data (1:5000 mapping at 10 m intervals) and the need to smooth the small-scale terraced features (1–3 m) which are not accurately mapped. It should be noted that this smoothing reflects an assumption that the dominant topographic control on subsurface water movement operates at the hillslope scale.

EXPERIMENTAL METHOD

Two model tests were undertaken. The first was a "lumped" catchment parameterisation, where soil hydraulic properties are taken to be spatially uniform, and the second was a distributed parameterisation in which hydraulic properties are identified as a function of the dominant land use. This has been done to represent the difference between the high vertical hydraulic conductivity of the headwater forest areas and the low vertical hydraulic conductivities in the *khet*, which results from surface compaction due to rolling before planting, and the development of a hydraulic and fine sediment seal under the standing water. The model was applied with an hourly time step to a one-month (August) climatic sequence during the summer monsoon of 1992. The model was then evaluated on the basis of a comparison between the observed and simulated runoff, and also upon distributed maps of soil moisture deficit. No model validation exercise is shown here. Full details of model validation exercises are given in Brasington (1997).

RESULTS

Model parameters have been determined by manual calibration, based on a combination of visual comparison and the minimisation of the sum of squares as an objective function, where:

$$F = \sum_{i=1}^{n} (Qobs_i - Qpred_i)^2 \tag{12}$$

Whilst parameters have been manipulated to achieve the best fit between the predicted and observed data, care has been taken to ensure that physically based parameters remain within "realistic" field values.

In addition to the sum of squared errors, the model efficiency (Nash and Sutcliffe, 1970), which is related to the regression coefficient of determination, was also computed to examine model performance. Efficiency is defined as:

$$Eff = 100 \times \left[\frac{(F_0 - F)}{F_0} \right] \tag{13}$$

where

$$F_0 = \sum_{i}^{n} (Qobs_i - \bar{Q})^2 \tag{14}$$

In the case of error-free prediction, an efficiency of 100% is returned, whereas a negative efficiency indicates that the sum of the squared model residuals exceeds the variance of the streamflow record.

LUMPED PARAMETERIZATION

The results of the simulation are shown in Figure 19.4 and a summary is given in Table 19.1.

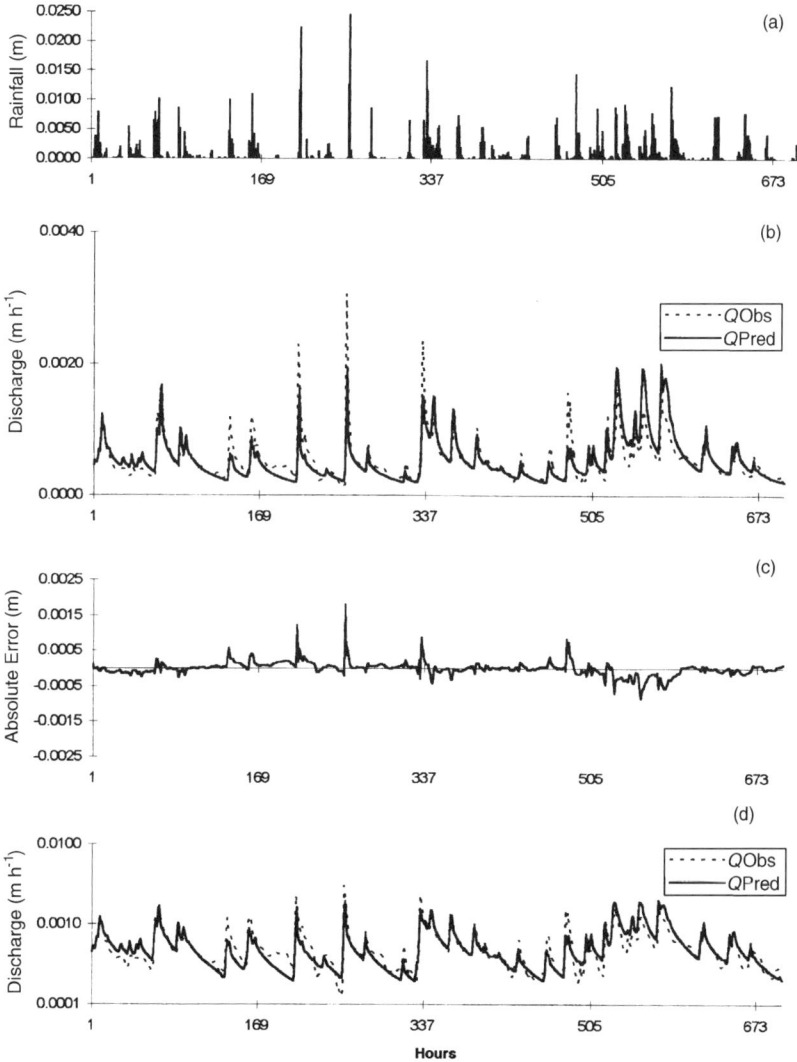

Figure 19.4   Lumped parameterisation results. (a) Rainfall. (b) Observed and predicted flows. (c) Absolute error series. (d) Log observed and predicted flows

Table 19.1   Simulation results from the lumped parameterisation

	Observed	Predicted
Sum rain (m)	0.662	
Sum discharge (m)	0.375	0.379
Sum actual *Et* (m)	–	0.079
Sum losses to deep groundwater (m)	–	0.234
Sum overland flow (m)	–	0.034
Sum subsurface flow (m)	–	0.345
Nash efficiency (%)	–	66.6

The model performs acceptably, simulating flow volumes and flood timing well. Two major sources of error are however evident in the simulation. Firstly, the model seriously underpredicts three storms during the middle of the month, these appear clearly on the absolute error time-series (Figure 19.4c) as distinct positive peaks. The second source of error results from a general over-prediction of streamflow during a series of consecutive storms towards the end of the month. Decomposition of the model predictions, as shown in Table 19.1, shows that subsurface flow dominates, with over 91% of the simulated flow delivered by this mechanism. Losses to groundwater are high, as would be expected, as it is this period which provides the bulk of recharge to deep groundwater that sustains perennial flows.

DISTRIBUTED PARAMETERISATION

Here parameters have been distributed on the basis of the land use pattern within the catchment. The land use distribution has been used rather than soil series data, which is less commonly available at fine resolutions. Empirical evidence from plot and infiltrometer experiments shows land use to be the dominant influence on soil characteristics. The use of this simplification in the parameterisation helps to reduce the total number of parameters and minimise computer memory requirements.

Vertical saturated hydraulic conductivities have been estimated from double-ring infiltrometer tests in the field. The remaining soil hydraulic parameters have been kept constant, with values taken from the secondary sources which, as Sivapalan and Wood (1986) pointed out, is a reasonable assumption, due to the overwhelming sensitivity of hydraulic conductivity estimates on unsaturated zone processes. Distributed interception storage capacities have been applied, again with values taken from secondary sources, although this could be estimated from remote sensed imagery, based on leaf area index (LAI) measurements (e.g. Herwitz, 1985). The distributed parameters and the values used in this simulation are shown in Table 19.2, in which a comparison is given with the values used in the lumped distribution.

Results from the distributed simulation are presented in Figure 19.5, and are summarised in Table 19.3. Again the results are acceptable, with Nash Efficiency (Nash and Sutcliffe, 1970) above 65%. The timing and volume of flows is reasonable, and the match to peak is better than the lumped simulation, although the recession limb is slightly more linear (see the log arithmetic plot of discharges, Figure 19.5d). The decomposition of flows shows a greater volume of runoff generated from

Table 19.2   Parameters used in the lumped and distributed parameterisations

	Lumped	Forest	*Khet*	Maize
Int. capacity (m)	0.002	0.007	0.003	0.004
Root zone capacity (m)	0.005	0.150	0.050	0.100
K_{vs} (m h^{-1})	0.410	0.640	0.008	0.120

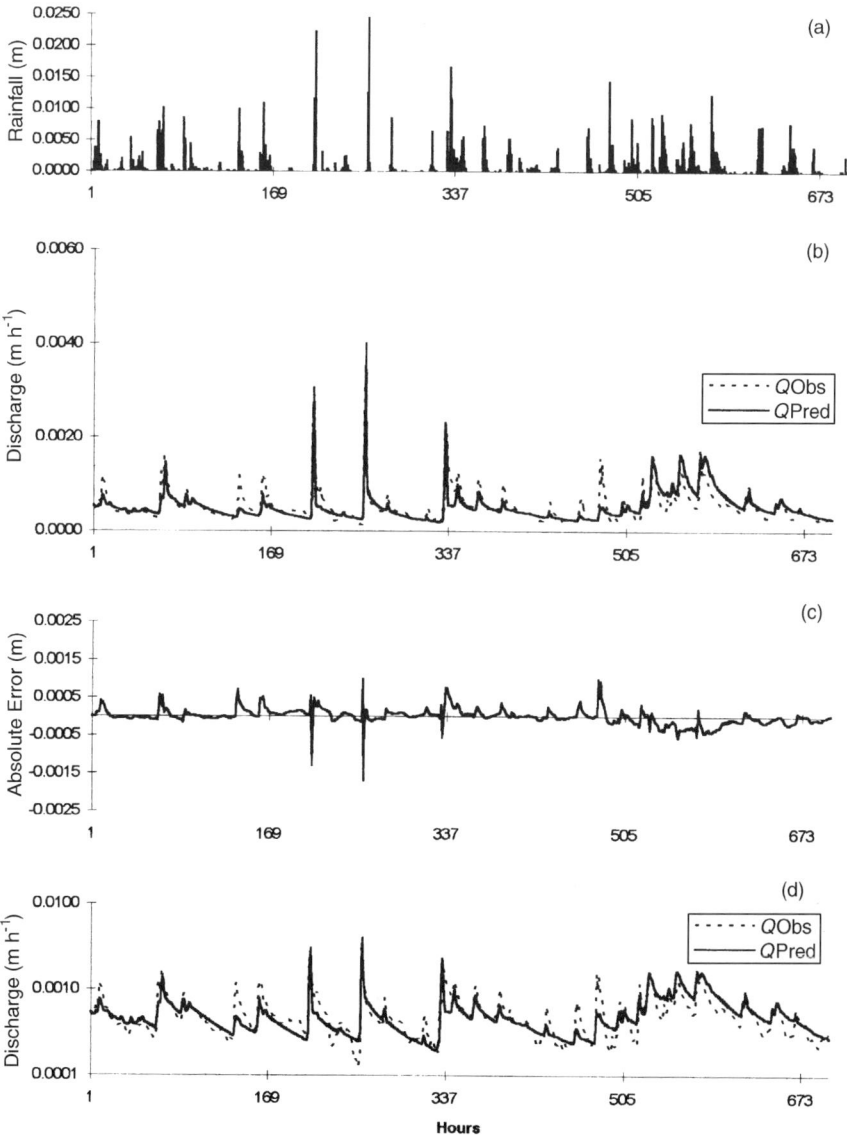

Figure 19.5   Distributed parameterisation results. (a) Rainfall. (b) Observed and predicted flows. (c) Absolute error series. (d) Log observed and predicted flows

Table 19.3   Simulation results from the distributed parameterisation

	Observed	Simulated
Sum rain (m)	0.662	
Sum discharge (m)	0.375	0.373
Sum actual Et (m)	–	0.093
Sum losses to deep groundwater (m)	–	0.220
Sum overland flow (m)	–	0.045
Sum subsurface flow (m)	–	0.328
Nash efficiency (%)	–	66.2

overland flow, comprising 12% of the total, compared with 9% in the lumped parameter simulation.

DISCUSSION

It is apparent that both parameterisations of the model are reasonable (although far from perfect) simulators of the system under these conditions. Close examination of the observed record shows a number of inconsistencies between rainfall and runoff, which may result from potential measurement errors. In any rainfall–runoff modelling exercise, the model can only be as good as the input data, and sources of error within this data inevitably manifest themselves in the simulation. In these exercises, the arithmetic average of two rainfall gauges has been used, and yet despite their relative close proximity (less than 1 km, and 300 m in elevation) regression between the hourly rainfall records from the two gauges yields a coefficient of determination of less than 30%. As noted above, the localised nature of rainfall may complicate streamflow response even in small catchments. Significant advances may be made, therefore, through the application of spatially variable rainfall fields estimated from a distributed rainfall gauge network.

Differences between the simulations, in terms of the timing and total volumes of streamflow response, are not strongly discernible, especially when analysed using simple statistical indices of fit. This is despite clear differences in the spatial parameterisation (Table 19.3).

This similarity of response from different parameter sets has been termed the problem of model equifinality by Beven and Binley (1992). Such equifinality is a key problem facing distributed hydrological modelling. It arises from the large number of degrees of freedom in a fully distributed model, so that various combinations of parameters may be used to satisfy a simple objective function, such as those used here (e.g. equation 12). Theoretically, one would expect a more complex distribution of variables which reflect the inherent spatial variability of the landscape to perform better than a simple lumped distribution, at least until errors in the input data become limiting. Following this principle does, however, lead us to question the problem of the identifiability of the model parameters, as the problem of parameter interdependence and compensatory effects may result in multiple solutions to the same problem. One of the main uses of distributed, quasi-physically based models, is their application in determining the hydrological consequences of changes in catchment conditions, such as land use change (see Introduction). Questionable parameter identifiability clearly poses considerable problems when the model is applied in non-stationary conditions. It is likely that in many cases apparently equally suitable parameter sets identified during calibration may produce a broad spectrum of responses when applied outside the range of the calibration conditions. This is especially pertinent where different model processes are invoked to produce similar outputs and where a wide variation in the spatial distribution of model processes is evident between parameterisations. Such problems have clear implications when distributed predictions (such as flow generation rates, soil moisture deficits and evaporation rates) are used to drive secondary models for the prediction of, for example, erosion/depositional processes and land surface–climate interaction modelling.

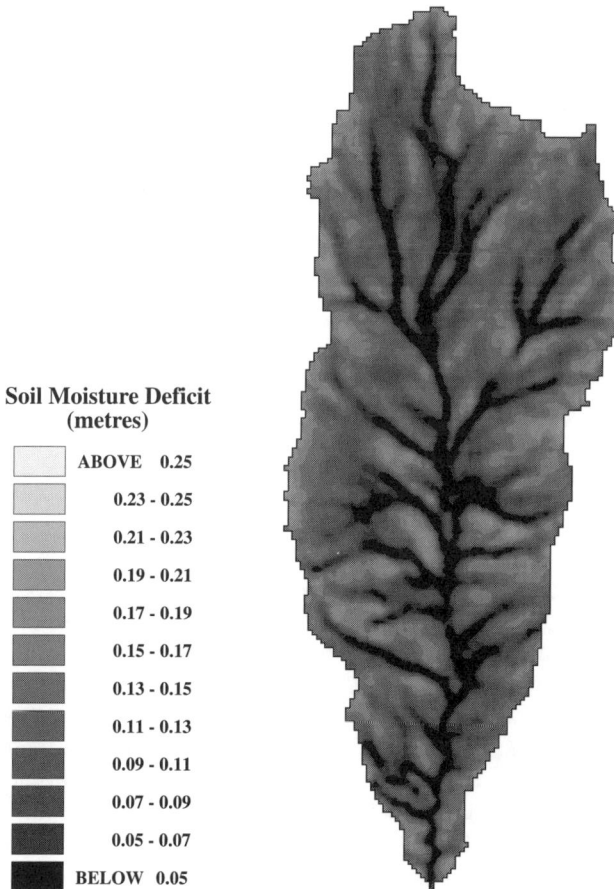

Soil Moisture Deficit (metres)

ABOVE 0.25
0.23 - 0.25
0.21 - 0.23
0.19 - 0.21
0.17 - 0.19
0.15 - 0.17
0.13 - 0.15
0.11 - 0.13
0.09 - 0.11
0.07 - 0.09
0.05 - 0.07
BELOW 0.05

Figure 19.6 Predicted soil moisture deficits for the wettest period of the simulation. (a) Lumped parameterisation.

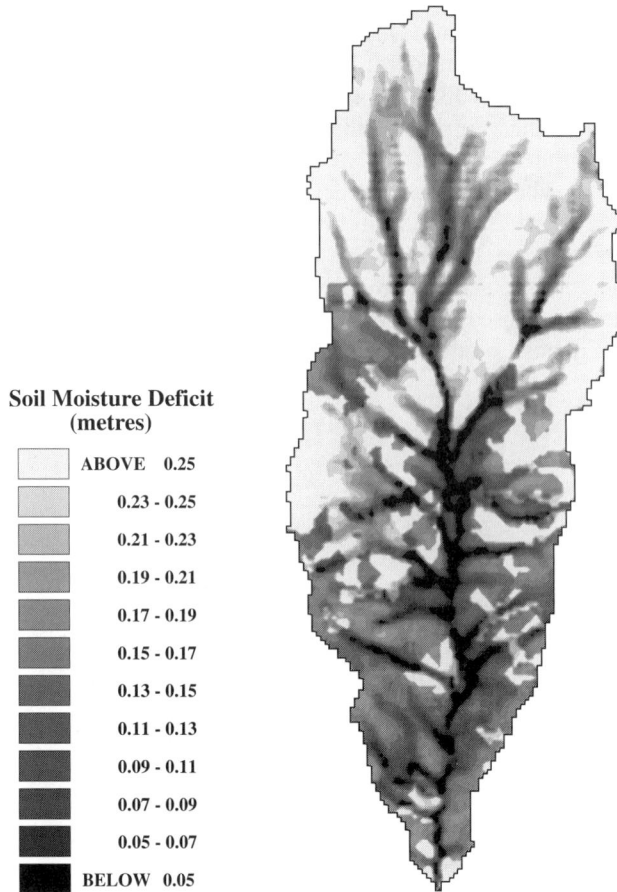

Soil Moisture Deficit (metres)

ABOVE 0.25
0.23 - 0.25
0.21 - 0.23
0.19 - 0.21
0.17 - 0.19
0.15 - 0.17
0.13 - 0.15
0.11 - 0.13
0.09 - 0.11
0.07 - 0.09
0.05 - 0.07
BELOW 0.05

Figure 19.6(*continued*)   (b) Distributed parameterisation

Whilst the two simulations described above reveal only minor differences on examination of the predicted streamflow response, the spatial predictions of the model, in terms of soil moisture deficit, do, however, show clear discrepancies. Figure 19.6 shows the predicted soil moisture deficits in the catchment during the wettest and driest periods of the record for the two parameterisations. In the case of the lumped parameterisation, the wettest soils are clearly related to the riparian areas of the catchment, highlighting the classic dentritic structure of the river network. High soil moisture deficits occur on the interfluves where hydraulic gradients are high and flow downslope is rapid. Examination of the distributed parameterisation, however, shows the dominant influence of the land use distribution. This creates a strongly polarised pattern of soil moisture deficits with dry forested headwaters and large areas of saturation extending into the *khet* terraces. The riverine structure of the catchment is still evident though the land use effects and the location of regularly saturated areas are clearly cross-correlated with this topographic effect. This pattern fits our perceptual model of the runoff system in the catchment. The vast majority of flows are thought to be subsurface-dominated, especially in this widely forested catchment. However, low infiltration rates under the *khet* terraces, coupled with the effects of topographic convergence of flows, have been witnessed to produce significant surface runoff generating areas close to the channels and ravines in the *khet* system. Flow generated here, in the lower reaches of the catchment, is quickly routed into a channel or gully system and therefore results in a rapid time of concentration. It has been suggested by some workers that *khet* systems may serve to attenuate peak flows, as drainage over the bund and the low gradients of the terrace should slow response. However, on steep slopes near channel systems these terraces have been observed to overflow, and high phosphate concentrations which could only be derived from the fertiliser inputs into the *khet* terraces have been observed in stormflow conditions (A. Jenkins, pers. comm.). The importance of the lower reaches of the catchment in controlling stormflow response is further illustrated though examination of the lower and upper rainfall gauge records and their comparison with the runoff series. A simple correlation shows that 50% of the runoff variance may be explained in terms of the lower gauge rainfall alone, compared to only 29% for the upper gauge.

CONCLUSIONS

The development of a distributed rainfall–runoff model has been described, and results from a simple test application within the Nepal Middle Hills presented. The computational model is based on a perceptual model of the hydrological system, and, where possible, processes within the model have been derived from accepted hydrological theory. Results from the test simulations show the model to perform acceptably, despite a complicated and potentially error-strewn rainfall–runoff record. Evaluation of the model with respect to streamflow prediction alone has identified the potential for equifinality of parameter sets, and a number of fears have been raised for the use of such models in the simulation of hydrological response under uncertainty. The importance of the nature of the model distributed predictions has been highlighted and it is contended that this type of study must become a future focus for distributed rainfall–runoff modelling. This is especially pertinent where distributed predictions are utilised as input data in secondary environmental models, when the possibility of compound errors becomes rife.

It is suggested that alternative data sources may aid the evaluation procedure. Distributed soil moisture and/or water-table data could be used to provide another source of information to assist the model validation procedure. These data may be used to confine the possible array of acceptable parameter sets to those which conform with the observed hydrological processes whilst remaining good rainfall–runoff filters. The use of "soft" data, as applied in this case, is also important. Here the modeller's qualitative impression of the catchment behaviour may be used to aid the choice of parameter sets when quantitative data are unavailable. This is particularly useful where comprehensive instrumentation of catchments is rare, as is common in areas throughout the tropics. It is believed that the use of new and more rigorous validation criteria may improve model parameterisations and eventually lead to the application of this and similar distributed models in scenario evaluation modelling.

ACKNOWLEDGEMENTS

This work was undertaken as part of the Royal Geographic Society/Institute of Hydrology Landuse Soil Conservation and Water Resource Management in the Nepal Middle Hills project, led jointly by Drs R. Gardener and A. Jenkins, and funded by the Overseas Development Adminstration and LandRover. JB was also in receipt of NERC studentship GT4/92/4/P. Professor Keith Beven is also thanked for his helpful comments upon the model.

REFERENCES

Anderson, M.G. and Burt, T.P. 1978. The role of topography in controlling throughflow generation. *Earth Surface Processes*, **3**, 331–344.

Anderson, M.G. and Burt, T.P. 1990. Subsurface Runoff. In *Process Studies in Hillslope Hydrology* (eds M.G. Anderson and T.P. Burt). Wiley, Chichester.

Barling, R.D., Moore, I.D. and Grayson, R.B. 1994. 'A quasi-dynamic wetness index for characterising zones of surface saturation and soil water content'. *Water Resources Research*, **30**, 1029–1044.

Beck, M.B., Jakeman, A.J. and McAleer, M.J. 1993. 'Construction and evaluation of environmental models'. In *Modelling Change in Environmental Systems* (eds A.J. Jakeman, M.B. Beck and M.J. McAleer), pp. 3–35. Wiley, Chichester.

Beven, K.J. 1984. 'Infiltration into a class of vertically non-uniform soils'. *Hydrological Sciences Journal*, **29**, 425–434.

Beven, K.J. 1986. 'Runoff production and flood frequency in catchments of order n: an alternative approach'. In *Scale Problems in Hydrology* (eds V.K. Gupta, I. Rodriguez-Iturbe and E.F. Wood), pp. 107–131. Reidel, Dordrecht.

Beven, K.J. 1991. 'Spatially distributed modelling: Conceptual approach to runoff prediction'. In *Recent Advances in the Modelling of Hydrologic Systems* (eds D.S. Bowes and P.E. O'Connell), pp. 373–387. Kluwer Academic, Dordrecht.

Beven, K.J. and Binley, A.M. 1992. 'The future of distributed models: model calibration and uncertainty prediction'. *Hydrological Processes*, **6**, 279–298.

Beven, K.J. and Kirkby, M.J. 1979. 'A physically based variable contributing area model of basin hydrology'. *Hydrological Sciences Bulletin*, **24**, 43–69.

Beven, K.J. et al. 1994. *TOPMODEL AND GRIDATB*. CRES Technical Report TR 110/94, Lancaster University, UK.

Beven, K.J., Lamb, R., Quinn, P.F., Romanoxvicz, R. and Freer, J. 1998. 'TOPMODEL'. In *Computer Models of Watershed Hydrology* (ed. V.P. Singh). Water Resource Publications (in press).

Blackie, J.R. and Eeles, C.W.O. 1985. 'Lumped catchment models'. In *Hydrological Forecasting* (eds M.G. Anderson and T.P. Burt), pp. 311–345. Wiley, Chichester.

Blaikie, P.M. 1985. '*The Political Economy of Soil Erosion in Developing Countries*'. Longman, Harlow.

Bonell, M. and Balek, J. 1993. 'Recent scientific developments and research needs in hydrological processes of the Humid Tropics'. In *Hydrology and Water Management in the Humid Tropics* (eds M. Bonell, M.M. Hufschmidt and J.S. Gladwell), pp. 167–260. UNESCO, CUP, Cambridge.

Boorman, D. and Gardner, R. 1995. Climate and hydrology. In Land use, soil conservation and water resource management in the Nepal Middle Hills. Unpublished report.

Brasington, J. 1997. Monitoring and modelling runoff response and sediment yield in heterogeneous highland catchments. Unpublished PhD Thesis, University of Cambridge.

Bruijnzeel, S.A. 1986. 'Environmental impacts of (de)forestation in the Humid Tropics: a watershed perspective'. *Wallaceana*, **46**, 3–13.

Bruijnzeel, S.A. 1989. (De)forestation and dry season flow in the tropics: a closer look. *Journal of Tropical Forest Science*, **1** 229–243.

Bruijnzeel, S.A. 1990. '*Hydrology of Moist Tropical Forests and Effects of Conversion: A State of Knowledge Review*'. UNESCO IHP, Humid Tropics Programme, Paris.

Bruneau, P., Gascuel-Odoux, C., Robin, P., Merot, Ph. and Beven, K.J. 1995. 'Sensitivity to space and time resolution of a hydrological model using digital elevation data'. *Hydrological Processes*, **9**, 69–81.

Cordova, J.R. and Rodriguez-Iturbe, I. 1983. Geomorphologic estimation of extreme flow probabilities. *Journal of Hydrology*, **65**, 159–173.

Dunne, T. and Black, R.D. 1970. 'Partial area contributions to storm runoff in a small New England watershed'. *Water Resources Research*, **6**, 1296–1311.

Eckholm, E. 1976. '*Losing Ground*'. Norton & Co, New York.

El-Hames, A., Brasington, J. and Richards, K.S. (in prep.). 'The development and evaluation of a fully distributed grid based hydrological model using digital elevation data'.

Gardner, R., Thapa, K. and Tripathi, R. 1995. 'Soil erosion from overland flow on hillslopes'. In Land use, soil conservation and water resource management in the Nepal Middle Hills. Unpublished report.

Garrotte, L. and Bras, R.L. 1995. 'A distributed model for real-time flood forecasting using digital elevation models'. *Journal of Hydrology*, **167**, 279–306.

Henderson-Sellers, A. and Gornitz, V. 1984. 'Possible climatic impacts of land cover transformations, with particular emphasis upon tropical deforestation'. *Climate Change*, **6**, 231–257.

Herwitz, S.R. 1985. 'Interception storage capacities of tropical rainforest canopy trees'. *Journal of Hydrology*, **77**, 237–252.

Hewlett, J.D. and Hibbert, A.R. 1967. 'Factors affecting the response of small watersheds to precipitation in humid areas'. In *Proceedings of the First International. Symposium on Forest Hydrology* (eds W.E. Sopper and H.W. Lull), pp. 275–290. Pergamon, Oxford.

Howes, S. and Anderson, M.G. 1988. 'Computer simulation in geomorphology'. In *Modelling Geomorphological Systems* (ed. M.G. Anderson), pp. 421–440. Unwin, London.

Hutchinson, M.F. 1989. A new procedure for gridding elevation data and stream line data with automatic removal of spurious pits. *Journal of Hydrology*, **106**, 211–232.

Ives, J.D. and Messerli, B. 1989. '*The Himalayan Dilemma: Reconciling Development and Conservation*'. Routledge. London.

Jenkins, A., Sloan, W.T. and Cosby, B.J. 1995. 'Stream chemistry in the middle hills and high mountains of the Himalayas, Nepal'. *Journal of Hydrology*, **166**, 61–79.

Kirkby, M.J. 1985. 'Hillslope hydrology'. In *Hydrological Forecasting* (eds M.G. Anderson and T.P. Burt), pp. 37–75. Wiley, Chichester.

Kirkby, M.J. 1997. 'TOPMODEL: a personal view'. *Hydrological Processes*, **11**, 1087–1097.

Klemes, V. 1993. The problems of the Humid Tropics – Opportunities for reassessment of hydrological methodology. In *Hydrology and Water Management in the Humid Tropics* (eds M. Bonell, M.M. Hufschmidt and J.S. Gladwell), pp. 45–52. UNESCO, CUP, Cambridge.

Lal, R. 1983. 'Soil erosion in the Humid Tropics with particular reference to agricultural land development and soil management'. In *Hydrology of Humid Tropical Regions with Particular Reference to the Hydrological Effects of Agriculture and Forestry Practice* (ed. R. Keller.) IASH Publ. 140, Wallingford.

Lal, R. 1987. '*Tropical Hydrology and Physical Edaphology*'. Wiley, Chichester.

Maidment, D.R., Olivera, J.F., Calver, A., Eatherall, A. and Fraczek, W. 1996. 'Unit hydrograph derived from a spatially distributed velocity field'. *Hydrological Processes*, **10**, 831–844.

Milly, P.C.D. 1986. 'An event-based simulation model of moisture and energy fluxes at a bare soil surface'. *Water Resources Research*, **22**, 1680–1692.

Moore, I.D. and Burch, G.J. 1986. 'Modelling erosion and deposition: topographic effects. *Transactions of the American Society of Agricultural Engineers*, **29**, 1624–1630.

Moore, I.D., Grayson, R.B. and Ladson, A.R. 1991. Digital Terrain Modelling I: review of hydrological, geomorphological and biological applications. *Hydrological Processes*, **5**, 3–30.

Myers, N. 1986. 'Environmental repercussions of deforestation in the Himalayas'. *Journal of World Forest Management*, **2**, 63–72.

Nash, J.E. and Sutcliffe, J.V. 1970. River flow forecasting through conceptual models. Part I – a discussion of principles. *Journal of Hydrology*, **83**, 307–335.

O'Loughlin, E.M. 1986. 'Prediction of surface saturation zones in natural catchments by topographic analysis'. *Water Resources Research*, **22**, 794–804.

Philip, J.R. 1957. 'The theory of infiltration'. *Soil Science*, **83, 84, 85**.

Quinn, P.F., Beven, K.J., Chevallier, P. and Planchon, O. 1991. 'The prediction of hillslope flow paths for distributed hydrological modelling using digital terrain models'. *Hydrological Processes*, **5**, 59–79.

Quinn, P.F., Beven, K.J. and Culf, A. 1995a. 'The introduction of macroscale hydrological complexity into land surface-atmosphere transfer models and the effect on planetary boundary layer development'. *Journal of Hydrology*, **166**, 421–444.

Quinn, P.F., Beven, K.J. and Lamb, R. 1995b. The $ln(a/tan\beta b)$ index: how to calculate it and how to use it within the TOPMODEL framework'. *Hydrological Processes*, **9**, 161–182.

Robson, A., Beven, K.J. and Neal, C. 1992. 'Towards identifying sources of subsurface flow: a comparison of components identified by a physically based runoff model and those determined by a chemical mixing model'. *Hydrological Processes*, **6**, 199–214.

Rundle, S.D., Jenkins, A. and Ormerod, S.J. 1993. 'Macroinvertebrates communities in streams in the Himalaya, Nepal'. *Freshwater Biology*, **30**, 169–180.

Sivapalan, M. and Wood, E.F. 1986. 'Spatial heterogeneity and scale in the infiltration response of catchments'. In *Scale Problems in Hydrology* (ed. V.K. Gupta), pp. 81–106. Reidel, Dordrecht.

Thompson, M., Warburton, M. and Hately, T. 1986. '*Uncertainty on a Himalayan Scale*'. Milton Ash, London.

Vertessy, R.A. and Wilson, C.J. 1992. 'Predicting erosion hazard areas using digital terrain analysis'. In *Research Needs and Applications to Reduce Erosion and Sedimentation in Tropical Steeplands*, pp. 298–308. IASH Publ. No. 192, Wallingford.

Ward, R.C. 1984. 'On the response to precipitation of headwater streams in humid areas'. *Journal of Hydrology*, **74**, 171–189.

Wheater, H.S., Jakeman, A.J. and Beven K.J. 1993. 'Progress and directions in rainfall–runoff modelling'. In *Modelling Change in Environmental Systems* (eds A.J. Jakeman, M.B. Beck and M.J. McAleer), pp. 101–132. Wiley, Chichester.

Wigmosta, M.S., Vail, L.W. and Lettenmaier, D.P. 1993. 'A distributed hydrology–vegetation model for complex terrain'. *Water Resources Research*, **30**, 1665–1679.

Wood, E.F., Sivapalan, M. and Beven, K.J. 1990. 'Similarity and scale in catchment storm response'. *Reviews of Geophysics*, **28**, 1–18.

Wu, K. and Thornes, J.B. 1994. 'Terrace irrigation of mountainous hill slopes in the Middle Hills of Nepal: stability or instability'. In *Water and the Quest for Sustainable Development in the Ganges Valley* (eds. G.P. Chapman and M. Thompson), pp. 41–63. Mansell, London.

Zhang, W. and Montgomery, D.R. 1993. Digital elevation model grid size, landscape representation and hydrological simulations. *Water Resources Research*, **30**, 1019–1028.

CHAPTER 20

Modelling Lake Level Changes: Examples from the Eastern Rift Valley, Kenya

MATTHEW STUTTARD
Remote Sensing Applications Consultants Ltd, Alton, UK

JULIAN B. HAYBALL
Earth Observation Sciences Ltd, Farnham, UK

GIOVANNI NARCISO
Institute for Soil, Climate and Water, Pretoria, South Africa

MAURO SUPPO
Aquater Eni Group, Pesaro, Italy

LUKA ISAVWA and AMBROSE ORODA
RCSSMRS, Nairobi, Kenya

INTRODUCTION

The work presented in this chapter concerns a methodology for monitoring water level changes on three lakes in the Kenyan Rift Valley. The primary objective of the project was to create a GIS database and use it to implement a hydrological model capable of accurately predicting lake level changes for lakes Naivasha, Nakuru and Elementeita. Vital considerations were that the method developed should be transferable to other lake basins, should use relatively simple data and should incorporate the effect of management practices. The results presented in this chapter focus on Lake Naivasha, but a full account of the project and the potential application of the results is contained in the final report (Stuttard et al., 1995).

Lakes Naivasha, Nakuru and Elementeita are the most elevated and centrally located members of a chain of lakes within the eastern branch of the Great African Rift Valley. The three lake catchments are enclosed by an area of 90 × 90 km, located between latitudes 0°15′ to 0°30′ S, and longitudes 36°0′ to 36°30′ E. This area is 70 km north-west of Nairobi and is confined by the Nyandarua (Aberdare) Mountains to the east (exceeding 3,960 m) and the Mau Escarpment to the west (exceeding 3,000 m).

The Sustainable Management of Tropical Catchments. Edited by David Harper and Tony Brown.
© 1998 John Wiley & Sons Ltd.

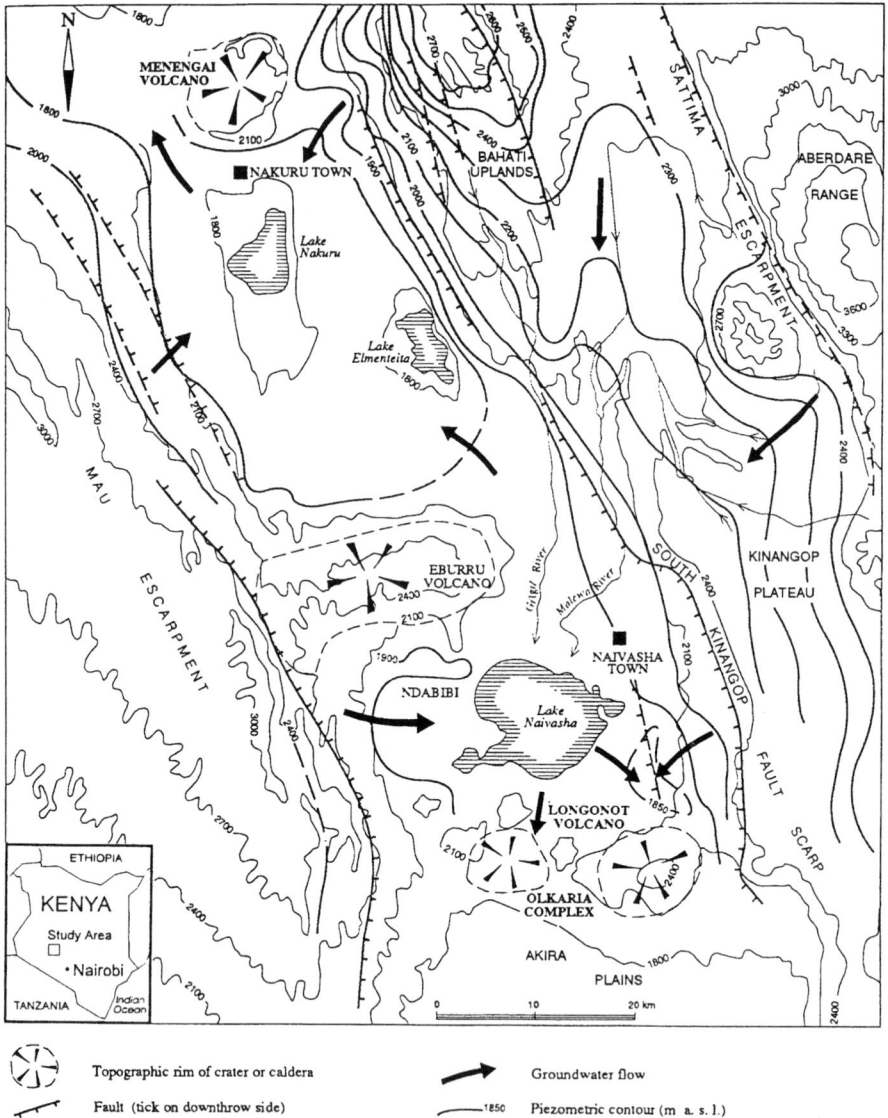

Figure 20.1   Location map of the project area

The valley width here is between 45 and 70 km. Figure 20.1 shows the general location of the lake catchments and includes both topographic and piezometric contours.

The lakes lie in the broad, level floor of the rift at an elevation around 1,800 m (1,200 to 2,000 m below the escarpments). Approximate surface areas of the lakes are: Naivasha, 133 km^2; Nakuru, 45 km^2; and Elementeita, 22 km^2. Nakuru and Elementeita are closed alkaline (soda) lakes, with no significant outflows. It has been suggested (Vareschi, 1982; Melack, 1988) that the levels of these two lakes are a direct response to the balance of rainfall and evaporation. The high soda content is a result of the contact of incoming water with the local volcanic rocks and subsequent concentration in the lake through evaporation. Between 1958 and 1974 the depth of Elementeita fluctuated between 0.3 m and 3.1 m.

Naivasha is unique in the Kenyan Rift Valley, as it is a freshwater lake. Though it has no surface outflow the lake water does not become alkaline, probably because of groundwater inputs and subsurface outflows (Ase, 1987). Naivasha's water level has always been subject to fluctuations. Regular measurements were started in 1908 and since then the lake level has varied by 9 m – the overall trend has been a reduction in level (Harper et al., 1990). Figure 20.2 shows lake level measurements covering the period used in the model simulation.

Rainfall in the area is highly variable, both spatially and temporally, as it is influenced not only by the tropical climate but also by the rain shadow of the surrounding highlands. At Naivasha and Elementeita, there is a discernible peak in April, whereas at Nakuru, there are lesser additional peaks in July/August and September/October. Naivasha experiences an average annual rainfall of 610 mm, and the wettest slopes of the Nyandarua (Aberdare) Mountains within the lake's catchment receive as much as 1,525 mm; the evaporation experienced by Naivasha is some 1,360 mm, so the runoff from the non-immediate catchment would seem to be broadly sufficient to sustain water level.

Wildlife, especially the huge flocks of flamingos on Elementeita and Nakuru, bring in significant tourist revenue, and the fresh water of Naivasha supports fisheries as well as high-value irrigated horticulture and agriculture; changes in lake level are thus intimately linked with both ecology and economy.

Whilst lake levels have fluctuated quite naturally since prehistoric times due to climatic change there is concern that recent trends are directly influenced by accelerating land use change and water demands in the area. In order to assess the potential impact of further environmental changes on lake level it is necessary to have

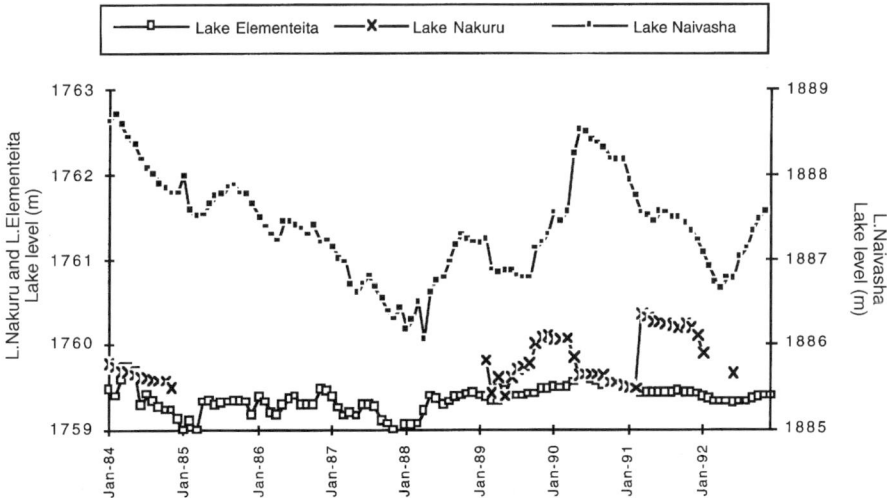

Figure 20.2   Measured lake level changes during the simulation period

a reliable model. The primary aim of the work was therefore to implement a hydrological model capable of accurately predicting lake level changes on a monthly basis, taking into account human influences. In order to calibrate and validate the model a time-series of data was assembled for 108 months between January 1985 and December 1993.

A key feature of the method developed in this project is the data model used to interface spatial datasets to the hydrological model. The model represents a series of entities, Land Reference Units (LRUs), with each one having a unique set of parameters defining the hydrology of the entity (e.g. area, slope, permeability). Additionally the dynamic data inputs: rainfall, number of rain days and actual evapotranspiration (AET) are represented by their contribution to the water flow within these LRUs on a monthly basis over the study period. This data model defines the outputs of spatial data integration and analysis, in terms of data requirements and agreed formats, as a series of tables which can be interfaced directly with the hydrological model.

The hydrological model simulates the total discharge to the lake on a monthly basis. In addition, by integrating river section boundaries (representing the catchments of hydrometric stations at which river discharge data were available) into the spatial data sets, the model is able to simulate the monthly discharge of individual rivers. This latter mode of operation was used in the validation phase of the work.

HYDROLOGICAL MODEL

As the desired output was the net monthly change in storage of each lake the hydrological model couples a catchment model with a reservoir water balance model. By using a volume/depth relationship it is then possible to calculate the predicted lake depth and compare this to actual lake level measurements in order to calibrate and validate the model. The model was developed mathematically and then implemented as a stand-alone programme on a PC.

In order to ensure that the model could be applied to a wide range of lake basins it was implemented in the form of a general mathematical scheme describing the hydrological processes acting within the lakes' basins (this is termed off-line general conceptualisation) and then a set of basin-specific switches were defined which related parameters in a basin to sub-sets of the process equations (on-line conceptualisation). The off-line conceptualisation is a general approach to the hydrology of lake basins, whereas the on-line conceptualisation is tailored to the particular topography, meteorology, hydrogeology, etc., of the lake basin under study.

The monthly discharge into the lake was calculated as the sum of contributing discharges from the LRUs within the lake basin. In other words the model is not 'distributed' because water is not routed from one LRU to the next within the catchment. The monthly discharge from each impermeable LRU is simply proportional to the volume of rainfall falling on that LRU. The monthly discharge of each land cover class is the sum of the discharges from all the (permeable) LRUs belonging to that class. In this scheme the land cover classes were rangeland, pasture and agriculture.

Following classical hydrological modelling, the monthly discharge from each LRU is the sum of three components: direct flow (surface runoff), hypodermic flow (throughflow) and groundwater flow (deep flow), equation (1). Each of these components is calculated from a series of conditional tests and equations which are illustrated in Figure 20.3 and set out formally below.

$$VBasin_t = VDF_t + VHyp_t + VGF_t \tag{1}$$

where $VBasin$ = total inflow into the lake, VDF_t, $VHyp_t$ and VGF_t are the total contributions from direct flow, hypodermic flow and ground flow respectively and the subscript represents the monthly time base.

Direct flow is passed to the lake within the same month but hypodermic and groundwater flow have a delay of some months depending on the memory of the system for each type of flow.

The direct flow regime within an LRU is one of three exclusive types: impermeable, Hortonian or Dunnian. Hortonian flow occurs on steeper slopes when rain falls faster than it can infiltrate the soil, whilst Dunnian flow occurs in low-lying swampy areas when the water storage capacity of the soil is exceeded. The formulae used to calculate direct flow are given in equations (2)–(4). These show the calculation of the direct flow contribution of a single LRU for a particular month. The total contribution to the lake from direct flow for that month is the sum of the contributions from all LRUs contained within the lake catchment.

$$vDF_t^I = Rain_t \times Coeff\_RXimp \times S \tag{2}$$

$$vDF_t^H = \left(\left(df1\_Co_j \times Slope \times \frac{Rain_t}{dd_t}\right) + df2\_Co_j\right) \times Rain_t \times S \tag{3}$$

if $Rain_t + W_{t-1} - AWC \leq 0$ then

$$vDF_t^D = 0 \text{ else} \tag{4}$$

$$vDF_t^D = \gamma \times (Rain_t + W_{t-1} - AWC) \times S$$

where vDF_t is the volume of direct flow from the LRU, with the superscripts I, H and D referring to the impermeable, Hortonian and Dunnian regimes respectively. $Rain_t$ is the amount of rain falling on the LRU, dd_t is the number of raindays in the month, $Slope$ is the average slope within the LRU, AWC the available water capacity of the soil class, W_{t-1} is the soil moisture content from the previous month and S is the area of the LRU. $Coeff\_RXimp$, $df1\_Co_j$, $df2\_Co_j$, and γ are coefficients obtained during calibration and in the Hortonian case are dependent on land use class as indicated by the index j.

The equations for hypodermic and ground flow are similar and both depend on the average soil moisture content of the LRU. For these equations a parameter is defined to be the threshold of soil moisture content above which hypodermic and ground flow begin to occur and is represented as a constant percentage of the available water capacity of the soil class. In order to make the calculation of hypodermic and ground flow, an estimate of the soil moisture available for the current month is required and is derived thus:

$$W_t^* = W_{t-1} + Rain_t - \nu DF_t - AET_t \tag{5}$$

where W_{t-1} is the soil moisture content from the previous month and AET_t is the actual evapotranspiration from that LRU. Note that in the Dunnian case, for the purposes of updating soil moisture, νDF_t is taken to be $\nu DF_t^D/\gamma$ to account for the fact that a portion of water released through soil saturation remains at the surface and does not become part of direct flow.

In the case of $W_t^* < \sigma$ contributions from both hypodermic and ground flow are zero. Otherwise:

$$\nu Hyp_t = Hyp\_Co_j \times (W_t^* - \sigma) \times S$$
$$\nu GF_t = \beta_d \times GF\_Co_j \times (W_t^* - \sigma) \times S \tag{6}$$

where Hyp_Co_j and GF_Co_j are calibration coefficients (dependent on land use class) and β_d represents the proportion of the ground flow from the LRU which is not lost between leaving the LRU and entering the lake.

As stated above, whilst direct flow is assumed to enter the lake instantaneously (within the context of a monthly time step) there is a time delay between hypodermic and ground flows leaving the LRUs and the water entering the lake. These time delays are characterised by transfer functions which apportion the flows derived from the current month between the discharges to the lake calculated for the succeeding months. In this way the calculation of the total discharge into the lake due to hypodermic and ground flows involves a summation of the contribution from all the LRUs over the current month and a number of preceding months.

Once the hypodermic and ground flows have been calculated the soil moisture content can be updated fully using equation (7) which is applied when $W_t^* > \sigma$ and hypodermic and ground flows have taken place.

$$W_t = W_t^* - \nu Hyp_t - \frac{VGF_t}{\beta_d} \tag{7}$$

Once all the contributions of discharge into the lake have been calculated this can be fed into the lake's reservoir sub-model so that the updated volume of the lake can be calculated, equation (8).

$$Vlake_t = Vlake_{t-1} + Vbasin_t + RLake_t + EVLake_t + Deep_t - Utilisations_t \tag{8}$$

where for month t, $Vlake$ is the volume of the lake, $Vbasin$ is the discharge into the lake, $RLake$ is the rainfall on the lake, $EVLake$ is the open water evaporation from the lake, $Deep$ is the groundwater outflow from the lake and $Utilisations$ is the extraction of water from the lake by the human population around the lake.

So far a general mathematical scheme (off-line concept) describing the hydrology of the basin has been presented (summarised in Figure 20.3). This has then to be applied to the specific cases of Lakes Naivasha, Nakuru and Elementeita; that is, an on-line

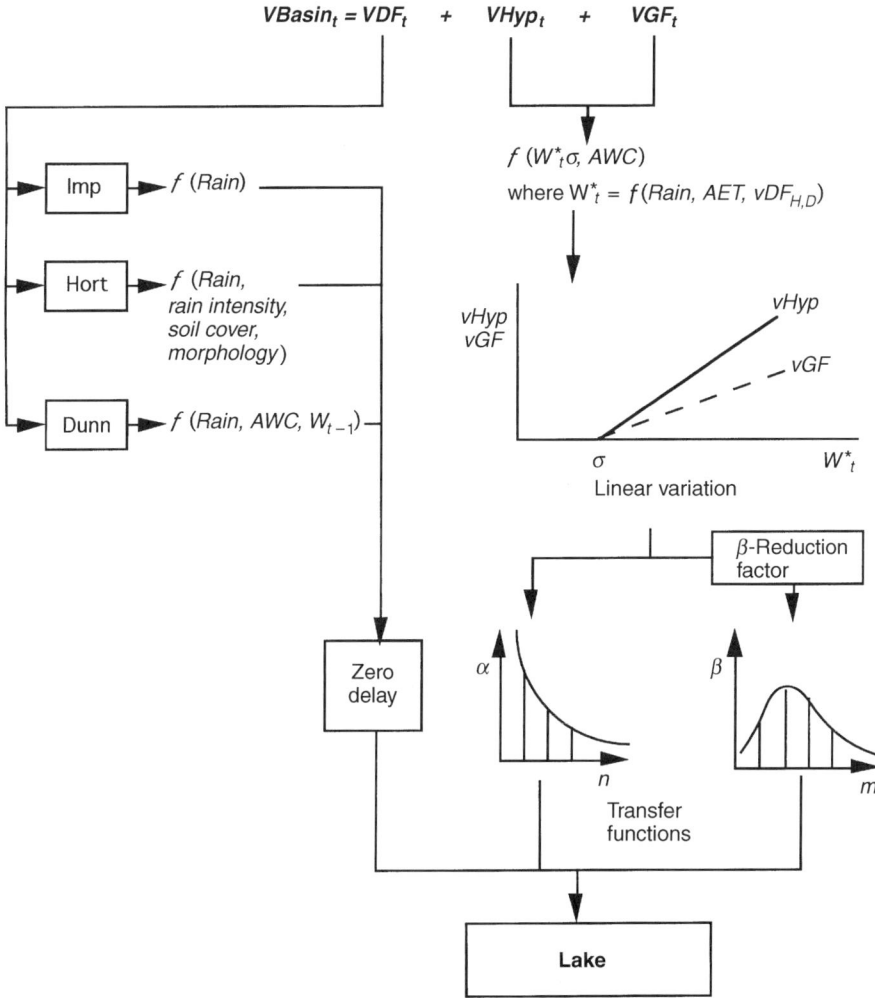

Figure 20.3   Schematic diagram of the hydrological model mathematical scheme

conceptualisation must be made. The results of this process can be summarised in a switch table which relates the geophysical parameters affecting the hydrology of a specific lake basin to the general concepts expressed above in mathematical form. The switch tables derived for the lakes studied in this project are shown in Table 20.1.

As an example, it can be seen in this table that if the land use class is 4, then the permeability switch is zero (impermeable) and the impermeable direct flow regime applies to that LRU.

In the specific case of the Rift Valley, altitude can be used to switch between the Hortonian and Dunnian direct flow regimes. This is due to the topography of the valley which has a flat floor (Dunnian regime) flanked by steep slopes (Hortonian regime) with the break between the two occurring at an altitude of about 2,100 m. In

Table 20.1 Model switch tables for Lake Naivasha (top) and Lakes Elementeita and Nakuru (bottom)

Land use	Perm	Alt	E/W Rift	W	R	W*	Aug peak	VDF	VHyp	VGF	Flow type
4	0	–	–	–	–	–	–	1	0	0	imperm.
1,2,3	1	0	–	0	–	0	–	1	0	0	Dunnian
1,2,3	1	0	–	0	–	1	–	1	1	0	Dunnian
1,2,3	1	0	–	1	0	0	–	0	0	0	Dunnian
1,2,3	1	0	–	1	0	1	–	0	1	0	Dunnian
1,2,3	1	0	–	1	1	0	–	1	0	0	Dunnian
1,2,3	1	0	–	1	1	1	–	1	1	0	Dunnian
1,2,3	1	1	0	–	–	0	–	1	0	0	Hortonian
1,2,3	1	1	0	–	–	1	0	1	0	0	Hortonian
1,2,3	1	1	0	–	–	1	1	1	1	0	Hortonian
1,2,3	1	1	1	–	–	0	–	1	0	0	Hortonian
1,2,3	1	1	1	–	–	1	0	1	0	1	Hortonian
1,2,3	1	1	1	–	–	1	1	1	1	1	Hortonian

Land use	Perm	Alt	E/W Rift	W	R	W*	Aug peak	VDF	VHyp	VGF	Flow type
4	0	–	–	–	–	–	–	1	0	0	imperm.
1,2,3	1	0	–	0	–	0	–	1	0	0	Dunnian
1,2,3	1	0	–	0	–	1	–	1	1	1	Dunnian
1,2,3	1	0	–	1	0	0	–	0	0	0	Dunnian
1,2,3	1	0	–	1	0	1	–	0	1	1	Dunnian
1,2,3	1	0	–	1	1	0	–	1	0	0	Dunnian
1,2,3	1	0	–	1	1	1	–	1	1	1	Dunnian
1,2,3	1	1	–	–	–	0	–	1	0	0	Hortonian
1,2,3	1	1	–	–	–	1	0	1	0	1	Hortonian
1,2,3	1	1	–	–	–	1	1	1	1	1	Hortonian

Land use = land use class (1, 2, 3, 4 = forest, pasture, agriculture and impermeable respectively)
Perm = permeability (0 = impermeable, 1 = permeable)
Alt = altitude (0 < 2100 m, 1 > 2100 m)
E/W = east or west Rift Valley slopes (0 = east, 1 = west)
W = soil moisture (0 ⇒ W < AWC, 1 ⇒ W > AWC)
R = rainfall effectiveness (0 ⇒ R < AWC − W, 1 ⇒ R > AWC − W)
W* = soil water content (0 ⇒ W* < σ, 1 ⇒ W* > σ)
Aug peak = August rainfall peak (1 = active, 0 = not active)
VDF = direct flow type (0 = not active, 1 = active)
VHyp = hypodermic flow type (0 = not active, 1 = active)
VGF = groundwater flow type (0 = not active, 1 = active)
Flow type = type of direct flow regime

this way an altitude switch was used to control the application of the different direct flow regimes. In an area of more varied topography, the switch might be based on average slope of the LRU instead of altitude.

The E/W switch is used in the Naivasha basin to differentiate the hydrogeological pattern. Studies have shown that the eastern rift (the Nyandarua Mountains) and the valley floor do not contribute groundwater to Lake Naivasha whereas the western rift (the Mau Escarpment and Eburru Mountain) does contribute (Geotermica Italiana,

1988). The E/W switch is therefore used to control which LRUs contribute ground-water flow to the lake.

Groundwater and hypodermic flows from each LRU depend on the water balance between incoming rainfall and losses from evapotranspiration or direct flow. From an analysis of rainfall/runoff relationships, it was seen that the amount of hypodermic flow depended on antecedent rainfall, thus a switch was introduced to reflect the rainfall regime of each LRU. The switch was based on whether or not the average annual rainfall had an August peak, as this indicates continuity of rainfall between the main spring and autumn rains. In this way geophysical and meteorological parameters were used to control the action of the model. It is clear that the construction of the switch table requires a detailed study of the lake basin and can incorporate the unique characteristics of the hydrological system to be modelled. What is also shown here is that spatial information in the form of process switches is critical for controlling the model and that the model requires spatial information which defines LRUs and populates them with data.

LINKING THE MODEL AND GIS

Once the model was stated mathematically it was possible to identify all those elements having a spatial expression. It was then possible to design the GIS data structure and its interface to the hydrological model. Some elements of the model are static (e.g. slope) and others are dynamic on a monthly time base (e.g. evapotran-spiration); this distinction has a significant effect on the data structure.

Figure 20.4 shows the flow of data between the GIS and the hydrological model. The model receives files from the GIS which are then processed to obtain the lake level

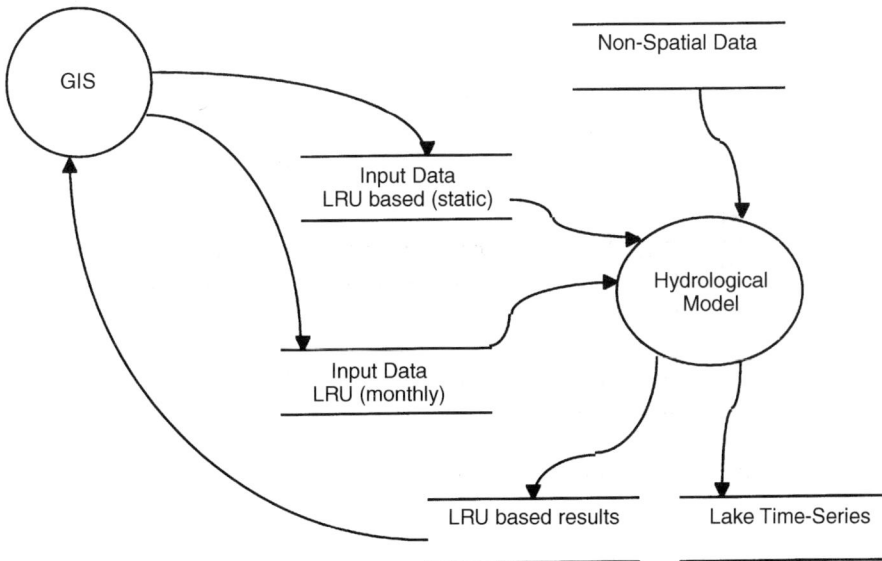

Figure 20.4   Data flow between the GIS and the hydrological model

at each month. The model can also output tables allowing maps to be made of monthly water flow in each LRU.

The LRU is the key to the interface between the information contained in the GIS and the hydrological model. It allows the model to treat each LRU as a discrete entity, with a single hydrological regime applied to a set of static and dynamic inputs. As the LRUs are hydrologically homogeneous they can therefore be represented as records in a table. Figures 20.5 and 20.6 show the table structures (as entity relationship diagrams) for the static and dynamic components respectively.

Referring to Figure 20.5, the LRU table holds attributes for most of the static components. These include the unique code for each LRU (lru_num), the LRU area, the lake catchment that the LRU falls in (basin_no), sub-catchment for river sections (sec_no), soil categories (soils_cl), Voronoi (Thiessen polygon) class for each rain gauge (voro_cl), detailed and simplified land use class (luse_det and luse), slope (average for the LRU) and switches for altitude (Dunnian or Hortonian regime), impermeable/permeable unit, August rainfall peak and lake/non-lake. The other static tables are linked in a one-to-many relationship with the LRU table. They contain data required by the model for each basin, soil class and basin/land use combination. By normalising the tables in this way, data redundancy is eliminated. For example many LRUs have the same soil class and therefore the same available water capacity (AWC), so the AWC values for each soil class are stored just once in the available water table. The basin/land use (bas_lu) table is not generated using the GIS, it contains calibration coefficients. The section table allows the model to compute discharge on gauged river sections (sec_no) as well as discharge to the lake, which is useful for calibration.

Referring to Figure 20.6, all the tables which contain dynamic variables (except Voro_raindays) are linked to the LRU table in a one-to-one relationship. The rainfall and AET (actual evapotranspiration) tables are input to the model and contained 108

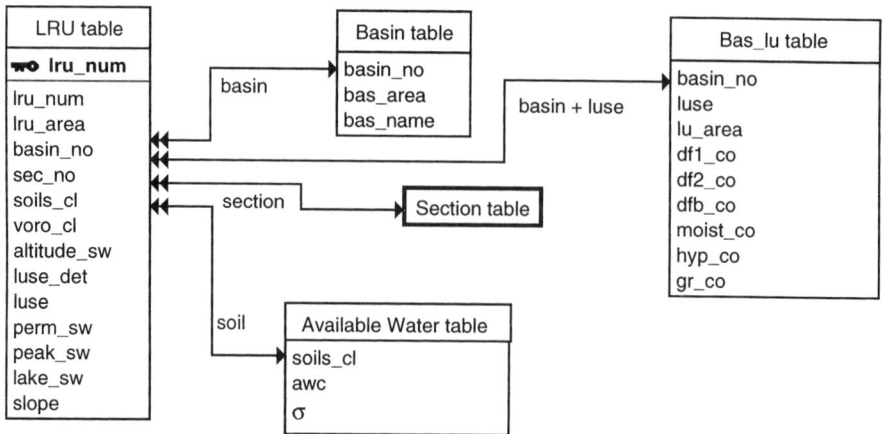

Relationship key: single arrow = one, double arrow = many.

Figure 20.5   Entity-relationship diagram of static data input to the model

LRU table

lru_num lru_num

LRU_Rainfall DF table

Voro_Raindays | lru_num | HY table
voro_cl | voro_cl | lru_num
Jan85 | | Jan85
Feb85 | | Feb85
Mar85 | | Mar85
..etc.. | | ..etc..
Nov93 | | Nov93
Dec93 | | Dec93

lru_num lru_num

LRU_AET GF table

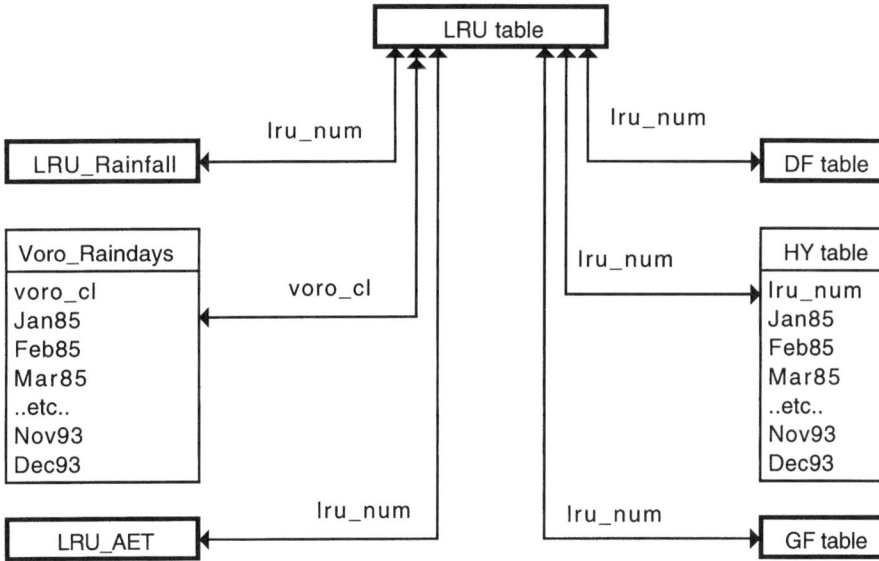

Figure 20.6 Entity-relationship diagram of dynamic data input/output between the model and the GIS

monthly values for each LRU. The raindays table is also a model input and it contains the number of raindays in each month for each rain gauge in the study area. As the Voronoi (Thiessen) polygon tessellation of the rain gauges is one of the LRU components it is unnecessary (and redundant) to store the raindays value for every LRU. The DF, HY and GF tables are output from the model and contain the predicted values for direct, hypodermic and groundwater flow contributions of every LRU for each month in the simulation period.

DATA COLLECTION AND INTEGRATION

The model is strongly influenced by the need to work with relatively simple data. Existing topographic maps of Kenya are good, so these were vector-digitised to obtain catchment boundaries, lake shoreline and contours. A digital elevation model was generated from the digitised contours and a slope map was derived from this. Available water capacity was estimated from the exploratory soil map of Kenya (Sombröek et al., 1982) using soil depth and texture.

Data for rainfall and raindays were obtained from a mixture of daily records on disk and monthly paper records. Spreadsheet software was ideal for integrating the data. Eventually, monthly datasets were assembled for 53 rain gauges over the 108-month period, but 14% of rainfall and 11% of raindays figures were missing. Missing data were reconstructed using a separate stepwise multiple regression model for each month. In each case the best combination was found of long-term averages, antecedent and subsequent monthly values, height, latitude and longitude. Interestingly, the regression approach was found to provide a better estimate at a missing data point than did spatial interpolation because of the high spatial variability of rainfall in the project area.

A land use map was derived using geometrically rectified satellite images (Landsat MSS (1976) and TM (1989)) covering the study area. Supervised classification was carried out with image processing software using ground data collected from field survey. A thematic map containing 10 spectrally distinct classes was produced. This was then simplified to five classes (Forest, Pasture, Agriculture, Urban, Water) by merging classes from the more detailed map. Both maps were easily transferred to the GIS from the image processing software.

The most problematic dataset to assemble was for the components of actual evapotranspiration (AET). AET can be calculated as open water evaporation multiplied by a transpiration coefficient. On a monthly time base, the transpiration coefficient is determined by energy balance, thus it was estimated from albedo (Bøgh and Søgaard, 1993). Open water evaporation was estimated for the rainfall stations and the transpiration coefficient was based on the albedo of different land use classes. Firstly, monthly evaporation was obtained from pan evaporimeter measurements available from four meteorological stations in or near the area. This dataset was spatially redistributed over 53 rain gauge sites using cluster analysis to assign each gauge to the most similar meteorological station. Secondly, albedo estimates were obtained on an average monthly basis using a time-series of images from the AVHRR sensor on the NOAA-11 satellite (Saunders, 1990) The albedo maps were combined with the land use map to obtain average monthly albedo per land use class. All these datasets were stored as a combination of maps and associated tables in the GIS.

SPATIAL ANALYSIS

Having constructed the dataset, a series of spatial analysis operations was used to create the set of tables for input to the model.

CREATION OF THE LRU MAP

The LRU map was made by overlaying separate maps of lake basin, river section, rain gauge site, Voronoi class (i.e. Thiessen polygon), land use class, altitude switch, soil class and lake identity. Special effort was required to ensure that common boundaries in the different maps were identical so that slivers (spurious new regions) were minimised. This overlay operation produced a map of unique combinations arising from the union of classes on the input maps and it was further processed so that each topologically distinct region had a unique identifier. In order to eliminate very small regions a 3 × 3 modal filter was applied resulting in a final set of 8310 LRUs.

ADDITION OF STATIC DATA TO THE LRU TABLE

Having created the map and calculated the area of each LRU, the LRU centroids were used to extract class codes from the source maps (lake basins, river section sub-catchments, soils, Voronoi, land use) and append them as new fields in the LRU table (see Figure 20.5). The rainfall peak switch was flagged as "on" if the average August rainfall in an LRU exceeded both the July and September figures by a factor greater than 1.2; this required only a simple operation on the table of long-term average rainfall. The permeable switch was flagged as "off" if the land use was urban. The slope in each LRU was obtained by calculating an area weighted average from the slope map.

CALCULATING RAINFALL FOR EACH LRU

To calculate rainfall in each LRU for all 108 months, each month of rainfall data was interpolated as a continuous surface from the rain gauge (point) data, then an area weighted average was computed for each LRU (Figure 20.7) using the GIS.

Following a comparative trial of interpolation accuracy and efficiency for four different methods of rainfall interpolation (Thiessen, linear, non-linear and distance decay), the linear interpolation method was selected as it provided a stable surface which passed through the available data points with good interpolation accuracy. The trial procedure examined accuracy at known data points by splitting the rain gauge dataset into active and control sets, interpolating on the active set and measuring accuracy on the control set.

CALCULATING RAINDAYS FOR EACH LRU

The nature of the raindays data is such that any form of weighted interpolation is not physically realistic because, in this area, there is little correlation between the number

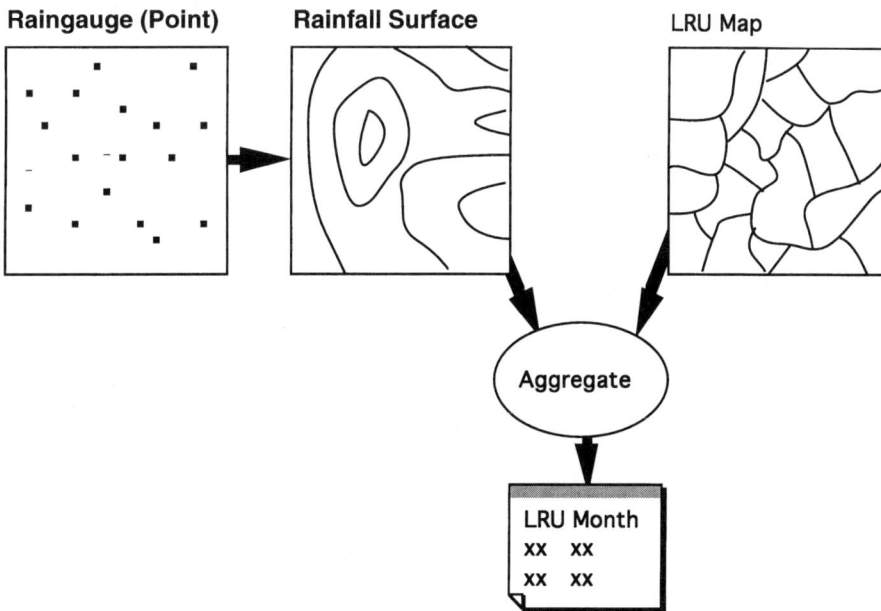

Figure 20.7   Rainfall aggregation on Land Reference Units (LRUs)

of raindays at any given site and the sites closest to it. Therefore a spatial estimation of raindays was obtained by using the Voronoi map derived from the locations of the rain gauge sites. All LRUs falling within a particular Voronoi polygon were assigned the same number of monthly raindays that occurred at the gauge located within the polygon.

As the Voronoi map was used as one input to the LRU map, each LRU fell entirely within just one Voronoi polygon. Hence the raindays value for each LRU was derived by looking up the raindays value for the matching rain gauge.

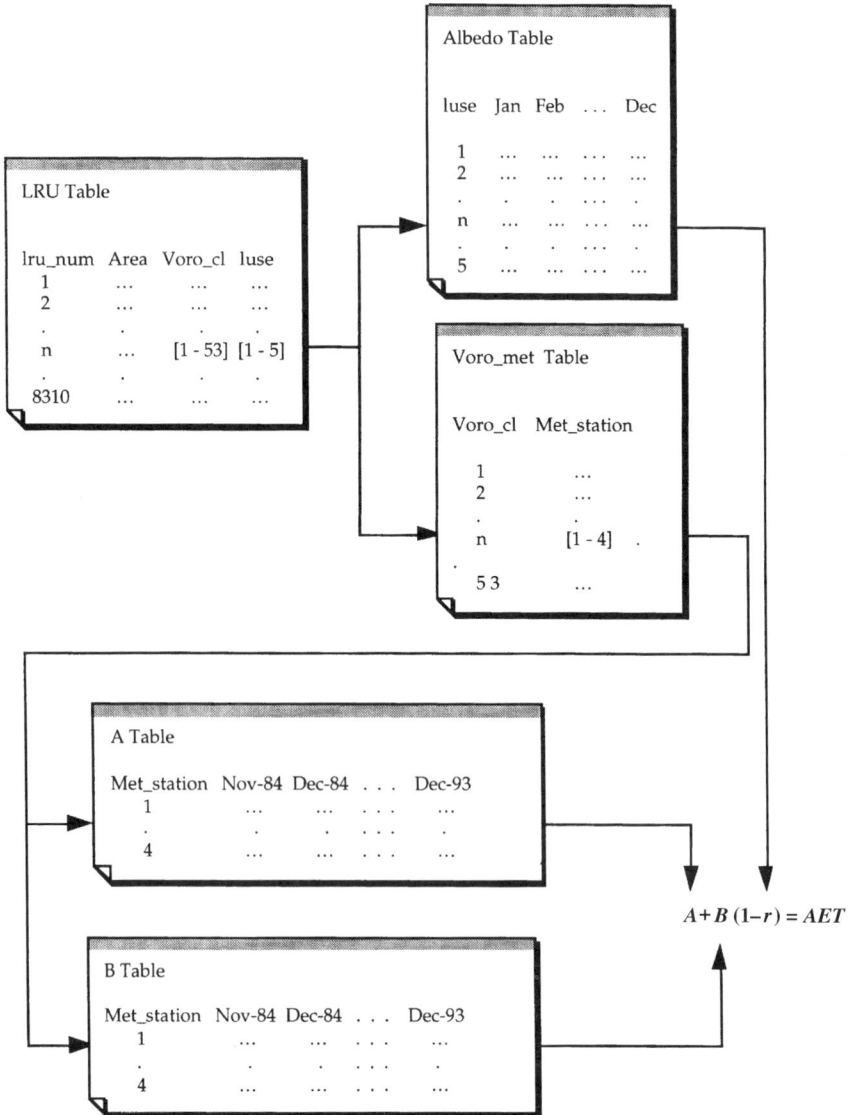

Figure 20.8   Calculation of actual evapotranspiration (*AET*) on each LRU by table look up

DETERMINING AET FOR EACH LRU

Evaporation from each LRU was obtained by a relational look-up table of monthly evaporation values for the meteorological station associated with the rain gauge site corresponding to the Voronoi class of the LRU. The albedo value was found by a look up on the table of albedo values for the month of the year and the land use class of the LRU. Using these values monthly *AET* was calculated for each LRU in the study area. The procedure is shown in Figure 20.8.

DETERMINING VOLUME/DEPTH RELATIONSHIP FOR LAKES

The GIS was also used to obtain the lake volume/depth relationship. A gridded depth model for each lake was created from bathymetric contours. Volume and surface area summations for different lake depths were carried out using the depth grid and a power function was fitted to the data by least squares.

CALIBRATION AND VALIDATION

For this phase of the project the data was split into two sets, one for calibration and one for validation. This process was hampered because of problems with the completeness and quality of the available lake level data (see Figure 20.2). So the exact time-series chosen for each lake basin and each analysis was made on an individual basis with regard to the data available for the basin. Indeed for Lake Nakuru the lake level data available was over such a restricted time period that a rigorous calibration and validation was not possible.

The calibration procedure was split into two phases; in the first, all the aspects relating to climatic, morphological and hydrogeological characteristics of the system were examined and probable ranges for the coefficients were established. Using these ranges as a basis, a trial and error procedure was adopted to assign exact values to the model's coefficients. The trial and error procedure involved varying a single coefficient within its assumed range while holding all other coefficients constant. The results of the simulations were then compared with the measured lake levels over the calibration period and various goodness-of-fit indicators were calculated. These indices were then used to establish the optimum value for the coefficient in question. The calibrated parameters that corresponded to the best fit are shown in Table 20.2. These gave a correlation of 0.95 and an explained variance of 87% between measured and modelled lake height in the 33-month calibration period.

In general the calibration was sensitive to variations in the parameters concerned with direct flow, but less so for parameters linked to hypodermic and ground flow, including the system delay parameters (transfer functions). The effect of varying initial values for soil water content was found to be very limited. For the choice of terms relating to storage volume balance of the lake (Coeff AET: pan evaporation coefficient, Deep: deep losses from the lake) reference was made to the available literature (Ase, 1987). It was found that minor variations in these parameters affected

Table 20.2   Best-fit model coefficients and deep loss terms, calibrated for Lake Naivasha

	Regime	Coefficient	Range	Imperme-able	Forest	Pasture	Agriculture
Direct Flow	Impermeable	Coeff_RX		0.2			
	Hortonian	DF1_CO			1.5×10^{-3}	1.5×10^{-3}	1.5×10^{-3}
		DF2_CO			0.01	0.02	0.03
	Dunnian	γ	0.7				
Hypo flow		HYP_CO			0.65	0.65	0.65
Ground flow		GF_CO			0.2	0.2	0.2
		β_d	0.4				
AET (lake)		COEFF_AET	0.9				
Deep losses		Deep $(10^6 \, \text{m}^3)$	3.75				
%AWC $= \sigma$		PERC_AWC	0.3				
Start Moisture		W_{t-1}	1				

the accumulation of errors over time. Importantly it was found that coefficients which depend on land use class had a significant effect on the results.

In the validation phase of the study an extensive analysis of the simulations was made, in order to characterise the operation of the model. Comparisons were made between the simulated results and the recorded values for the levels of all three lakes and also the discharge at five river gauging stations. Additionally an analysis of the sensitivity of the model to various input parameters was made. Perturbations to the rainfall, AET, AWC data and land use classes were made and the model re-run with the new data. The results were compared with the simulation using the unperturbed data and this indicated that the model behaved robustly over a range of inputs.

RESULTS

The results of the simulation are shown in Figure 20.9, along with the recorded lake level data. It can be seen that with a few notable exceptions monthly changes in lake level are modelled with a reasonable degree of accuracy. It will be noted, however, that the errors that are introduced in each estimation of lake volume change accumulate over time and hence the simulated and recorded data tend to diverge. Since the model is not intended to be a precise predictive tool but was designed to be an aid to water resource management in the catchment, the accumulation of errors is acceptable because, whilst it is not absolutely accurate, it is still physically realistic. The model can therefore be used to make comparisons between successive simulations under different catchment conditions.

In order to make a judgement on the quality of the output from the model a simple statistical hydrological model was implemented, based on one developed previously for Lake Naivasha (Ase et al., 1986). River discharge data and precipitation data aggregated over river sections was used to derive rainfall/runoff relationships allowing an estimate of the discharge into the lakes to be made from the catchment-wide rainfall levels. In this model all the discharge into the lake was assumed to originate from the rivers Malewa and Gilgil. Values for the precipitation over the lake and the

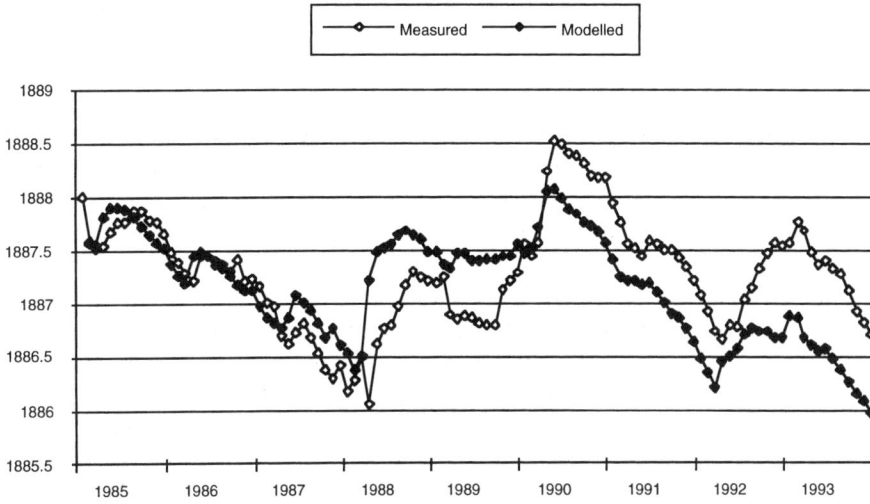

Figure 20.9   Results from the hydrological model

evaporation from the lake surface were taken from the same values used in the project model. Groundwater losses were estimated by assuming constant outflow and calibrating against the same time-series used for calibrating the project model. In Figure 20.10 the results of the two models along with the measured lake level data are shown.

These results indicate that although the statistical model can reproduce the general trend in lake level, the project model is superior in modelling the more detailed month-to-month changes. For example, the peak in lake level occurring in January 1993 is accurately estimated by the project model whereas the statistical model barely registers this event. This indicates that the project model more nearly represents the

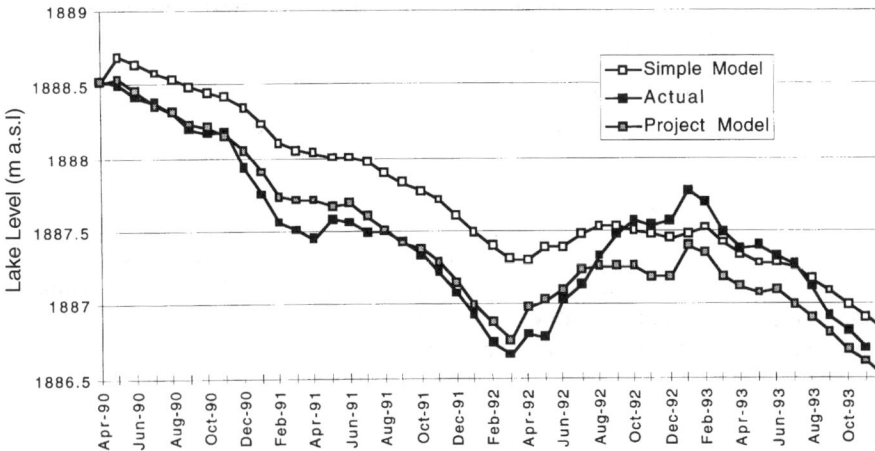

Figure 20.10   Comparison of level of Lake Naivasha predicted by the project model with the statistical model and with actual data over the validation period

real situation as regards the actual flow of water in the Naivasha basin. The statistical model also relies on the calibrated groundwater balance factor to cover systematic errors in the estimation of inputs and outputs to the lake water balance. Thus, from calibration, a figure of $-8.6 \times 10^6 \, \text{m}^3$, obtained for the groundwater balance factor in the statistical model probably takes into account systematic underestimation of evaporation from the lake and poor modelling of the discharge into the lake through the rivers. The equivalent groundwater calibration figure for the project model was $-3.75 \times 10^6 \, \text{m}^3$ which compares quite well to available estimates of actual groundwater loss.

Perhaps the main disadvantage of the statistical "black-box" model is that it is not possible to carry out scenario analysis. This is because "black-box" models are driven by purely empirical (statistical) relationships (in this case between rainfall and runoff) which cannot be used to evaluate the response of the basin system when changes are made to its state (e.g. changed land use or introduction of localised water abstraction). Whilst a simple statistical model is less costly (easier, quicker) to implement than an LRU-based model, it cannot be used to assess the impact of different management scenarios. Therefore it is of little value as a decision support tool for environmental managers and this is what is needed.

USING THE MODEL

Management decisions concerning land resources, especially water resources and land use, are widely assumed to have an impact on the overall discharge of the basin thus influencing, to a certain extent, the level of water in the lakes. The model that has been developed in this project can be applied to the study of these interactions. Land and water use practices can potentially affect all aspects of hydrological and ecological systems and an integrated use of GIS and modelling facilities can effectively simulate new "scenarios".

As an example of the use of the model as a tool for analysing the effects of water management policies in terms of lake level fluctuations, two hypothetical situations were considered with reference to a gauged section of the Malewa river:

- a water diversion with a variable monthly release rate; and
- a water reservoir with a given monthly release into the river.

FIRST SCENARIO – WATER DIVERSION

The amount of the applied diversion was computed by analysing the discharge diagram and applying a monthly 20% diversion reduced by $0.03 \times 10^6 \, \text{m}^3$ (roughly corresponding to the minimum flow in the analysed period) in order to guarantee a constant free water flow into the river section.

The resulting reduced river flow was applied to the whole of the Lake Naivasha system and the results are shown in Figure 20.11, where the modified value can be compared with the undisturbed situation. The overall result is a generalised reduction in lake level (a few centimetres) with respect to the undisturbed trend and no significant change in the average slope of that same trend.

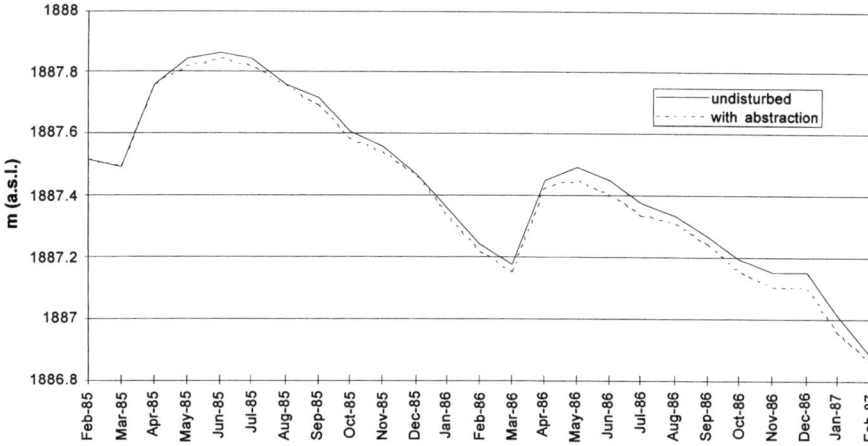

Figure 20.11   Predicted effect of water diversion upon Lake Naivasha level

SECOND SCENARIO – WATER RESERVOIR

As a second example a hypothetical reservoir was created, with a water storage capacity around $13-14 \times 10^6 \, m^3$ and an imposed release into the river of $1 \times 10^6 \, m^3$ for the months of January, February, March, June, July and December and of $2 \times 10^6 \, m^3$ for the remaining months. By means of this regulation the residual discharge into the river is lower than the natural one during the wet season but largely greater in the dry season. The overall effect of the reservoir is a relative reduction in lake level (within a range of 2 to 14 cm) which still follows the undisturbed pattern of fluctuation (Figure 20.12).

Based on these results it becomes possible to decide whether the calculated decrease of the water level is significant with reference, for example, to the natural environment or the economy of the surrounding agricultural areas.

CONCLUSIONS

The technical outputs of this project, comprising an environmental information system and a specially designed hydrological model structure, form a powerful management tool to support development of water and land management policies.

The best use of the model is to answer questions such as:

- what would have happened to the lake level if a specific water diversion had been implemented?
- what would have happened to the lake level if different land use practices had been introduced?

Whilst many other institutional and political factors tend to govern major decisions on watershed management, decision-makers can employ this type of information as

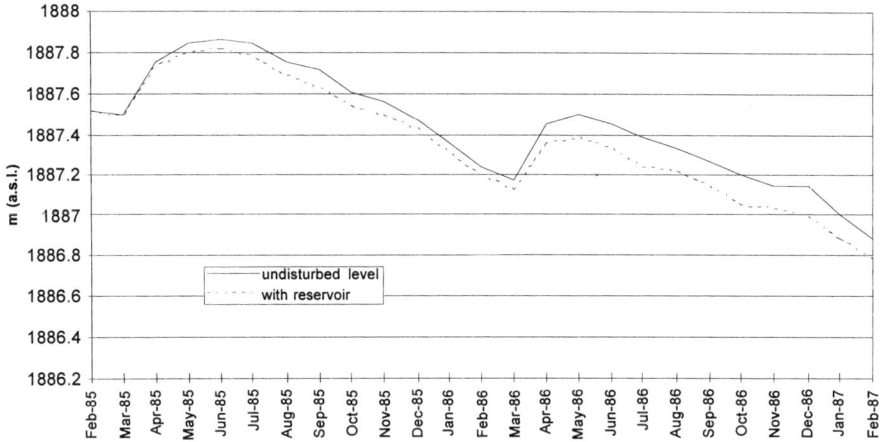

Figure 20.12   Predicted effect of reservoir construction upon Lake Naivasha level

an "objective" analysis element in the framework of sustainable socio-economic and environmental development.

As it stands, the model concept and structure is not suited to forecasting what the lake level will be in the future. This kind of problem requires a different approach mainly based on stochastic analysis but by adding further meteorological analysis (e.g. expanding from 9 to 50 years of historical data) it would also become possible to examine the likely effects of different rainfall patterns (e.g. phases of low rainfall with a given recurrence period or probable changes resulting from longer term climate trends).

The basic environmental data which has been integrated into a homogeneous and accessible form also has great potential for a wider range of applications in the project area. These include land use planning, wildlife and vegetation studies, and wetland conservation planning.

There is scope for further primary research and also for immediate application arising from this work. None of the components are new, in terms of science, but the project has demonstrated how an integrated approach embracing a relatively simple physical model, GIS and remote sensing technology can result in an effective decision support tool.

Increasing conflict over use of water resources is widely observed, notably on shared lakes in the tropics. Major opportunities therefore exist to apply this method to other lake basins in the East African region. This includes lakes both within the Rift Valley, such as lakes Baringo, Bogoria, Magadi, Natron and Turkana, and outside it, such as lakes Chala, Jipe, Amboseli and particularly Lake Victoria.

SOFTWARE NOTES

The software used for this project was as follows: SPANS for digitising and GIS analysis, ERDAS for image processing and SciStat for statistical analysis. Spreadsheets (Quattro Pro and Excel) were used for entering, reorganising and analysing tabular data. For speed and convenience the hydrological model was written in the 'C' programming language running on a PC and was designed to interface directly with SPANS tabular files. The hydrological model could be implemented in most programming languages or even a spreadsheet macro. Several GIS packages that would be suitable for the spatial work are now available on low-cost PCs; however, the GIS must be able to overlay multiple layers of gridded data.

ACKNOWLEDGEMENT

The work reported here was partially financed from the Commission of the European Communities, Directorate-General XII for Science, Research and Development; under the programme of Life Sciences and Technologies for Developing Countries (STD3).

REFERENCES

Ase, L.E. 1987. 'A note on the water budget of Lake Naivasha, Kenya – Especially the role of *Salvinia molesta* Mitch. and *Cyperus papyrus* L'. *Geografiska Annaler*, **69A**, 415–429.

Ase, L.E., Sernbo, K. and Syren, P. 1986. 'Studies of Lake Naivasha, Kenya, and its drainage area'. *Naturgeografiska Institutionen Stockholms Universitet, Forskningsrapport*, **63**, 1–75.

Bøgh, E. and Søgaard, H. 1993. 'Interaction between landforms, vegetation and waterbalance in the Sahel studied by use of GIS, satellite images and data from HAPEX-Sahel'. *Proceedings from the Seminar on Satellite Remote Sensing in Developing Countries*, pp. 73–90. Institute of Geography, University of Copenhagen.

Geotermica Italiana srl, 1988. '*Integrated Geophysical Survey in the Menengai and Suswa-Longonot Areas of the Kenya Rift Valley. Final Report 1 Synthesis of Results*', TCD CON 7/85-KEN 82/002. Ministry of Energy and Regional Development, Republic of Kenya, Nairobi.

Harper, D.M., Mavuti, K.M. and Muchiri, S.M. 1990. 'Ecology and management of Lake Naivasha, Kenya, in relation to climatic change, alien species' introductions, and agricultural development'. *Environmental Conservation*, **17**, 328–336.

Melack, J.M. 1988. 'Primary producer dynamics associated with evaporative concentration in a shallow, equatorial soda lake (Lake Elementeita, Kenya)'. *Hydrobiologia*, **158**, 1–14.

Saunders, R.W. 1990. 'The determination of broad surface albedo from AVHRR visible and near infrared radiances'. *International Journal of Remote Sensing*, **11**, 741–752.

Sombroek, W.G., Braun, H.M.H. and van der Pouw, B.J.A. 1982. '*Exploratory Soil Map and Agro-climatic Zone Map of Kenya 1980*'. Exploratory Soil Survey Report E1, Kenya Soil Survey, Nairobi.

Stuttard, M.J., Narciso, G., Oroda, A., Hayball, J.B., Suppo, M., Isavwa, L., Baraza, J., Carizi, G. and Catani, R. 1995. '*Monitoring Lakes in Kenya: An Environmental Analysis Methodology for Developing Countries – Final Report to European Commission*'. Contract No. TS3-CT92-0016, European Commission, Brussels.

Vareschi, E. 1982. 'The ecology of Lake Nakuru (Kenya). III. Abiotic factors and primary production'. *Oecologia*, **55**, 81–101.

CHAPTER 21

The Influence of Tropical Catchments upon the Coastal Zone: Modelling the Links between Groundwater and Mangrove Losses in Kenya, India/Bangladesh and Florida

JURGEN TACK

and

PHILIP POLK

Laboratory of Ecology and Systematics, Free University of Brussels, Belgium

INTRODUCTION

Mangrove forests are the only protection in the tropics against coastal erosion, are nursery grounds for a variety of marine animals and provide a focus for sedimentation of river-borne materials. Mangrove distribution in the tropics and sub-tropics is often linked with the presence of estuaries and creeks (Macnae, 1968; Barth, 1982; Blasco, 1991). There is a consensus that the brackish water microenvironment which is the key factor for the development of mangroves is caused by river discharges into the oceans (Snedaker, 1982). The influence of the catchment area on the mangrove area through the runoff of the river is clear. However, all over the world there are mangrove areas where no rivers or estuaries are in the immediate neighbourhood. This chapter suggests an explanation for those exceptions, making use of a mathematical groundwater flow model. This links the catchment area again with the mangrove area, but through groundwater rather than surface flow.

The use of mathematical models can be a powerful tool in predicting unknown variables, e.g. the impact of a changing groundwater flow pattern on the ecology. The model is able to explain the distribution of mangroves in three different areas (Kenya,

The Sustainable Management of Tropical Catchments. Edited by David Harper and Tony Brown.
© 1998 John Wiley & Sons Ltd.

Bangladesh and the Florida Everglades, USA). Modelling also shows its usefulness in predicting mangrove destruction in these areas whose groundwater flow pattern is not known, e.g. the impact of a changing groundwater flow pattern on the mangrove ecosystem. This study describes the groundwater flow of three different regions of the world by a mathematical groundwater model (Ituli, 1984; Dapaah-Siakwan, 1986).

MODEL DEVELOPMENT

Groundwater flow depends on change in piezometric heads (if all other factors are constant). This means that by solving the model equation for piezometric heads and by knowing their distribution in the study areas we are able to obtain an idea about the groundwater flow pattern. The model consists of a set of mathematical differential equations with their boundary and initial conditions. It is designed to simulate in one or two dimensions the response of a phreatic, semi-confined or confined aquifer to an imposed stress. The model allows homogeneous or heterogeneous aquifers with irregular boundaries. The model also allows constant or variable recharge (e.g. variable areal net precipitation), constant or variable discharge (e.g. variable total transpiration) and surface inflow or outflow. Horizontal flow is considered in the model. Because of the large surface of the study areas the aquifers vary in geological composition from place to place. Transmissivity varies with the geological composition of the aquifer.

To simulate groundwater flow through porous media the model considers Darcy's Law and the Law of Conservation of Mass. It considers a steady-state two-dimensional, horizontal flow through a non-homogeneous isotropic aquifer of thickness d. Including source and sink terms, respectively precipitation and withdrawal, the governing equation of groundwater flow becomes elliptic and is known as Poisson's equation.

To formulate the governing equation for groundwater flow the continuity equation is applied to an elemental volume within the aquifer:

$$\frac{\partial q_x}{\partial x} \Delta x (d\Delta y) + \frac{\partial q_y}{\partial y} \Delta y (d\Delta x) = R(x,y)\Delta x \Delta y - Q(x,y)\Delta x \Delta y \qquad (1)$$

where x and y are the Cartesian co-ordinates, q_x is the flux in the x-direction, q_y is the flux in the y-direction, d is the thickness of the aquifer, $R(x,y)$ is the recharge to the elemental volume, and $Q(x,y)$ is the surface outflow.

By application of Darcy's Law in the two directions of flow we can substitute the values of q_x and q_y into equation (1). Darcy's Law states:

$$q_x = -K\frac{\partial h}{\partial x} \qquad (2)$$

$$q_y = -K\frac{\partial h}{\partial y} \qquad (3)$$

where K is the hydraulic conductivity, and h is the piezometric head.

Equation (1), after substitution of the values of q_x and q_y, and after dividing it through $\Delta x \Delta y$ becomes:

$$\frac{\partial}{\partial x}\left(-Kd\,\frac{\partial h}{\partial x}\right) + \frac{\partial}{\partial y}\left(-Kd\,\frac{\partial h}{\partial y}\right) = R(x, y) - Q(x, y) \tag{4}$$

If the transmissivity is defined as $T = Kd$ for a homogeneous medium or

$$T = \int_0^d Kdz \tag{5}$$

for a non-homogeneous medium equation (4) can be written as:

$$\frac{\partial}{\partial x}\left(T\,\frac{\partial h}{\partial x}\right) + \frac{\partial}{\partial y}\left(T\,\frac{\partial h}{\partial y}\right) + R(x, y) - Q(x, y) = 0 \tag{6}$$

This is the Poisson equation which is used in the model. The regional aquifer is divided into zones with a hydraulic conductivity that can be considered as uniform. These zones also reflect the geological units in the areas under study.

The model considers two zones, a recharge zone and a discharge zone. In the recharge zone, if the piezometric head, h, is less than the topographic level, h_t, the surface outflow, $Q(x, y)$, is taken to be equal to zero: $Q(x, y) = 0$ for $h < h_t$. Equation (6) becomes:

$$\frac{\partial}{\partial x}\left(T\,\frac{\partial h}{\partial x}\right) + \frac{\partial}{\partial y}\left(T\,\frac{\partial h}{\partial y}\right) + R(x, y) = 0 \tag{7}$$

In the discharge zone, if the piezometric head, h, is equal to or greater than the topographic level, h_t, the surface outflow, $Q(x, y)$, is taken to be greater than zero. Here the piezometric head is taken to be equal to the topographic level: $Q(x, y) > 0$ for $h = h_t$. Equation (6) then becomes:

$$\frac{\partial}{\partial x}\left(T\,\frac{\partial h}{\partial x}\right) + \frac{\partial}{\partial y}\left(T\,\frac{\partial h}{\partial y}\right) + Q(x, y) = 0 \tag{8}$$

If surface outflows occurs, the model considers that the water flows downwards to the adjoining areas with lower topographic levels. The water is there added to the normal recharge input in these areas.

The flow equation now becomes:

$$\frac{\partial}{\partial x}\left(T\,\frac{\partial h}{\partial x}\right) + \frac{\partial}{\partial y}\left(T\,\frac{\partial h}{\partial y}\right) + R(x, y) - Q(x, y) + Q_s(x, y) = 0 \tag{9}$$

where Q_s is the surface inflow from the neighbouring areas.

Excessive withdrawals from wells can have adverse effects on the groundwater storage in the aquifer. This effect can be studied when the well is located at a point,

(x, y). If the well is pumped at a rate of $Q_w(x, y)$ and this withdrawal is incorporated in the flow equation it becomes the following:

$$\frac{\partial}{\partial x}\left(T\frac{\partial h}{\partial x}\right) + \frac{\partial}{\partial y}\left(T\frac{\partial h}{\partial y}\right) + R(x, y) - Q_w(x, y) - Q(x, y) + Q_s(x, y) = 0 \quad (10)$$

where Q_w is the withdrawal from the well.

In order to solve a mathematical model the boundary and the initial conditions of the model have to be specified. A groundwater flow domain can be defined by several types of boundary conditions. In the regions studied two boundary conditions were used, potential and no-flow boundary conditions.

In potential boundary conditions the piezometric heads are known: $h = h^*(x, y)$ where h^* is a known piezometric head for all points along the boundary. In the model this type of boundary represents that part of the aquifer where the piezometric head would not change with time. In natural conditions, such boundary conditions occur as recharge boundaries or areas beyond the influence of hydraulic stresses and are defined by known equi-potential lines.

No-flow boundary conditions are defined by a line across which no flow is occurring, thus:

$$\frac{\partial h}{\partial x} = 0 \quad (11)$$

or

$$\frac{\partial h}{\partial y} = 0 \quad (12)$$

This means that in the x- or y-direction perpendicular to the boundary, no flow is occurring. This kind of boundary can be defined in nature by two situations: the existence of an outcropping of impervious rock or water divide.

The groundwater flow equation is in a steady-state condition so no initial condition is required. However, the topographic levels are required to calculate the piezometric heads in the study area. This enables the study area to be divided into discharge and recharge zones.

A number of assumptions were made in the model used:

- Darcy's Law is valid and piezometric head gradient is the only significant driving mechanism for the groundwater flow.
- Horizontal groundwater flow prevails in the aquifer system.
- Gradients of the density of water, viscosity and temperature do not affect the velocity distribution.
- The aquifer is isotropic.
- No chemical reactions occur that affect the water properties or the aquifer properties.
- Vertical variations in head and concentration are negligible.

The central finite-difference approximation method was used to solve the partial differential equation describing the groundwater flow. The model equations describing the regional groundwater flow were solved by a computer program written in FORTRAN IV and originally created for the simulation of regional groundwater flow with solute transport in the lower Athi–Tana Basin, Kenya (Dapaah-Siakwan, 1986). The programme was adapted for use in the areas studied. The basin characteristics which serve as inputs for the computer to solve the model equations were the transmissivity values, T, the areal net precipitation, R, the topographic levels, h_t, the aquifer thickness, d, and the porosity of the aquifer material, n.

The transmissivity values were obtained by the product of hydraulic conductivity and the thickness of the aquifer, $T = KD$, where K is the hydraulic conductivity and d is the thickness of the aquifer. The flow domains were divided into several zones having different transmissivity values. Those zones correspond to the geological units distinguished in the flow domains. Similarly the flow domains were divided into zones of equal transmissivity and of equal net precipitation.

The areal net precipitation was estimated from:

$$R = P - E_t \tag{13}$$

where R is the net precipitation which may be available for infiltration, P is the mean annual precipitation, and E_t is the mean annual evapotranspiration.

The topographic levels were obtained by averaging the topographic levels on each element of a 33×38 grid system imposed on a topographic map of the study areas. The square grids measured respectively 10×10 km for the Athi–Tana river basin, 12.5×12.5 km for the Hugli–Ganges basin and 7.9×7.9 km for Southern Florida. A thickness of 80 m and a porosity of 0.4 were assumed for the aquifer.

All the calculated parameters were transformed into graphical output.

MODEL OUTPUTS

KENYA COAST

Figure 21.1 shows the mangrove areas along the Kenyan coast. Most of the mangrove areas are in the vicinity of one or more rivers. However, the rivers north of Mombasa are perennial while the rivers south of Mombasa are seasonal, drying up after the rainy season. There are also a few mangrove areas where no rivers are in the immediate neighbourhood.

Figure 21.2 shows the graphical output of the model. To give a clearer view of what is happening the groundwater flow is only considered along the coastal line (Figure 21.3), which shows those areas along the Kenyan coast with groundwater flow higher than $1 \, \text{m}^2 \, \text{day}^{-1}$. High discharge of fresh groundwater into the sea supports the assertion that, close to the shores in some points, seepage of freshwater occurs (Isaac and Isaac, 1968; Knutzen and Jasuund, 1979; Ruwa and Polk, 1986). Figures 21.1 and 21.3 show a very clear linkage between groundwater flow and the distribution of the mangroves along the coast; in all cases mangroves are growing in areas with a high groundwater flow. The majority of coastal towns and villages are also found in those areas, suggesting that the Swahili population detected the correlation between mangroves and freshwater many decades ago. The only village outside an area with

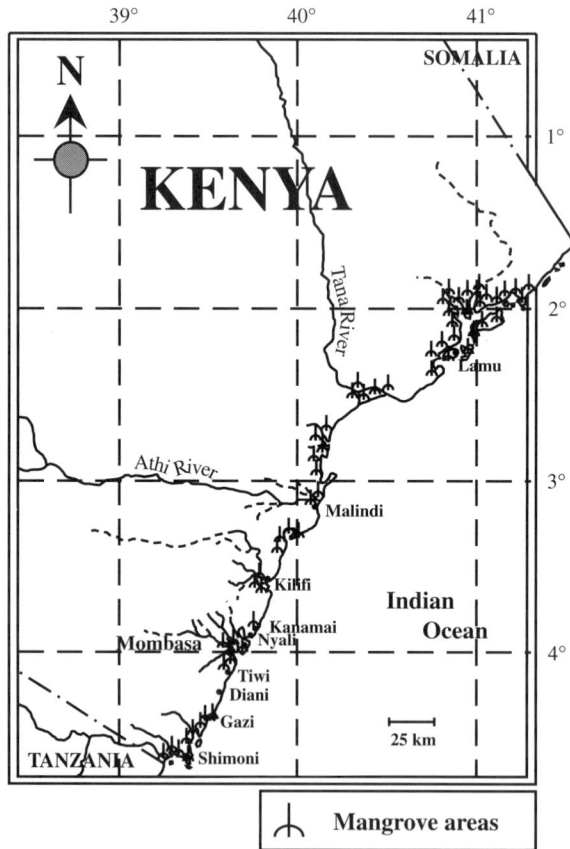

Figure 21.1   Distribution of mangrove areas along the Kenyan coast

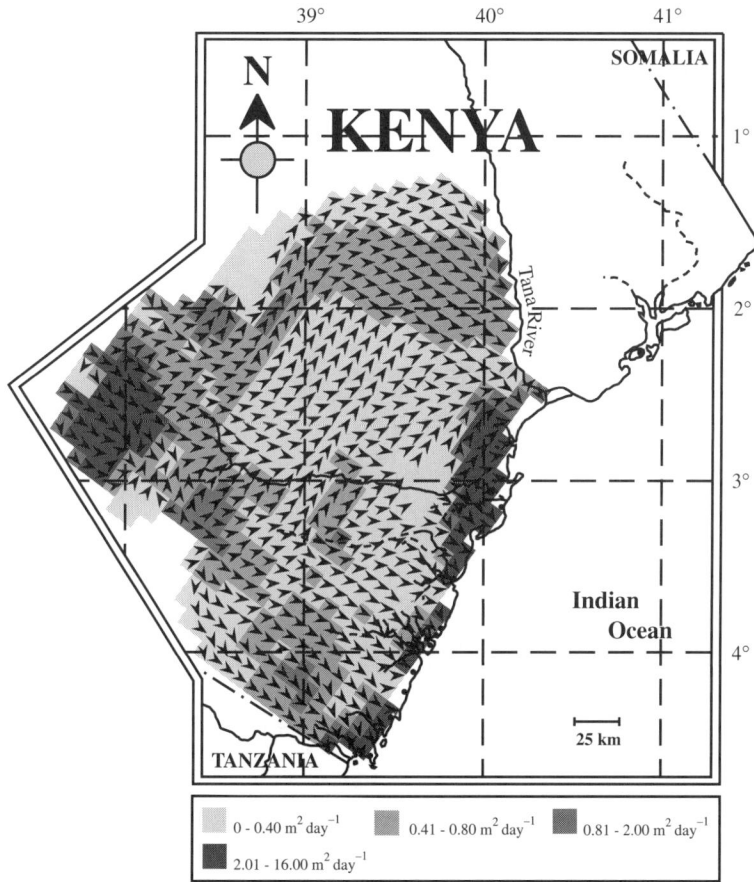

Figure 21.2　Groundwater flow in Kenya: graphical output of the model

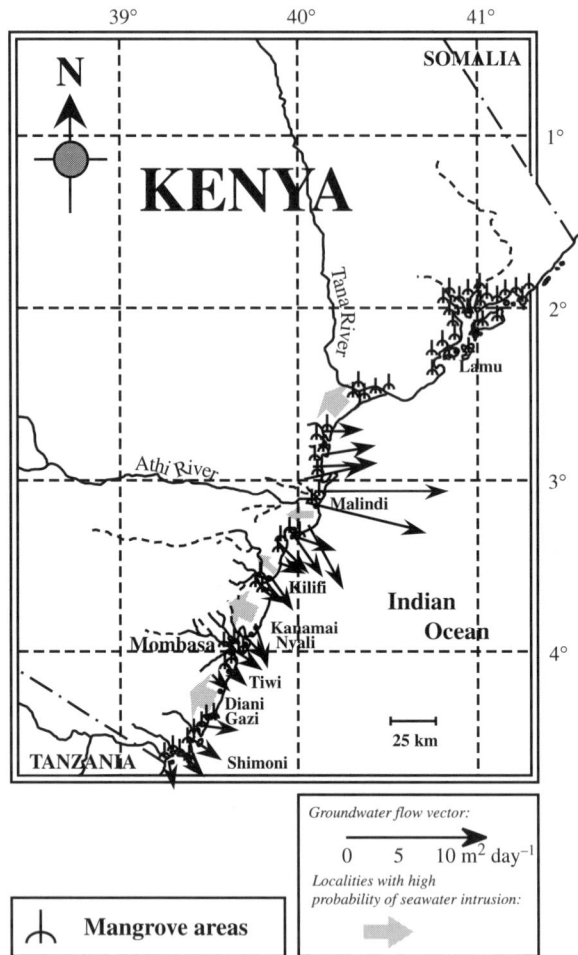

Figure 21.3   Groundwater flow: graphical output of the actual situation of the Kenyan coast between the Tanzanian border and Tana River. Only those areas with groundwater flow higher than $1\,m^2\,day^{-1}$ and regions with a high probability of seawater intrusion are indicated

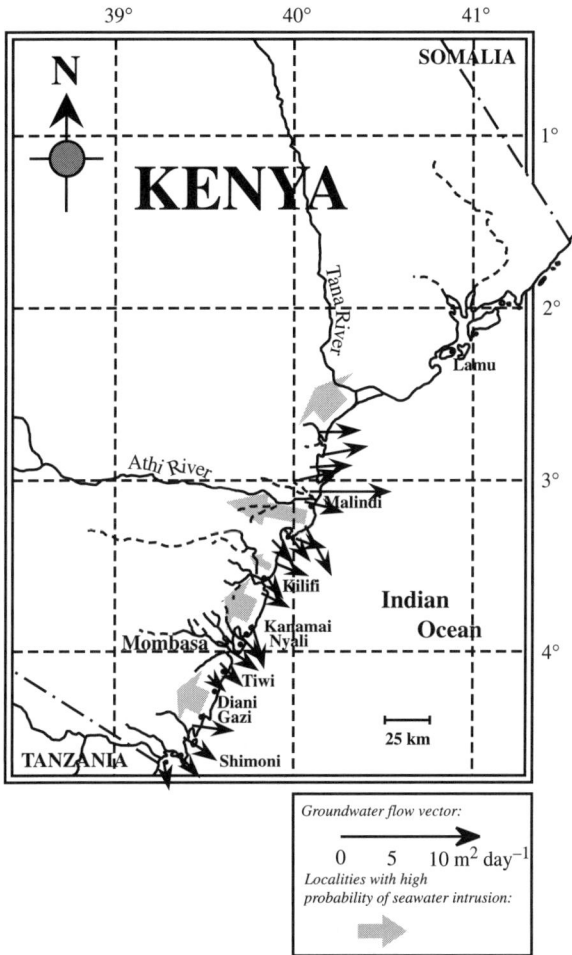

Figure 21.4 Groundwater flow: graphical simulation of the groundwater flow when $10^6 \, \mathrm{m^3 \, day^{-1}}$ are pumped at Mzima springs

high groundwater flow is Diani. However, this village was built more recently, at a time when drinking water was provided by pipeline.

Problems of seawater intrusion have been reported (Ituli, 1984) and areas which are threatened by seawater intrusion can be identified by the model. Areas with a groundwater flow lower than $0.22\,m^2\,day^{-1}$ are highly susceptible to seawater intrusion, whereas the areas with a groundwater flow above $1\,m^2\,day^{-1}$ are less susceptible to seawater intrusion. Salinity measurements of groundwater samples confirm the results of this model.

To solve the drinking water problem of Mombasa, plans exist to pump more water from Mzima Springs, a natural spring inside Tsavo National Park in the catchment of the Athi river collecting runoff from the porous Chuyli Hills, between Nairobi and Mombasa. Pumping up huge amounts of water will lead to an alteration of the regional groundwater flow along the coast (Figure 21.4). Compared with Figure 21.2 a clear decrease in groundwater flow and an increase in seawater intrusion is predicted. The southern part of the coast is geographically closest to Mzima Springs, but it is especially the area around Malindi further north which is affected by the groundwater flow alterations.

THE INDIAN–BANGLADESHI COAST

Simulation of the groundwater flow of the coastal area of India and Bangladesh between the Hugli river and the Ganges river shows very low groundwater flows in the area; so low that seawater intrusion would potentially occur along the whole coast (Figure 21.5). However, this process can be balanced by a sufficient surface water flow

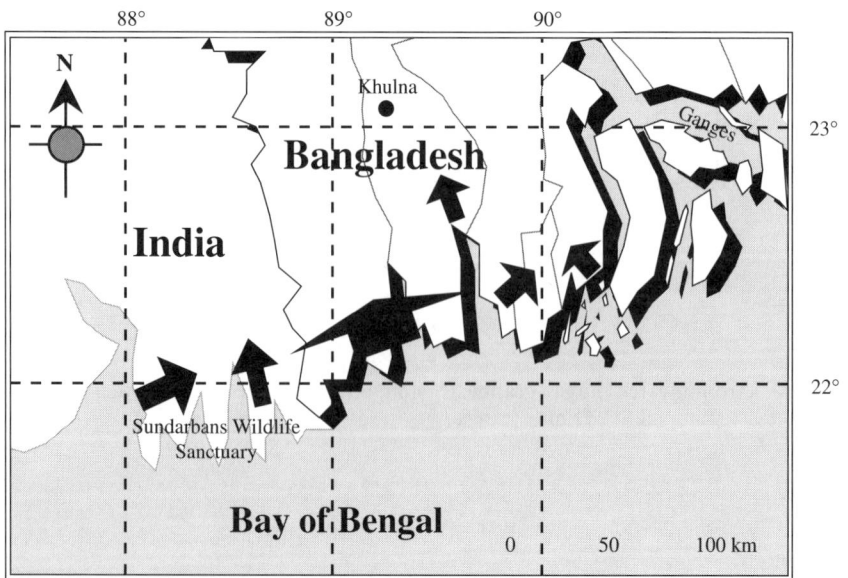

Figure 21.5 India–Bangladesh: regions with high probability of seawater intrusions. Groundwater flow is lower than $1\,m^2\,day^{-1}$ in the whole coastal area

provided by the river Ganges. This has always been the case up to the building of the Farakka barrage in India near the border of Bangladesh. Since then the runoff of the Ganges has diminished and at the same time seawater intrusion has started. This is leading to the destruction of vast areas of mangrove forest in the area.

THE FLORIDA EVERGLADES

Figure 21.6 shows the application of the model on the southern part of Florida and on the Everglades, around 1900. Recently a number of publications dealing with the

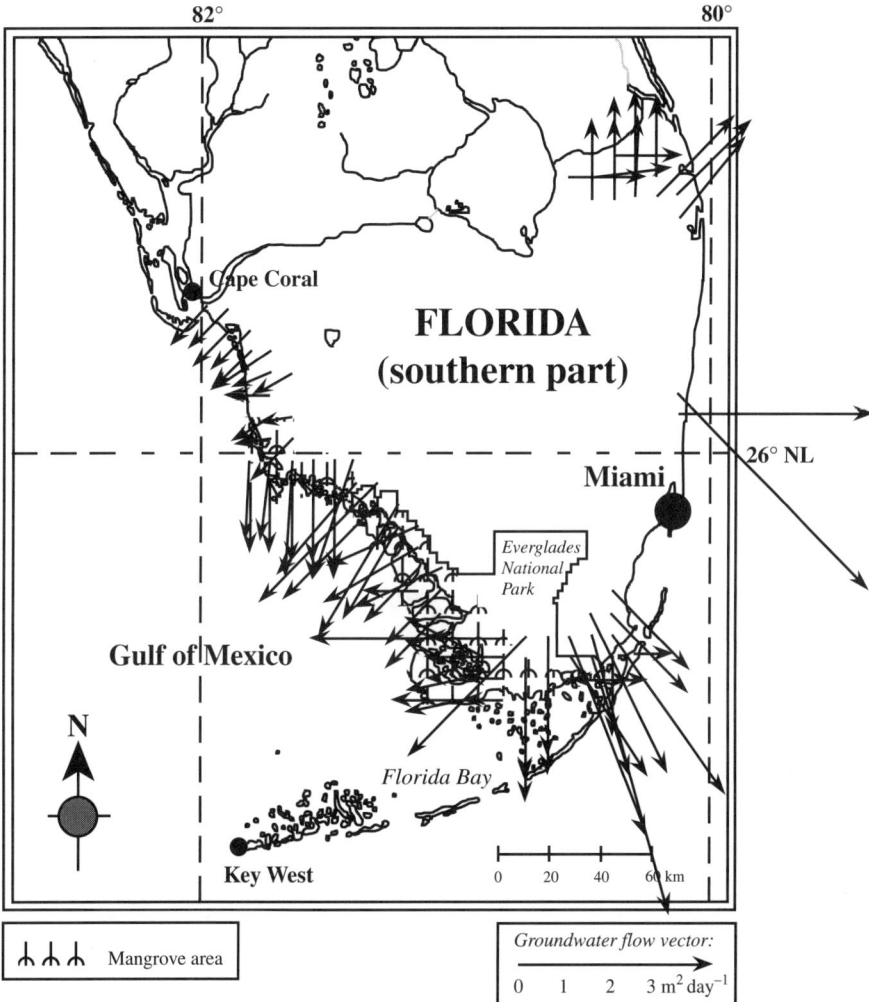

Figure 21.6 Groundwater flow in the southern part of Florida in the beginning of the twentieth century. In the western part of the study region groundwater flow was high in the direction of the Gulf of Mexico. In the eastern part there were only a few areas with high groundwater flow

hydrological problems (mostly surface water) of the Everglades have been published (Holloway, 1994; Mairson, 1994). The present-day situation (Figure 21.7) shows a serious decrease of groundwater flow along the coastal line. The complete absence of groundwater flow in the south-eastern part of the Everglades National Park, together with the diminished surface water flow can explain the doubling of salinity in Florida Bay and the sea grass die-off in the bay. The absence of groundwater flow in the

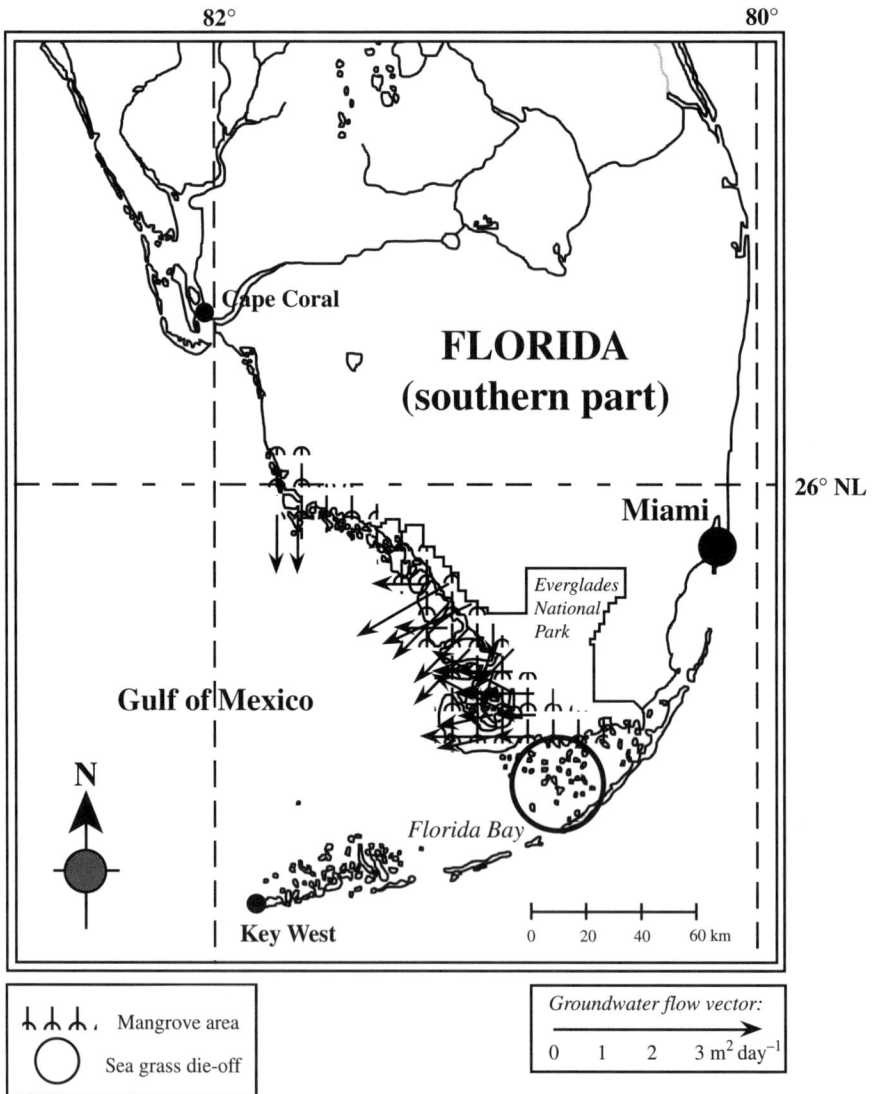

Figure 21.7   Groundwater flow in the southern part of Florida based on data for February 1994. The figure also shows places with sea grass die-off because of the high salinity in Florida Bay

northern part of the mangrove area indicates that there is a serious risk that the mangrove stock in this area will decline in the near future. To save the Everglades for posterity it will be necessary to go much further than the restoration of the surface water flow.

CONCLUSIONS

It is clear that tropical catchments, sometimes several hundreds of kilometres inland, influence the mangrove areas along the coastal lines of the continents. Changes in the catchment area sometimes several hundreds of kilometres away, either caused by human interference or by nature, can have influences. It is clear, therefore, that research on tropical catchments and/or impact assessment of dam building has to take a truly catchment-scale approach, as well as consider fully the groundwater flow consequences, if it is to avoid such disasters as are happening in the Everglades and in the Sundarbans National Parks.

ACKNOWLEDGEMENTS

We thank R. Hauspie and F. De Smedt (Hydrology – Free University of Brussels) for their assistance, the Water Resources Research Center (Florida) for their hydrological data, the Kenya Marine and Fisheries Research Institute and Dr Okemwa for their co-operation in Kenya, the Belgian National Science Foundation and the Kenya Belgium Project in Marine Sciences for their financial support.

REFERENCES

Barth, H. 1982. 'The biogeography of mangroves'. In *Tasks for Vegetation Science 2* (eds. D.N. Sen and K.S. Rasjpurohit), pp. 35–59. Dr. W. Junk Publishers, The Hague.

Blasco, F. 1991. 'Les mangroves'. *La Recherche*, **231**, 444–453.

Dapaah-Siakwan, S. 1986. 'Simulation of regional groundwater flow with solute transport in the lower Athi–Tana Basin, Kenya'. Unpublished MSc Thesis, Free University of Brussels.

Holloway, M. 1994. 'Nurturing nature'. *Scientific American*, **270**, 76–84.

Isaac, W.E. and Isaac, F.M. 1968. 'Marine botany of the Kenya coast 3: general account of environment, flora and vegetation'. *Journal of the East African Natural History Society*, **27**, 7–12.

Ituli, J.T. 1984. 'A regional groundwater flow model for the lower Athi–Tana catchment basin, Kenya'. Unpublished MSc Thesis, Free University of Brussels.

Knutzen, J. and Jasuund, E. 1979. 'Note on the littoral algae from Mombasa, Kenya'. *Journal of the East African Natural History Society*, **168**, 1–4.

Macnae, W. 1968. 'A general account of the flora and fauna of mangrove swamps and forests in the Indo-West pacific region'. *Advances in Marine Biology*, **6**, 73–270.

Mairson, A. 1994. 'The Everglades: dying for help'. *National Geographic*, **185**, 2–35.

Ruwa, R.K. and Polk, P. 1986. 'Additional information on mangrove distribution in Kenya: some observations and remarks'. *Kenya Journal of Sciences Series B*, **7**, 41–45.

Snedaker, S.C. 1982. 'Mangrove species zonation: why?'. In *Tasks for Vegetation Science 2* (eds D.N. Sen and K.S. Rasjpurohit), pp. 111–125. Dr. W. Junk Publishers, The Hague.

Index

A horizon, 63, 79, 104
Aberdare, 22, 32, 35, 41, 45, 51, 164
Aberdares, 41, 51, 164, 239
Acacia, 136, 182, 190, 216, 221, 244, 268, 269, 270
Acacia mangium, 136, 182
Aceros corrugatos. See wrinkled hornbill
Achnanthes, 192
acidity, 63, 100, 105, 155, 167
acrisols, 32, 143, 153
aerial deposition, 168
aerial photography, 89, 263
Africa, 8, 11–14, 16–17, 36, 47, 50, 58, 61, 71, 73–74, 141, 148, 154, 165, 200, 204, 225–6, 234, 237, 241, 249, 255, 257–9, 262, 271
African groundnut scheme, 89
Agathis, 173, 177
aggregates, 81, 87–8, 104, 111–18
aggregation, 77, 80, 88, 104, 253, 266
agile gibbon, 173
Agricultural and Horticultural Society of India, 6
agroecological, 32, 35, 41, 47, 63, 75, 79–80, 107
Agroecological Zones Programme, 90
agroecosystems, 61
agroforestry, 11, 159, 225, 255
agro-silvo-pastoralism, 222
AIm index, 112
Alestes, 247
alkaline (soda) lakes, 338
Alkalinity, 64
Alnus, 190
alpha diversity, 200
Alseodaphne coriacea. See gemur tree
aluminium, 61, 72, 153, 180
Amazonia, 19, 89, 91–3, 95, 100, 104–7, 258
Amblycipitidae, 196

Anas undulata, 260
anastomosing channels, 172
anatase, 100
Anderson soils, 144, 154
Andes mountain, 299
andisoils, 77, 80
andosols, 32, 63, 87
Andropogon, 94
Anguillidae, 196
Anisoptera, 195
Annapurna, 191, 201
Ants, 151
aquifer, 14, 180, 253, 360–3
Argentina, 297–9, 311
Artocarpus heterophyllus. See jackfruit
as flash floods, 299
Athi River, 242, 254
AVHRR, 348, 357
avulsions, 212
Awash, 25, 26

Baetidae, 195
Bagaridae, 196
Balag (irrigated forest), 216
balags, 217
Bangladesh, 14, 359, 368–9
Bantu, 212
Barbus tanensis, 248
barley, 110, 117
basalt, 32, 52, 62
Basement Complex, 63, 67–8
beans, 63, 110, 116–17
bedload, 189
Beja, 210
Bella, 13
Bench terracing, 54
Benchmark Soil Project, 90
Berrending, 227, 236
beta diversity, 200

The Sustainable Management of Tropical Catchments. Edited by David Harper and Tony Brown.
© 1998 John Wiley & Sons Ltd.